Spectral Clustering and Biclustering

Spectral Clustering and Biclustering

Learning Large Graphs and Contingency Tables

Marianna Bolla

Institute of Mathematics
Budapest University of Technology and Economics, Hungary

This edition first published 2013

Registered office
John Wiley & Sons Ltd, The Atrium, Southern Gate, Chichester, West Sussex, PO19 8SQ, United Kingdom

For details of our global editorial offices, for customer services and for information about how to apply for permission to reuse the copyright material in this book please see our website at www.wiley.com.

Library of Congress Cataloging-in-Publication Data

Bolla, Marianna.
Spectral clustering and biclustering : learning large graphs and contingency tables / Marianna Bolla.
pages cm
Includes bibliographical references and index.
ISBN 978-1-118-34492-7 (hardback)
1. Multivariate analysis. 2. Contingency tables. 3. Graph theory. I. Title.
QA278.B65 2013
515′.35–dc23

2013011891

A catalogue record for this book is available from the British Library.

ISBN: 978-1-118-34492-7

Set in 10/12pt Times by Thomson Digital, Noida, India

1 2013

To my son and students

Contents

Preface

When dealing with networks, statisticians want to explore the structure of the actual data set whatever it is. Meanwhile, we realized that statistical methods are not immediately applicable for graphs, as graph theory is mainly concerned with special structures which are too sensitive to minor changes. Therefore, we decided to bridge, at least partially, the gap between statistics and graph theory with ideas on how graphs and contingency tables can be considered as statistical data and how they can be treated by methods reminiscent of classical and modern techniques of multivariate statistical analysis. We want to provide answers to the demanding questions which arise when statisticians are confronted with large weighted graphs or rectangular arrays like microarrays. In contemporary data analysis, experts are striving to recover the structure of communication, social, and biological networks, or in other situations, to compare networks of different sizes and classify them based on a training sample. In the case of large data sets, random effects and measurement errors also have to be treated.

We plan to use existing knowledge on graph spectra and experiences in analyzing real-world data in the form of hypergraphs, weighted graphs, and contingency tables. At the end of the 1980s, together with my PhD advisor, Gábor Tusnády, we used spectral methods for a binary clustering problem, where the underlying matrix turned out to be the generalization of the graphs' Laplacian to hypergraphs. Then we defined the Laplacian for multigraphs and edge-weighted graphs, and went beyond the expanders by investigating gaps within the spectrum and used eigenvectors corresponding to some structural eigenvalues to find clusters of vertices. We also considered minimum multiway cuts with different normalizations that were later called ratio- and normalized cuts. In the 1990s, spectral clustering became a fashionable area and a lot of papers in this topic appeared, sometimes redefining or modifying the above notions, sometimes having a numerical flavor and suggesting algorithms without rigorous mathematical explanation.

At the turn of the millennium, thanks to the spread of the World Wide Web and the human genome project, there was a rush to investigate evolving graphs and random situations different from the classical Erdős–Rényi one. Graph theorists, for example, László Lovász and co-authors, considered convergence of graph sequences and testable graph parameters, bringing – without their volition – statistical concepts into this discrete area. They also started to consider noisy graph and contingency table sequences, and testability of some balanced versions of the already defined minimum multiway cut densities. Meanwhile, physicists introduced other measures of characterizing a community structure of networks. In this book we discuss some penalized versions of the Newman–Girvan modularity and regular cuts which are present in situations when the network has clusters of vertices with a homogeneous information flow within or between them.

A wide array of classical and modern statistical methods are presented and adapted to weighted graphs and contingency tables. We recommend algorithms for recovering the structure of large networks burdened with random noise, further to clustering and classifying them. Through representations, we establish a common outline structure for the contents of each algorithm along with various fundamental elements which are combined via unified notions and principles. We only give a general treatment to the problems exposed; for more examples we refer to papers and also the supporting website (`www.wiley.com/go/spectral`) of the book. Hence, the book is a resource for professionals mining large datasets related to communication, social, biological, or industrial networks, and for students learning statistical methods applicable to networks.

Throughout the book we use advanced linear algebra, probability, and multivariate statistics. A collection of frequently used facts in these fields is to be found in the Appendices. However, a basic knowledge of graph theory and computer sciences suffices as we only consider statistical graph properties which are unaffected by minor perturbations of the vertex- or edge-weights, and we just touch upon the computational complexity of the proposed algorithms. We also discuss possible applications to real-life networks. However, the main contribution of this book is the well-motivated presentation and comparison of a great variety of statistical methods applicable to large networks or network sequences. Well-known facts from former books concerning graph spectra are not proven (references are given), but all other propositions and theorems are proven with mathematical precision. Therefore the book is suitable for a one-semester graduate course on this topic. For those who are interested only in the algorithms, these proofs can be skipped.

Marianna Bolla

Budapest, December 2012

Acknowledgements

I wish to express gratitude to persons whose books, papers, and lectures on spectral graph theory and multivariate statistics turned my interest to this field: F. Chung, D. M. Cvetkovic, M. Fiedler, A. J. Hoffman, B. Mohar, and C. R. Rao.

I am indebted to my PhD advisor, Gábor Tusnády with whom I have worked on several projects and applied statistical methods in real-life problems. I am grateful to László Lovász for valuable discussions on the spectral graph and graph convergence topics. I wish to thank many colleagues for their useful comments on the related research and applications: László Babai, András Benczur, Endre Boros, Imre Csiszár, Villő Csiszár, Zoltán Füredi, Ferenc Juhász, János Komlós, Vilmos Komornik, Tamás Lengyel, Nathan Linial, András Lukács, Péter Major, György Michaletzky, Dezső Miklós, Tamás F. Móri, Dénes Petz, Tomaz Pisanski, András Prékopa, András Recski, Lídia Rejtő, Tamás Rudas, András Simonovits, Miklós Simonovits, Vera T. Sós, Domokos Szász, Gábor J. Székely, Endre Szemerédi, András Telcs, László Telegdi, Bálint Tóth, György Turán, and Katalin Vesztergombi. I am particularly indebted to Katalin Friedll, András Krámli, and János Tóth for promotive discussions, to Ákos Csizmadia for helping me to present the references of this book in a unified form; further, to Ibolya Barna, Miklós Gergi, and Péter István for their technical help.

I would like to thank my students at the Budapest University of Technology and Economics, Central European University, and Budapest Semester of Mathematics, who worked on the spectral clustering topic and helped me with questions, calculations, and processing computer programs: Andrea Bán, Erik Bodzsár, Brian Bullins, Sorathan Chaturapruek, Shiwen Chen, Ahmed Elbanna, Max Del Giudice, Haris Khan, Tamás Kói, Anett Lovas, Gábor Molnár–Sáska, Ildikó Priksz, Viktor Szabó, Zsolt Szabó, Joan Wang, and László Nagy who actually assisted me with preparing some figures for this book.

I am grateful to Károly Simon for encouraging me to write this book and letting me go on sabbatical two years ago when I assembled the pieces of this material together. The related research was partly supported by the Hungarian National Research Grant OTKA-KTIA 77778; and by the TÁMOP-4.2.2.C-11/1/KONV-2012-0001 project, sponsored by the European Union.

List of abbreviations

Notational abbreviations

General

$\mathbb{C}$ set of complex numbers

$\mathbb{R}$ set of real numbers

$a, b, c, x, y, z, \alpha, \beta, \theta, \ell, \ldots$ scalars (real or complex numbers)

$\mathbf{a}, \mathbf{b}, \mathbf{c}, \mathbf{x}, \mathbf{y}, \mathbf{z}, \boldsymbol{\alpha}, \boldsymbol{\beta}, \boldsymbol{\theta}, \ldots$ vectors of real coordinates (column vectors)

$\mathbf{0}$ vector of all 0 coordinates

$\mathbf{1}$ vector of all 1 coordinates

$\boldsymbol{A}, \boldsymbol{B}, \boldsymbol{C}, \boldsymbol{\Lambda}, \boldsymbol{\Delta}, \ldots$ matrices of real entries

$\boldsymbol{A} \leq \boldsymbol{B}$ means that $\boldsymbol{B} - \boldsymbol{A}$ is positive semidefinite ($\boldsymbol{A}$ and $\boldsymbol{B}$ are symmetric matrices)

$\boldsymbol{O}$ matrix of all zero entries

$\boldsymbol{I}_n$ $n \times n$ identity matrix

$\mathbf{a}^T, \boldsymbol{A}^T$ transpose of the vector $\mathbf{a}$ or matrix $\boldsymbol{A}$

$\boldsymbol{A} \oplus \boldsymbol{B}$ Kronecker-sum of the matrices $\boldsymbol{A}$ and $\boldsymbol{B}$

$\boldsymbol{A} \otimes \boldsymbol{B}$ Kronecker-product of the matrices $\boldsymbol{A}$ and $\boldsymbol{B}$

$\boldsymbol{A} \circ \boldsymbol{B}$ Schur or Hadamard product of the matrices $\boldsymbol{A}$ and $\boldsymbol{B}$

$\|\mathbf{a}\|$ Euclidean norm of the vector $\mathbf{a}$

$\|\boldsymbol{A}\|$ spectral norm of the matrix $\boldsymbol{A}$ (its maximal singular value)

$\|\boldsymbol{A}\|$ determinant of the matrix $\boldsymbol{A}$

$\|\boldsymbol{A}\|_2$ Frobenius norm of the matrix $\boldsymbol{A}$ (squareroot of the sum of the squares of its singular values)

$\|\boldsymbol{A}\|_4$ Schatten 4-norm of the matrix $\boldsymbol{A}$ (fourth root of the sum of the fourth powers of its singular values)

$\|\boldsymbol{A}\|_\square$ cut-norm of the matrix $\boldsymbol{A}$

$\mathrm{tr}(\boldsymbol{A})$ trace of the matrix $\boldsymbol{A}$

$o(f(n))$ 'little o' of $f(n)$, i.e., a function of n such that $\lim_{n\to\infty} \frac{o(f(n))}{f(n)} = 0$

$\mathcal{O}(f(n))$ 'big O' of $f(n)$, i.e., a function of n such that $\mathcal{O}(f(n)) \leq Cf(n)$ with some constant C, independent of n

$\Omega(f(n))$ a function of n such that $\Omega(f(n)) \geq cf(n)$ with some constant c, independent of n

$\Theta(f(n))$ is $\mathcal{O}(f(n))$ and $\Omega(f(n))$ at the same time

$\mathcal{P}_k$ set of the k-partitions of $\{1, \dots, n\}$ (into disjoint, non-empty subsets)

Probability and statistics

$X, Y, Z, \dots$ real-valued random variables

$\mathbf{X}, \mathbf{Y}, \mathbf{Z}, \dots$ random vectors (column vectors)

$\mathbb{E}(X)$ expectation of the random variable X (scalar)

$\mathbb{E}(\mathbf{X})$ expectation of the random vector $\mathbf{X}$ (column vector)

$\text{Cov}(X, Y)$ covariance between the random variables X and Y

$\text{Corr}(X, Y)$ correlation between the random variables X and Y

$\text{Var}(X)$ variance of the random variable X (nonnegative scalar)

$\text{Var}(\mathbf{X})$ covariance matrix of the random vector $\mathbf{X}$ (symmetric, positive semidefinite matrix)

$\mathcal{N}_p(\boldsymbol{\mu}, \boldsymbol{C})$ p-dimensional normal distribution with expectation (vector) $\boldsymbol{\mu}$ and covariance matrix $\boldsymbol{C}$

$S_k^2(\boldsymbol{X})$ k-variance of the row vectors of matrix $\boldsymbol{X}$

$\tilde{S}_k^2(\boldsymbol{X})$ weighted k-variance of the row vectors of matrix $\boldsymbol{X}$

Graphs

$G = (V, \boldsymbol{W})$ edge-weighted graph on vertex-set V and symmetric edge-weight matrix $\boldsymbol{W}$ of nonnegative entries (especially, for simple G, $\boldsymbol{W}$ is the adjacency matrix)

$H = (V, E)$ hypergraph with vertex-set V and set of hyper-edges E

$e(S)$ number of edges emanating from the vertex-subset U of a simple graph

$e(S, T)$ number of edges connecting the vertex subsets U and T of a simple graph

$w(S, T)$ weighted cut between the vertex subsets U and T of an edge-weighted graph

d_i generalized degree of vertex i of an edge-weighted graph $G = (V, \boldsymbol{W})$

$\mathbf{d}$ degree vector comprised of the generalized degrees of vertices of an edge-weighted graph

$\sqrt{\mathbf{d}}$ $=(\sqrt{d_1}, \dots, \sqrt{d_n})^T$

$\boldsymbol{D} = \text{diag}(\mathbf{d})$ degree matrix, i.e., diagonal matrix containing the generalized degrees in its main diagonal

$\boldsymbol{L} = \boldsymbol{D} - \boldsymbol{W}$ Laplacian of the edge-weighted graph $G = (V, \boldsymbol{W})$

$\boldsymbol{L}_D = \boldsymbol{I} - \boldsymbol{D}^{-1/2}\boldsymbol{W}\boldsymbol{D}^{-1/2}$ normalized Laplacian of the edge-weighted graph $G = (V, \boldsymbol{W})$

$\boldsymbol{M} = \boldsymbol{W} - \mathbf{d}\mathbf{d}^T$ modularity matrix of the edge-weighted graph $G = (V, \boldsymbol{W})$, when $\|\sqrt{\mathbf{d}}\| = 1$

$\boldsymbol{M}_D = \boldsymbol{D}^{-1/2}\boldsymbol{W}\boldsymbol{D}^{-1/2} - \sqrt{\mathbf{d}}\sqrt{\mathbf{d}}^T$ normalized modularity matrix of the edge-weighted graph $G = (V, \boldsymbol{W})$, when $\|\sqrt{\mathbf{d}}\| = 1$

$\text{Vol}(U) = \sum_{i \in U} d_i$ Volume of the vertex subset U of an edge-weighted graph

$g_k(G)$ k-way partition cut of the edge-weighted graph $G = (V, \boldsymbol{W})$

$f_k(G)$ k-way normalized cut of the edge-weighted graph $G = (V, \boldsymbol{W})$

$M_k(G)$ k-way Newman–Girvan modularity of the edge-weighted graph $G = (V, \boldsymbol{W})$

$BM_k(G)$ k-way balanced Newman–Girvan modularity of the edge-weighted graph $G = (V, \boldsymbol{W})$

$NM_k(G)$ k-way normalized Newman–Girvan modularity of the edge-weighted graph $G = (V, \boldsymbol{W})$

$W_\square$ cut-norm of the graphon W

Contingency tables

Row, Col row- and columnn-sets of a contingency table $\boldsymbol{C}$ of nonnegative real entries

$c(R, C)$ cut between the subsets R and C of the rows and columns of a contingency table

$d_{row,i}$ ith row-sum of the $m \times n$ contingency table $\boldsymbol{C}$

$d_{col,j}$ jth column-sum of the $m \times n$ contingency table $\boldsymbol{C}$

$\boldsymbol{D}_{row} = \text{diag}(d_{row,1}, \ldots, d_{row,m})$ diagonal row-sum matrix

$\boldsymbol{D}_{col} = \text{diag}(d_{col,1}, \ldots, d_{col,n})$ diagonal column-sum matrix

$\boldsymbol{C}_{corr} = \boldsymbol{D}_{row}^{-1/2}\boldsymbol{C}\boldsymbol{D}_{col}^{-1/2}$ normalized contingency table or correspondence matrix

$\text{Vol}(S) = \sum_{i \in S} d_{row,i}$ $\text{Vol}(T) = \sum_{j \in T} d_{col,j}$ volumes of the row- and column-subsets U and T of a contingency table

$\nu_k(\boldsymbol{C})$ normalized k-way bicut of the contingency table $\boldsymbol{C}$

$C_\square$ cut-norm of the contingon C

Verbal abbreviations

c.d.f. cumulative distribution function

dist: distance between vectors and/or subspaces

i.d. identically distributed (random variables)

i.i.d. independent, identically distributed (random variables)

p.d.f. probability density function

s.t. such that

ANOVA Analysis of variance

MANOVA Multivariate analysis of variance

SD spectral decomposition (of a symmetric matrix)

SVD singular value decomposition (of a matrix)

Introduction

'Happy families are all alike; every unhappy family is unhappy in its own way.'
Leo Tolstoy: *Anna Karenina*

Graphs with a small spectral gap can be very different by their nature, indeed. Those with a large spectral gap have been frequently studied since Metropolis *et al.* (1953); Nash-Williams (1959) and Cheeger (1970), establishing a lot of equivalent or near equivalent advisable features of them. There are beautiful results about the relation between this gap and the expanding properties of the graph (see e.g., Hoory, Linial and Widgerson (2006)), including a random walk view of Azran and Ghahramani (2006); Diaconis and Stroock (1991) and connectivity (the first results are due to Fiedler (1973) and Hoffman (1970)). Roughly speaking, graphs with a large spectral gap are good expanders; the random walk goes through them very quickly with a high mixing rate and short commute time, see for example Lovász (1993); Lovász and Winkler (1995); they are also good magnifiers as their vertices have many neighbors; in other words, their vertex subsets have a large boundary compared to their volumes characterized by the isoperimetric number: see Mohar (1988); equivalently, they have high conductance and show quasirandom properties discussed in Chung, Graham and Wilson (1989) and Chung and Graham (2008). Because of these favorable characteristics, they are indispensable in communication networks; further, there is basically one eigenvalue responsible for this property.

However, less attention has been paid to graphs with a small spectral gap, when several cases can occur: among others, the graph can be a bipartite expander (see Alon (1986)) or its vertices can be divided into two sparsely connected clusters, but the clusters themselves can be good expanders (see Ng *et al.* (2001) and Newman (2003)). In the case of several clusters of vertices the situation is even more complicated. The pairwise relations between the clusters and the within-cluster relations of vertices of the same cluster show a great variety; not surprisingly, there is more than one eigenvalue responsible for these versatile properties. Depending on the number and sign of the so-called structural eigenvalues of the normalized modularity matrix, defined in Bolla (2011), we help enable the reader to decide the number of underlying clusters and the type of connection between them.

Given a graph – simple or edge-weighted – that does not belong to the well manageable family of expanders, our task is to survey the numerous possibilities for this graph according to which it may have a structure. We wouldn't say that every such graph has its own structure, but there are types of relations between its vertices according to which they may form clusters. We prove and illustrate through random examples that based on the structural eigenvalues of the normalized modularity matrix, how the main types of intra- and inter-cluster relations can be established and the clusters themselves recovered via the corresponding eigenvectors. Our examples are mainly from the class of generalized random graphs. We deal with dense graphs as a statistical sample. Sparse graphs (like path, grid, or hypercube) are frequently used as examples where the minimum spectral gap shows up; however there are no sensible ways to divide their vertices into any number of clusters. Our methods are rather applicable to real-life situations when the number of edges is proportional to the square of the number of vertices; or in the edge-weighted graph setup, at least a linear number of weights in each row of the weight matrix differ from zero.

Spectra of graphs together with their application was discussed by several authors, for example, Biggs (1974); Chung (1997); Cvetkovic, Doob and Sachs (1979); Spielman (2007); von Luxburg (2007). There are quick numerical algorithms to find spectral or singular value decomposition (SD or SVD) of the graph or contingency table based matrix with normalization adopted to the objective function. Most of these algorithms use the k smallest (normalized or unnormalized) Laplacian eigenvalues which are capable to reveal a sparse k-way cut (see Lee *et al.* (2012); Louis *et al.* (2011)). We do not confine ourselves to the bottom of the Laplacian spectrum, but take into consideration the top eigenvalues as well, which together form the large absolute value (structural) eigenvalues of the normalized modularity matrix, introduced just for this convenience. As a statistician, I believe there must be a structure in each network. Theoretical results (like the Szemerédi's regularity lemma, see Komlós *et al.* (2002)) also support this, though sometimes with an enormously large number of clusters. There is a lot of knowledge about how some specific structures can be characterized by means of spectra, but little is known about how to recognize the structure itself. Sometimes the same continuous relaxation is used to optimize different types of cuts (see Ding *et al.* (2003)), sometimes it is noted that the result of the Laplacian-based clustering depends on the graph itself (see von Luxburg *et al.* (2008)). I plan to approach this problem from the opposite direction: first to be acquainted with the graph so as to apply the best possible clustering to it. This means that based on the spectrum (gaps between the structural and the bulk eigenvalues and the sign of the structural ones) we use the most appropriate eigenvectors for representation and clustering. We consider the graph as statistical data itself, find out which structure is expected, and then select the method.

Though spectral graph theory seems to be a versatile endless tale, we speak about it in the framework of a moderate length book. Of course, we cannot discuss all the details, but what we discuss is compatible with unified notation, and we just touch upon subjects (e.g., random walk view) that have been thoroughly discussed by other authors. However, there is reality behind every tale, and facing it helps us to unveil the problem. We learn, for example, that the reality behind the diabolic kernel trick – when using reproducing kernel Hilbert spaces – is the century old Riesz–Fréchet representation theorem, according to which linear functionals over a Hilbert space are in one-to-one correspondence with the space itself; here it is extended to non-linear functionals that are related to the elements of a more complicated Hilbert space (not the original one). Knowing this, the trick becomes a simple tool as to how the selection of an appropriate kernel helps us in discovering non-linearities in our data and separate clusters that cannot otherwise be separated by simply applying the k-means algorithm to it. We start the sections with a brief exposition of the problems and roughly explain the solution, but later on we encounter all the definitions and facts belonging to the issue. We elucidate everything with full precision and also point in new directions. The following explains the contents of the strongly interwined chapters.

Chapter 1 is devoted to multivariate analysis techniques for representing edge-weighted graphs and contingency tables. Different kinds of graph-based matrices are introduced together with spectra and spectral subspaces which optimize different quadratic placement problems subject to different constraints. Our purpose here is to develop tools for the unique treatment of the multiway cut problems of Chapter 2. These methods are reminiscent of classical methods of multivariate statistical analysis, like principal component analysis, correspondence analysis, or analysis of variance. As a result, we get low dimensional representation of the graph's vertices or rows and columns of the contingency table so that the representation somehow favors the classification criteria of Chapter 2. Non-linearities are treated by mapping the data into a feature space (reproducing kernel Hilbert space), which is a useful technique if we build a graph on given data points. We also generalize the notion of representation for joint distributions, and use it in Chapter 4 when convergence of normalized spectra is discussed.

In Chapter 2, minimum ratio and normalized multiway cut problems are presented together with modularity cuts. Since the optima of the corresponding objective functions are taken on partition vectors belonging to the hidden clusters, via Chapter 1, they are related to Laplacian, normalized Laplacian, or modularity spectra, whereas the precision of the estimates depends on the distance between the subspaces spanned by the corresponding eigenvectors and partition vectors. By an analysis of variance argument, this distance is the sum of the inner variances of the underlying clusters, the objective function of the k-means clustering. Therefore, instead of speaking about continuous relaxation of the discrete optimization tasks, we rather solve the combinatorial problems by using the same linear algebra machinery as in Chapter 1. In this way, estimates for minimal or maximal multiway cuts are obtained by relating the overall minima or maxima to those taken on partition vectors.

Chapter 3 applies the results of Chapters 1 and 2 for spectral clustering of large networks. Networks are modeled either by edge-weighted graphs or contingency tables, and usually subject to random errors due to their evolving and flexible nature. Asymptotic properties of SD and SVD of the involved matrices are discussed when not only the number of the graph's vertices or that of the rows and columns of the contingency table tends to infinity, but the cluster sizes also grow proportionally with them. Mostly, perturbation results for the SD and SVD of blown-up matrices—burdened with a Wigner-type error matrix—are investigated. Conversely, given a weight-matrix or rectangular array of nonnegative entries, we look for the underlying block-structure. We show that under very general circumstances, clusters of a large graph's vertices and simultaneously those of the rows and columns of a large contingency table can be identified with high probability. In this framework, so-called volume regular cluster pairs are also considered with homogeneous information flow within the clusters and between pairs of them; and their pairwise discrepancies are related to spectra.

In Chapter 4, the theory of convergent graph sequences and graphons, elaborated by graph theorists, is used for vertex- and edge-weighted graphs and contingency tables. The convergence can be formulated in terms of the cut-distance, and limit objects are defined. Testable parameters are nonparametric statistics defined on graphs and contingency tables that can be consistently estimated based on a smaller sample selected randomly from the underlying huge network. Real-word graphs or rectangular arrays are sometimes considered as samples from a large network, and we deduce the network's parameters from the same parameters of its smaller parts. The theory guarantees that this can be done if the investigated parameter is testable. We prove that certain balanced multiway cut densities are indeed testable, and the increasing noisy graph sequences of Chapter 3 converge in this sense too.

Chapter 5 presents parametric and nonparametric statistical methods to find the underlying clusters of a given network or to classify graphs or contingency tables given some distinct prototypes. In fact, the minimum multiway cuts or bicuts and maximum modularities of Chapter 2 are nonparametric statistics that are estimated by means of spectral methods. Algorithms for the representation-based spectral clustering are reviewed together with recommendations about the choice of the eigenvectors which are best capable to reveal the underlying structure. A special algorithm for finding disjoint clusters of the edges, and not necessarily disjoint clusters of the vertices of a hypergraph, is also presented. We also discuss parametric models and give maximum likelihood estimation for the parameters. Eventually, we discuss possible applications of discriminant analysis for classifying graphs and contingency tables.

Frequently used facts about linear operators, random matrices, and multivariate statistical methods are collected in the Appendices. We have relatively few illustrations as believe that instead of visualization, the step by step comprehension, rather than the delusive view, can lead to a deeper understanding of the topic.

References

Alon N. (1986) Eigenvalues and expanders. *Combinatorica* **6**, 83–96.

Azran A. and Ghahramani Z. 2006 Spectral methods for automatic multiscale data clustering. Proc. IEEE Computer Society Conference on Computer Vision and Pattern Recognition (CVPR 2006), *New York NY*, (eds Fitzgibbon A., Taylor C.J. and Lecun Y.), IEEE Computer Society, Los Alamitos, California, pp. 190–197.

Biggs N.L. 1974 *Algebraic Graph Theory*, Cambridge University Press, Cambridge.

Bolla M. 2011 Penalized versions of the Newman–Girvan modularity and their relation to multi-way cuts and k-means clustering. *Phys. Rev. E* **84**, 016108.

Cheeger J. 1970 A lower bound for the smallest eigenvalue of the Laplacian, in *Problems in Analysis* (ed. R. C. Gunning), Princeton Univ. Press, Princeton NJ, pp. 195–199.

Chung F. 1997 *Spectral Graph Theory*, CBMS Regional Conference Series in Mathematics 92. American Mathematical Society, Providence RI.

Chung F, Graham R.L. and Wilson R.K. 1989 Quasi-random graphs. *Combinatorica* **9**, 345–362.

Chung F. and Graham R. 2008 Quasi-random graphs with given degree sequences, *Random Struct. Algorithms* **12**, 1–19.

Cvetkovic D.M., Doob M. and Sachs H. 1979 *Spectra of Graphs*, Academic Press, New York.

Diaconis P. and Stroock D. 1991 Geometric bounds for eigenvalues of Markov chains. *Ann. Appl. Probab.* **1**, 36–62.

Ding C., He X., Zha H., Gu M. and Simon H.D. 2003 A Minmax Cut Spectral Method for Data Clustering and Graph Partitioning. Technical Report 54111, Lawrence Berkeley National Laboratory.

Fiedler M. 1973 Algebraic connectivity of graphs. *Czech. Math. J.* **23** (98), 298–305.

Hoffman A.J. 1970 On eigenvalues and colorings of graphs. In *Graph Theory and its Applications* (ed. Harris B), pp. 79–91. Academic Press, New York.

Hoory S., Linial N. and Widgerson A. 2006 Expander graphs and their applications. *Bull. Amer. Math. Soc. (N. S.)* **43** (4), 439–561.

Komlós J., Shokoufanden A., Simonovits M. and Szemerédi E. 2002 Szemerédi's Regularity Lemma and its Applications in Graph Theory, in *Lecture Notes in Computer Science*, Springer, Berlin, vol 2292, pp. 84–112.

Lee J.R., Gharan S.O. and Trevisan L. 2012 Multi-way spectral partitioning and higher-order Cheeger inequalities, in *Proc. 44th Annual ACM Symposium on the Theory of Computing (STOC 2012)*, New York NY, pp. 1117–1130.

Louis A., Raghavendra P., Tetali P. and Vempala S. 2011 Algorithmic extension of Cheeger's inequality to higher eigenvalues and partitions, in *Approximation, Randomization and Combinatorial Optimization. Algorithms and Techniques. Lecture Notes in Computer Science* (eds Goldborg LA, Jansen K, Ravi R and Rolim JDP), Vol. 6845, Springer, pp. 315–326.

Lovász L. 1993 Random walks on graphs: a survey, in *Combinatorics, Paul Erdős is Eighty. János Bolyai Society, Mathematical Studies*, Keszthely, Hungary, vol. 2, pp. 1–46.

Lovász L. and Winkler P. 1995 Exact mixing in an unknown Markov chain. *Electron. J. Comb.* **2**, Paper R15, 1–14.

Metropolis N., Rosenblut A., Rosenbluth M., Teller A. and Teller E. 1953 Equation of state calculation by fast computing machines. *J. Chem. Physics* **21**, 1087–1092.

Mohar B. 1988 Isoperimetric inequalities, growth and the spectrum of graphs. *Linear Algebra Appl.* **103**, 119–131.

Nash-Williams CStJA 1959 Random walks and electronic currents in networks. *Proc. Cambridge Phil. Soc.* **55**, 181–194.

Newman M.E.J 2003 Mixing patterns in networks. *Phys. Rev. E* **67**, 026126.

Ng A.Y. Jordan M.I. and Weiss Y. 2001 On spectral clustering: analysis and an algorithm. Proc. 14th Neural Information Processing Systems Conference (NIPS 2001) (Dietterich T.G., Becker S. and Ghahramani Z. eds), MIT Press, Cambridge, MA, pp. 849–856.

Spielman D.A. 2007 Spectral graph theory and its applications. *Proc. 48th Annual IEEE Symposium on Foundations of Computer Science (FOCS 2007), Providence, RI*, IEEE Computer Society, Los Alamitos, California, pp. 29–38.

von Luxburg U 2007 A tutorial on spectral clustering. *Stat. Comput.* **17** (4), 395–416.

von Luxburg U., Belkin M. and Bousquet O. 2008 Consistency of spectral clustering. *Ann. Stat.* **36** (2), 555–586.

1

Multivariate analysis techniques for representing graphs and contingency tables

Graph spectra have been used for about 40 years to recover the structure of graphs. Different kinds of spectra are capable of finding multiway cuts corresponding to different optimization criteria. While eigenvalues give estimates for the objective functions of the discrete optimization problems, eigenvectors are used to find clusters of vertices which approximately solve the problems. These methods are reminiscent of some classical methods of multivariate statistical analysis (see Appendix C). Throughout this chapter, methods of principal component analysis and correspondence analysis are used to solve quadratic placement tasks on weighted graphs and contingency tables. As a result, we get low dimensional representation of the graph's vertices or rows and columns of the contingency table by means of linear methods so that the representation somehow favors our classification criteria. Non-linearities are treated by mapping the data into a feature space (reproducing kernel Hilbert space). We also generalize the notion of representation for joint distributions.

1.1 Quadratic placement problems for weighted graphs and hypergraphs

In multivariate statistical analysis, principal component analysis and factor analysis (see Section C.1) are crucial methods for reducing the dimensionality of the data via representing them by a smaller number of independent factors. The representation also

Spectral Clustering and Biclustering: Learning Large Graphs and Contingency Tables, First Edition.
Marianna Bolla.

gives rise to clustering the data in the factor space. Given an $n \times n$ positive definite covariance matrix $\boldsymbol{C}$ of a random vector, the principal components are determined successively, and they are the maximum variance normalized linear combinations of the components of the underlying random vector conditioned on the uncorrelatedness (in the Gaussian case, independence) of them.

Now, our data matrix corresponds to a graph. First, let $G = (V, E)$ be a simple graph on the vertex-set V and edge-set E with $|V| = n$ and $|E| \le \binom{n}{2}$. Thus, the $|E| \times n$ data matrix $\boldsymbol{B}$ has 0-1 entries, the rows correspond to the edges, the columns to the vertices, and b_{ij} is 1 or 0 depending on whether the edge i contains the vertex j as an endpoint or not. The Gram-matrix (see Definition A.3.6) $\boldsymbol{C} = \boldsymbol{B}^T\boldsymbol{B}$ is the non-centralized covariance matrix based on the data matrix $\boldsymbol{B}$, and is both positive definite (provided there are no multiple edges) and a Frobenius-type matrix with nonnegative entries to which Theorem A.3.30 is applicable. Sometimes the matrix $\boldsymbol{C}$ is called signless Laplacian (see Cvetković *et al.* (1997)) and its eigenspaces are used to compare cospectral graphs. It is easy to see that $\boldsymbol{C} = \boldsymbol{D} + \boldsymbol{A}$, where $\boldsymbol{A} = (a_{ij})$ is the usual *adjacency matrix* of G (it is symmetric and a_{ij} is 1 or 0 depending on whether vertices i and j are connected or not; $a_{ii} = 0$, $i = 1, \dots, n$), while $\boldsymbol{D}$ is the so-called degree-matrix, that is, diagonal matrix, containing the vertex degrees in its main diagonal. $\boldsymbol{A}$ being a Frobenius-type matrix (see Theorem A.3.30), its maximum absolute value eigenvalue is positive, and by Proposition A.3.33, it is at most the maximum vertex-degree, and apart from the trivial case – when there are no edges at all – it is indefinite, as the sum of its eigenvalues, that is, the trace of $\boldsymbol{A}$, is zero.

Instead of the positive definite matrix $\boldsymbol{C}$, for optimization purposes, as will be derived below, the Laplacian matrix $\boldsymbol{L} = \boldsymbol{D} - \boldsymbol{A}$ is more suitable, which is positive semidefinite, and always has a zero eigenvalue, since the row sums are zeros. This $\boldsymbol{L}$ is sometimes called combinatorial or difference Laplacian, whereas we will introduce the so-called normalized Laplacian, $\boldsymbol{L}_D$ too. If our graph is regular, then $\boldsymbol{D} = d\boldsymbol{I}$ (where d is the common degree of the vertices and $\boldsymbol{I}$ is the identity matrix) and therefore, the eigenvalues of $\boldsymbol{C}$ and $\boldsymbol{L}$ are obtained from those of $\boldsymbol{A}$ by adding d to them or subtracting them from d, respectively.

There are too many matrices around, not to mention the modularity and normalarized modularity matrices, the latter one closely related to the normalized Laplacian, akin to the so-called transition probability matrix $\boldsymbol{D}^{-1}\boldsymbol{A}$, or $\boldsymbol{I} - \boldsymbol{D}^{-1}\boldsymbol{A}$ which is sometimes called the random walk Laplacian (this matrix is not symmetric, still it has real eigenvalues). Our purpose is to clarify in which situation which of these matrices is the best applicable. The whole story simplifies if we use edge-weighted graphs, and all these matrices come into existence naturally, while solving some optimization problems.

1.1.1 Representation of edge-weighted graphs

From now on, we will use the more general framework of an edge-weighted graph. A simple graph is a special case of it with 0-1 weights. Let $G = (V, \boldsymbol{W})$ be a graph on n vertices, where $V = \{1, \dots, n\}$ and the $n \times n$ symmetric *edge-weight matrix* $\boldsymbol{W}$ has nonnegative real entries and zero diagonal. Here w_{ij} is the similarity between vertices

i and j, where 0 similarity means no connection (edge) at all. If G is a simple graph, $\boldsymbol{W}$ is its adjacency matrix. Since $\boldsymbol{W}$ is symmetric, the weight of the edge between two vertices does not depend on its direction, that is, our graph is *undirected*. In this book, we will mostly treat undirected graphs, except in Subsection 3.3.3, where a non-symmetric $\boldsymbol{W}$ corresponds to a directed edge-weighted graph.

Let the row-sums of $\boldsymbol{W}$ be

$$d_i = \sum_{j=1}^{n} w_{ij}, \quad i = 1, \ldots, n$$

which are called *generalized vertex-degrees* and collected in the main diagonal of the diagonal *degree-matrix* $\boldsymbol{D} = \mathrm{diag}(\mathbf{d})$, where $\mathbf{d} = (d_1, \ldots, d_n)^T$ is the so-called *degree-vector*.

For a given integer $1 \le k \le n$ we are looking for k-dimensional representatives $\mathbf{r}_1, \ldots, \mathbf{r}_n \in \mathbb{R}^k$ of the vertices such that they minimize the objective function

$$Q_k = \sum_{i<j} w_{ij} \|\mathbf{r}_i - \mathbf{r}_j\|^2 \ge 0 \tag{1.1}$$

subject to

$$\sum_{i=1}^{n} \mathbf{r}_i \mathbf{r}_i^T = \boldsymbol{I}_k \tag{1.2}$$

where $\boldsymbol{I}_k$ is the $k \times k$ identity matrix. When minimized, the objective function Q_k favors k-dimensional placement of the vertices such that vertices connected with large-weight edges are forced to be close to each other.

Let us put both the objective function and the constraint in a more favorable form. Denote by $\boldsymbol{X}$ the $n \times k$ matrix of rows $\mathbf{r}_1^T, \ldots, \mathbf{r}_n^T$. Let $\mathbf{x}_1, \ldots, \mathbf{x}_k \in \mathbb{R}^n$ be the columns of $\boldsymbol{X}$, for which fact we use the notation $\boldsymbol{X} = (\mathbf{x}_1, \ldots, \mathbf{x}_k)$. Because of the constraint (1.2), the columns of $\boldsymbol{X}$ form an orthonormal system, hence, $\boldsymbol{X}$ is a suborthogonal matrix. Therefore, the constraint (1.2) can be formulated as $\boldsymbol{X}^T\boldsymbol{X} = \boldsymbol{I}_k$. With this notation, the objective function (1.1) is rewritten in the symmetrized form

$$\begin{aligned} Q_k &= \frac{1}{2}\sum_{i=1}^{n}\sum_{j=1}^{n} w_{ij}\|\mathbf{r}_i - \mathbf{r}_j\|^2 = \sum_{i=1}^{n} d_i \|\mathbf{r}_i\|^2 - \sum_{i=1}^{n}\sum_{j=1}^{n} w_{ij}\mathbf{r}_i^T\mathbf{r}_j \\ &= \sum_{\ell=1}^{k} \mathbf{x}_\ell^T(\boldsymbol{D} - \boldsymbol{W})\mathbf{x}_\ell = \mathrm{tr}[\boldsymbol{X}^T(\boldsymbol{D} - \boldsymbol{W})\boldsymbol{X}]. \end{aligned} \tag{1.3}$$

Definition 1.1.1 *The matrix $\boldsymbol{L} = \boldsymbol{D} - \boldsymbol{W}$ is called the Laplacian corresponding to the edge-weighted graph $G = (V, \boldsymbol{W})$.*

For simple graphs, we get back the usual definition of the Laplacian. About the physical meaning of it, see papers by Biggs (1974); Chung (1997); Cvetković *et al.* (1979); Mohar (1988).

The Laplacian is always positive semidefinite, as can easily be seen from the $Q_1 \geq 0$ relation, and it always has a zero eigenvalue, since its rows sum up to zero. It can be shown, that the multiplicity of 0 as an eigenvalue of $\boldsymbol{L}$ is equal to the number of connected components of $G = (V, \boldsymbol{W})$, that is, the maximum number of disjoint subsets of V such that there are no edges connecting vertices of distinct subsets (where no edge means an edge with zero weight). In terms of $\boldsymbol{W}$, the number of connected components of G is the maximum number of the diagonal blocks which can be achieved by the same permutation of the rows and columns of $\boldsymbol{W}$. For simple graphs, the proof is in von Luxburg (2006), among others, and it is analogous for the edge-weighted case. Consequently, if G is connected, then 0 is a single eigenvalue with corresponding unit-norm eigenvector $\mathbf{u}_0 = \frac{1}{\sqrt{n}}\mathbf{1}$, where $\mathbf{1}$ denotes the all 1's vector. In the sequel, we will assume that G is connected (or equivalently, $\boldsymbol{W}$ is irreducible).

Theorem 1.1.2 (Representation theorem for edge-weighted graphs) *Let $G = (V, \boldsymbol{W})$ be a connected edge-weighted graph with Laplacian matrix $\mathbf{L}$. Let $0 = \lambda_0 < \lambda_1 \leq \cdots \leq \lambda_{n-1}$ be the eigenvalues of $\mathbf{L}$ with corresponding unit-norm eigenvectors $\mathbf{u}_0, \mathbf{u}_1, \ldots, \mathbf{u}_{n-1}$. Let $k < n$ be a positive integer such that $\lambda_{k-1} < \lambda_k$. Then the minimum of (1.1) subject to (1.2) is*

$$\sum_{i=0}^{k-1} \lambda_i = \sum_{i=1}^{k-1} \lambda_i$$

and it is attained with the optimum representatives $\mathbf{r}_1^*, \ldots, \mathbf{r}_n^*$, the transposes of which are row vectors of $\boldsymbol{X}^* = (\mathbf{u}_0, \mathbf{u}_1, \ldots, \mathbf{u}_{k-1})$.

Proof. Using (1.3), our objective function is

$$Q_k = \mathrm{tr}[\boldsymbol{X}^T \boldsymbol{L} \boldsymbol{X}]$$

and it is to be minimized under $\boldsymbol{X}^T \boldsymbol{X} = \boldsymbol{I}_k$. Then Proposition A.3.12 guarantees that the solution is the required one.

Definition 1.1.3 *The vectors $\mathbf{r}_1^*, \ldots, \mathbf{r}_n^*$ giving the optimum in Theorem 1.1.2 are called optimal k-dimensional representatives of the vertices, while the eigenvectors $\mathbf{u}_0, \mathbf{u}_1, \ldots, \mathbf{u}_{k-1}$ of $\mathbf{L}$ are called vector components taking part in the optimal k-dimensional representation.*

We remark the following.

- The dimension k does not play an important role here; the vector components can be included one after the other up to a k such that $\lambda_{k-1} < \lambda_k$.
- We remark that the eigenvectors can be arbitrarily chosen in the eigenspaces corresponding to possible multiple eigenvalues, under the orthogonality conditions. Further, the representatives can also be rotated in $\mathbb{R}^k$. Indeed, neither the objective function nor the constraint is changed if we use $\boldsymbol{R}\mathbf{r}_i$'s instead of

$\mathbf{r}_i$'s, or equivalently, $\boldsymbol{XR}$ instead of $\boldsymbol{X}$, where $\boldsymbol{R}$ is an arbitrary $k \times k$ orthogonal matrix.

- Since the eigenvector $\mathbf{u}_0$ has equal coordinates, the same first coordinates of the vertex representatives do not play an important role in the representation, especially when the representatives are used for clustering purposes. Therefore, $\mathbf{u}_0$ can be omitted, and an optimum $(k-1)$-dimensional representation is performed based on the eigenvectors $\mathbf{u}_1, \ldots, \mathbf{u}_{k-1}$.
- For the time being, we assume that $\boldsymbol{W}$ has zero diagonal. We can also see that in the presence of possible loops (some or all diagonal entries of $\boldsymbol{W}$ are positive) the objective function and the Laplacian remains the same, hence, Theorem 1.1.2 is applicable to this situation too.

1.1.2 Representation of hypergraphs

For hypergraphs, the minimum placement problem is formulated in terms of the representatives of vertices and hyperedges, but it will turn out that we can always assign an edge-weighted graph to our hypergraph so that the two quadratic placement problems are equivalent in terms of the vertices.

Let $H = (V, E)$ be a hypergraph with vertex-set $V = \{v_1 \ldots v_n\}$ and edge-set $E = \{e_1 \ldots e_m\}$. H is uniquely defined by its $m \times n$ incidence matrix $\boldsymbol{B}$ (0-1 data matrix) such that $b_{ij} = \mathcal{I}(v_j \in e_i)$, where $\mathcal{I}(v_j \in e_i)$ is 1 if the vertex v_j is contained in the hyperedge e_i, and 0, otherwise.

For a fixed integer k ($1 \le k \le n$), we are looking for k-dimensional representatives $\mathbf{r}_1, \ldots, \mathbf{r}_n$ of the vertices and $\mathbf{q}_1, \ldots, \mathbf{q}_m$ of the edges subject to

$$\sum_{i=1}^{n} \mathbf{r}_i \mathbf{r}_i^T = \boldsymbol{I}_k \tag{1.4}$$

so that the following sum of the costs of edges in this representation is minimized:

$$Q_k = \sum_{i=1}^{m} C(e_i), \tag{1.5}$$

where the cost of the edge e_i is

$$C(e_i) = \sum_{j=1}^{n} b_{ij} \|\mathbf{r}_j - \mathbf{q}_i\|^2.$$

The construction of the above objective function forces the representatives of the vertices to be close to those of the hyperedges they are contained in, while the constraint keeps them at a distance. As a compromise, the representatives of the vertices which are together in many hyperedges will be close to each other.

To minimize the objective function (1.5), let $\bar{\mathbf{r}}(e)$ denote the barycenter of the vertex-representatives contained in the hyperedge e:

$$\bar{\mathbf{r}}(e) = \frac{1}{|e|} \sum_{j=1}^{n} \mathcal{I}(v_j \in e)\mathbf{r}_j.$$

Let $\boldsymbol{X}$ and $\boldsymbol{Y}$ denote the $n \times k$ and $m \times k$ matrices containing the vertex- and edge-representatives as row vectors, respectively. Further, let $\boldsymbol{D}_v$ and $\boldsymbol{D}_e$ be the $n \times n$ and $m \times m$ vertex- and edge-valence matrices: they are diagonal matrices, with diagonal entries that are the column- and row-sums of the incidence matrix $\boldsymbol{B}$, respectively. Assume that $\boldsymbol{D}_e$ is not singular (there are no empty hyperedges).

With this notation, $C(e)$ is decreased by means of the Steiner's inequality:

$$C(e) \geq \sum_{j=1}^{n} \mathcal{I}(v_j \in e)\|\mathbf{r}_j - \bar{\mathbf{r}}(e)\|^2, \quad e \in E. \tag{1.6}$$

The right-hand side only depends on the incidence relations $\mathcal{I}(v_j \in e)$ and on the representatives of the vertices comprising the row vectors of the matrix $\boldsymbol{X}$. Denoting the right-hand side of (1.6) by $Q(e, \boldsymbol{X})$, with an easy calculation we get the following formula for it:

$$Q(e, \boldsymbol{X}) = \frac{1}{2|e|} \sum_{i=1}^{n} \sum_{j=1}^{n} \mathcal{I}(v_i \in e)\mathcal{I}(v_j \in e)\|\mathbf{r}_i - \mathbf{r}_j\|^2, \quad e \in E. \tag{1.7}$$

With the notation $Q(\boldsymbol{X}) = \sum_{e \in E} Q(e, \boldsymbol{X})$, the inequality $Q_k \geq Q(\boldsymbol{X})$ holds with any representation $\boldsymbol{X}$ of the vertices. But $Q(\boldsymbol{X})$ can be rewritten as was done with weighted graphs:

$$Q(\boldsymbol{X}) = \sum_{i=1}^{n} \sum_{j=1}^{n} \left[\frac{1}{2} \sum_{e \in E} \mathcal{I}(v_i \in e)\mathcal{I}(v_j \in e)\frac{1}{|e|} \right] \|\mathbf{r}_i - \mathbf{r}_j\|^2 = \frac{1}{2} \sum_{i=1}^{n} \sum_{j=1}^{n} \ell_{ij}\mathbf{r}_i^T \mathbf{r}_j,$$

where

$$\ell_{ij} = \begin{cases} -\mathcal{I}(v_i \in e)\mathcal{I}(v_j \in e)\frac{1}{|e|} & \text{if} \quad i \neq j \\ d_{vi} - \mathcal{I}(v_i \in e)\frac{1}{|e|} & \text{if} \quad i = j, \end{cases} \tag{1.8}$$

where d_{vi} is the ith diagonal entry of the diagonal matrix $\boldsymbol{D}_v$.

It is easy to see that the $n \times n$ matrix of the entries ℓ_{ij} is $\boldsymbol{D}_v - \boldsymbol{B}^T \boldsymbol{D}_e^{-1} \boldsymbol{B}$.

Definition 1.1.4 *The matrix* $\tilde{\boldsymbol{L}} = \boldsymbol{D}_v - \boldsymbol{B}^T \boldsymbol{D}_e^{-1} \boldsymbol{B}$ *is called the Laplacian of the hypergraph* H.

For simple graphs, the above $\tilde{\boldsymbol{L}}$ is one-half of the usual Laplacian $\boldsymbol{L}$, as each edge has valence 2. Further, to any hypergraph $H = (V, E)$ an edge-weighted graph $G = (V, \boldsymbol{W})$ can be assigned such that between their Laplacians the relation $\tilde{\boldsymbol{L}} = \boldsymbol{L}$

holds, in the following manner the edge-weights are:

$$w_{ij} = \begin{cases} \sum_{e \in E} \mathcal{I}(v_i \in e)\mathcal{I}(v_j \in e)\frac{1}{|e|} & \text{if} \quad i \neq j \\ 0 & \text{if} \quad i = j. \end{cases} \tag{1.9}$$

Therefore, $\tilde{\boldsymbol{L}}$ is positive semidefinite, and the multiplicity of 0 as an eigenvalue of H is equal to the number of its connected components. The connected components of a hypergraph are spanned by disjoint subsets of its vertices such that there are no hyperedges containing vertices from more than one component. If vertices of the distinct components are colored with different colors, there are only monocolored hyperedges within the components. H is connected if it has only one connected component. In the sequel, only connected hypergraphs will be considered.

Theorem 1.1.5 (Representation theorem for hypergraphs) *Let $H = (V, E)$ be a connected hypergraph with Laplacian matrix $\tilde{\mathbf{L}}$. Let $0 = \lambda_0 < \lambda_1 \leq \ldots \leq \lambda_{n-1}$ be the eigenvalues of $\tilde{\mathbf{L}}$ with corresponding unit-norm eigenvectors $\mathbf{u}_0, \mathbf{u}_1, \ldots, \mathbf{u}_{n-1}$. Let $k < n$ be a positive integer such that $\lambda_{k-1} < \lambda_k$. Then the minimum of the cost function (1.5) subject to (1.4) is*

$$\sum_{j=0}^{k-1} \lambda_j = \sum_{j=1}^{k-1} \lambda_j$$

and it is attained with the optimum representatives $\mathbf{r}_1^*, \ldots, \mathbf{r}_n^*$ of the vertices, the transposes of which are row vectors of $\boldsymbol{X}^* = (\mathbf{u}_0, \mathbf{u}_1, \ldots, \mathbf{u}_{k-1})$. Further, the optimum representatives of the edges are row vectors of the matrix

$$\boldsymbol{Y}^* = \boldsymbol{D}_e^{-1}\boldsymbol{B}\boldsymbol{X}^*.$$

Proof. It is easy to see that

$$Q(\boldsymbol{X}) = \text{tr}[\boldsymbol{X}^T \tilde{\boldsymbol{L}} \boldsymbol{X}]$$

and it is to be minimized subject to $\boldsymbol{X}^T\boldsymbol{X} = \boldsymbol{I}_k$. We again use the linear algebra fact of A.3.12 and the following easy observation: the relation (1.6) implies that the optimum representative of an edge e is the barycenter of the representatives of its vertices.

From the defining formula (1.8) of the Laplacian, it can easily be seen that the loop-edges ($|e| = 1$) do not give any contribution to it. Akin to weighted graphs, $(k-1)$-dimensional representatives will as well do after eliminating the first (trivial) coordinate.

We first defined the Laplacian of hypergraphs in a binary clustering problem: see Bolla (1989, 1993) and the algorithm in Subsection 5.2.2. We remark that Dhillon (2001) treats a similar problem in the framework of bipartite graphs, also related to contingency tables, see Section 1.2.

1.1.3 Examples for spectra and representation of simple graphs

Here we give some examples for the smallest Laplacian eigenvalues and the representation based on the corresponding eigenvectors in a straightforward dimension of basic simple graphs. Their adjacency matrix $\boldsymbol{A}$ can also be considered as a 0-1 weight matrix of an edge-weighted graph, and they are also hypergraphs with all edge-valences two. The reader can find many examples for adjacency spectra in Biggs (1974); Cvetković *et al.* (1979); Hoffman (1969) and Lovász (1993). In most of our examples the underlying graph is nearly regular, therefore there is an asymptotic relation between their Laplacian and adjacency spectra, at least for large n, the object of Chapter 3.

Note that neither the adjacency nor the Laplacian spectrum is affected by the labeling of the vertices, in other words, isomorphic graphs have the same spectrum. However, the converse is not true: there are graphs with the same spectrum, though they are not isomorphic (e.g., Cvetković *et al.* (1979) discussed such cospectral graphs in details). This is not surprising, since it is the SD (spectrum and eigenvectors or eigenspaces together) which uniquely characterizes a symmetric matrix, and not the spectrum itself. However, Kelmans (1967) found a class of graphs that is characterized by the Laplacian spectrum. For example, it can be shown that almost all trees are cospectral; see also Mohar (1991).

(a) The adjacency matrix $\boldsymbol{A}$ of the *complete graph* K_n on n vertices is comprised of entries $a_{ij} = 1$ for $i \neq j$ and $a_{ii} = 0$ for $i = 1, \dots, n$. Consequently, its corresponding Laplacian is $\boldsymbol{L} = (n-1)\boldsymbol{I}_n - \boldsymbol{A}$ and $\mathrm{tr}(\boldsymbol{L}) = n(n-1)$. For symmetry reasons, the spectrum of $\boldsymbol{L}$ is nothing else but

$$\lambda_0 = 0, \quad \lambda_1 = \cdots = \lambda_{n-1} = \frac{n(n-1)}{n-1} = n.$$

The unique eigendirection corresponding to the eigenvalue zero is designated by the vector $\mathbf{1}$, while the eigenspace corresponding to the multiple eigenvalue n is $\mathbf{1}^{\perp}$, the orthogonal complementary subspace of $\mathbf{1}$ in $\mathbb{R}^n$. Since there is only one positive eigenvalue (with multiplicity $n-1$), here only the trivial one- or the n-dimensional representation of K_n makes sense. The one-dimensional representatives of the vertices are the same points, whereas the n-dimensional representatives form a simplex on n vertices in the $(n-1)$-dimensional hyperplane $\mathbf{1}^{\perp}$ of $\mathbb{R}^n$. These are, in fact, zero- and $(n-1)$-dimensional representations, because of the trivial first eigenvector $\mathbf{1}$.

(b) The adjacency matrix $\boldsymbol{A}$ of the *path graph* P_n on n vertices is a tridiagonal matrix with zero diagonal and all 1's above and below the diagonal entries. Here the vertices are labeled in their natural succession. It is proved in Cvetković *et al.* (1979); Lovász (1993) that the adjacency spectrum of $\boldsymbol{A}$ consists of the numbers

$$2\cos\frac{i\pi}{n+1}, \quad i = 1, \dots, n.$$

The Laplacian spectrum of P_n (see Anderson and Morley (1985); Mohar (1991)) consists of the eigenvalues

$$\lambda_i = 4\sin^2\frac{i\pi}{2n} = 2\left(1-\cos\frac{i\pi}{n}\right), \quad i=0,1,\ldots,n-1.$$

Hence, the smallest positive Laplacian eigenvalue of P_n is $\lambda_1 = 2(1-\cos\frac{\pi}{n})$. For odd n (say, $n = 2\ell+1$), disregarding the trivial dimension, the one-dimensional representatives of the vertices, that is, the coordinates of $\mathbf{u}_1$, are the numbers

$$x_j = \sqrt{\frac{2}{n}}\sin j\frac{\pi}{n}, \quad j=-\ell,\ldots,-1,0,1,\ldots,\ell \tag{1.10}$$

forming a path, where the distances between representatives of neighboring vertices follow the sine rhythm of (1.10).

(c) The 2-dimensional *$m \times n$ grid $G_{m,n}$* is the Cartesian product (in other words, direct sum) of P_m and P_n, hence its adjacency eigenvalues (see Cvetković *et al.* (1979)) are the numbers

$$\alpha_{i,j} = 2\cos\frac{i\pi}{m+1} + 2\cos\frac{j\pi}{n+1}, \quad i=1,\ldots,m; \quad j=1,\ldots,n.$$

With the considerations of Fiedler (1973), a similar result holds for he Laplacian eigenvalues of the Cartesian product of the simple graphs G_1 and G_2: they are equal to all possible sums of eigenvalues of the two factors. Since $G_{m,n}$ is the Cartesian product of P_m and P_n, the Laplacian eigenvalues of the $m \times n$ grid are

$$\lambda_{i,j} = 4\sin^2\frac{i\pi}{2m} + 4\sin^2\frac{j\pi}{2n} = 2\left(1-\cos\frac{i\pi}{m}\right) + 2\left(1-\cos\frac{j\pi}{n}\right),$$
$$i=0,1,\ldots,m-1; \quad j=0,\ldots,n-1.$$

More generally, denote by $Grid_{d,\ell}$ the d-dimensional cubic grid ($d \geq 2$ is an integer) with $n = (2\ell+1)^d$ vertices, where the vertices are characterized by d-tuples of integers $-\ell,\ldots,-1,0,1,\ldots,\ell$ such that two vertices are adjacent if and only if their d-tuples differ in exactly one coordinate. $Grid_{2,\ell} = G_{2\ell+1,2\ell+1}$ and $Grid_{d,\ell}$ is the Cartesian product of d copies of $P_{2\ell+1}$. Using these facts, the adjacency eigenvalues of $Grid_{d,\ell}$ are the numbers

$$2\sum_{j=1}^{d}\cos\frac{i_j\pi}{2\ell+2}, \quad i_1,\ldots,i_d = 1,\ldots,2\ell+1.$$

In Lovász (1993), the adjacency spectrum of the d-dimensional hypercube is also derived via Cartesian products. With similar considerations, the Laplacian eigenvalues of $Grid_{d,\ell}$ are

$$\lambda_{i_1,\dots,i_d} = 4\sum_{j=1}^{d} \sin^2 \frac{i_j \pi}{2(2\ell+1)} = 2\sum_{j=1}^{d}\left(1 - \cos\frac{i_j\pi}{2\ell+1}\right)$$
$$i_1,\dots,i_d = 0,\dots,2\ell. \tag{1.11}$$

The smallest positive Laplacian eigenvalue of the d-dimensional cubic grid on $n = (2\ell+1)^d$ vertices is a λ with all but one subscripts 0, and the non-zero subscript is 1 in (1.11). As there are d choices for the non-zero index, the smallest positive Laplacian eigenvalue is

$$4\sin^2\frac{\pi}{2(2\ell+1)} = 2\left(1 - \cos\frac{\pi}{2\ell+1}\right)$$

with multiplicity d. The d-dimensional representatives of the vertices (after leaving out the trivial dimension) form a grid in a d-dimensional hyperplane of $\mathbb{R}^n$, its center of gravity being the origin, whereas the distances between the representatives of adjacent vertices follow the sine rhythm of (1.10).

(d) Let $K_{n_1,\dots,n_k}$ be the *complete k-partite graph*, where $n = \sum_{i=1}^{k} n_i$ is the number of its vertices. Let $V_1,\dots,V_k$ denote the non-empty, disjoint, independent sets of the vertices (called clusters), where $|V_i| = n_i$ $(i = 1,\dots,k)$.

Proposition 1.1.6 *The Laplacian spectrum of $K_{n_1,\dots,n_k}$ consists of a single 0, the numbers $n - n_i$ with multiplicity $n_i - 1$ $(i = 1,\dots,k)$ and the number n with multiplicity $k-1$. Further, in the $(k-1)$-dimensional representation of the vertices via any orthonormal set of $k-1$ eigenvectors corresponding to the largest eigenvalue n, the representatives of vertices of the same cluster coincide.*

Proof. Choose a labeling of the vertices such that the first n_1 vertices are contained in V_1, the next n_2 ones in V_2, etc. Since the complete multipartite graph is connected, 0 is a single eigenvalue with eigendirection $\mathbf{1}$. The adjacency matrix $\boldsymbol{A}$ of $K_{n_1,\dots,n_k}$ is a symmetric block matrix of $k \times k$ blocks, where the diagonal blocks are all zeros, and the off-diagonal ones have all 1 entries. The diagonal degree-matrix is $\boldsymbol{D} = \boldsymbol{D}_1 \oplus \cdots \oplus \boldsymbol{D}_k$, where $\boldsymbol{D}_i = (n - n_i)\boldsymbol{I}_{n_i}$. The eigenvalue–eigenvector equation for the Laplacian $\boldsymbol{L} = \boldsymbol{D} - \boldsymbol{A}$ of $K_{n_1,\dots,n_k}$ yields the system of equations

$$(n - n_i)x_j - \sum_{\ell \notin V_i} x_\ell = \lambda x_j, \quad j \in V_i;\ i = 1,\dots,k \tag{1.12}$$

where λ is an eigenvalue of $\boldsymbol{L}$ with corresponding eigenvector $\mathbf{x} = (x_1,\dots,x_n)^T$.

It is easy to see that with $\lambda = n$, the system (1.12) simplifies to

$$n_i x_j + \sum_{\ell \notin V_i} x_\ell = 0, \quad j \in V_i,\ i = 1, \ldots, k$$

which is solved by any piecewise constant vector $\mathbf{x}$ on the partition $(V_1, \ldots, V_k)$ such that it is orthogonal to the $\mathbf{1} \in \mathbb{R}^n$ vector at the same time. Indeed, let $x_j = y_i$ for $j \in V_i$ and because of the orthogonality condition, the numbers $y_1, \ldots, y_k$ satisfy $\sum_{i=1}^{k} n_i y_i = 0$. The subspace of $\mathbb{R}^n$ spanned by these $\mathbf{x}$'s is of dimension $k-1$. Therefore, the multiplicity of the eigenvalue n is $k-1$, and the $(k-1)$-dimensional representatives of the vertices – with vector components that form an orthonormal system within this eigenspace – yield k distinct points such that vertices of the same cluster are represented with the same point.

One can also verify that for any $i \in \{1, \ldots, k\}$, substituting $n - n_i$ for λ, the system (1.12) becomes:

$$\sum_{\ell \notin V_i} x_\ell = 0,$$

which is solved by any $\mathbf{x}$ such that $x_j = 0$ whenever $j \notin V_i$. For $j \in V_i$, we select the coordinates of $\mathbf{x}$ such that $\mathbf{x}$ is orthogonal to the $\mathbf{1}$ vector. This condition results in the restriction $\sum_{j \in V_i} x_j = 0$ for the non-zero coordinates of $\mathbf{x}$. This restriction also ensures the orthogonality to the piecewise constant vectors in the eigenspace corresponding to the eigenvalue n. Trivially, the subspace of such $\mathbf{x}$'s is of dimension $n_i - 1$, and hence, the multiplicity of the eigenvalue $n - n_i$ is $n_i - 1$, for $i = 1, \ldots, k$.

(e) Let S_d denote the *star graph* on $n = d + 1$ vertices. In fact, the star graph is a complete 2-partite graph, namely, $S_d = K_{1,d}$. Applying the result of (d), its Laplacian spectrum consists of a single 0, the number $n - d = 1$ (with multiplicity $d - 1$) and the single eigenvalue n. Making a d-dimensional representation, based on the d smallest Laplacian eigenvalues, the representatives of the d endpoints form the vertices of a simplex in the $(d-1)$-dimensional eigenspace corresponding to the eigenvalue 1, but they are not connected to each other, they are merely connected to the single vertex of degree d, the representative of which is the origin (the center of gravity of the simplex). This kind of representation is reminiscent of graph drawing and imitates the physical picture of the graph. However, a 1-dimensional representation is also possible based on the eigenvector corresponding to the largest single eigenvalue. In this structure-revealing representation, the representatives form two points on the real number line: one corresponds to the endpoints, and the other to the middle point.

(f) Let $S_{d,\ell}$ denote the subdivision graph of S_d, where each of the edges of S_d is divided into ℓ parts. We call $S_{d,\ell}$ a *spider graph* with d feet of ℓ sections. The number of its vertices is $n = d\ell + 1$. The Laplacian spectrum inherits features of that of the path and the star. Namely, the smallest positive Laplacian eigenvalue of $S_{d,\ell}$ is of multiplicity $d-1$ and is equal to $1-\cos\frac{\pi}{2\ell+1}$. The optimal d-dimensional (in fact, $(d-1)$-dimensional) representation of the spider is that of (e), where the feet of the spider are divided according to the sine rhythm of (1.10).

We remark that the star graph is the only tree with smallest positive Laplacian eigenvalue 1. Indeed, Maas (1987) and Merris (1987) independently proved that for a tree T, $\lambda_1(T) \le 1$, with equality if and only if T is a star.

Proposition 1.1.6 illustrates that in the case of empty clusters, the representation, based on eigenvectors corresponding to eigenvalues in the top of the Laplacian spectrum, is able to reveal this so-called anti-community cluster structure. This is also true for clusters (subsets) of vertices with sparse intra- and dense inter-cluster edge-densities (we will define these notions more precisely in Chapter 2). For example, hub authorities are of this type. Another example is a game of strategic substitutes, where an increase in other players' actions leads to relatively lower payoffs under higher actions of a given player. In strategic interaction games the agents are vertices of a graph and only agents connected with an edge influence each other's actions, see Ballester *et al.* (2006); Jackson and Zenou (forthcoming).

Analogously to the derivation of the Representation Theorem (Theorem 1.1.2), now the maximum of the quadratic form $Q_k = \mathbf{X}^T \mathbf{L} \mathbf{X}$ subject to $\mathbf{X}^T \mathbf{X} = \mathbf{I}_k$ is looked for, and hence, representatives of vertices connected with few, low-weight edges are stressed to be close to each other. As an easy consequence of Theorem A.3.12 in the Appendix, the maximum is the sum of the k largest Laplacian eigenvalues and it is attained with an $\mathbf{X}^*$ containing the corresponding unit-norm eigenvectors in its columns.

In Chapter 3 we will study more general cluster structures such that clusters are not necessarily dense or sparse subsets of the network, but rather homogeneous ones, as far as the intra- and inter-cluster relations of the vertices are concerned. In other words, vertices of the same cluster behave similarly with respect to each other and to vertices of any other cluster, like synopses of the brain. These types of clusters can be recovered via representation based on eigenvectors corresponding to some structural eigenvalues of the normalized modularity matrix to be introduced in Section 1.3. The so-called structure-revealing representation of Example (e) is also of this flavor. Bottom Laplacian eigenvalues and eigenvectors are mainly advisable for graph drawing purposes; for a detailed description see Liotta (2004). Even in the case of Laplacian spectra of chemical graphs we may select not necessarily consecutive eigenvalues together with eigenvectors for spatial representation of molecules, see Pisanski and Shawe-Taylor (2000) for details.

1.2 SVD of contingency tables and correspondence matrices

Now, more generally, our underlying objects will be contingency tables, that is, rectangular arrays with nonnegative, real entries. For example, keyword–document matrices or microarrays are such. In microarrays, rows correspond to genes and columns to different conditions, while the corresponding entries are expression levels of genes under specific conditions (a 0-1 matrix is a special case of it). Let $\boldsymbol{C}$ be a contingency table on row set $Row = \{1, \dots, m\}$, column set $Col = \{1, \dots, n\}$, where $\boldsymbol{C}$ is a $m \times n$ rectangular matrix of nonnegative real entries c_{ij}'s. Without loss of generality, we can assume that there are no identically zero rows or columns (otherwise they can be omitted). Here c_{ij} is some kind of association between the objects representing row i and column j, where 0 means no interaction at all. Usually, the entries of $\boldsymbol{C}$ are normalized, either with a uniform bound, say 1 (like probabilities), or the sum of the entries is 1 (reminiscent of a joint distribution). This normalization will have importance in Section 1.4; here it has no relevance, since the correspondence matrix to be introduced is invariant under scaling the entries of $\boldsymbol{C}$. Let the row-sums of $\boldsymbol{C}$ be

$$d_{row,i} = \sum_{j=1}^{n} c_{ij}, \quad i = 1, \dots, m \tag{1.13}$$

and the column-sums

$$d_{col,j} = \sum_{i=1}^{m} c_{ij}, \quad j = 1, \dots, n \tag{1.14}$$

which are collected in the main diagonal of the $m \times m$ diagonal matrix $\boldsymbol{D}_{row} = \text{diag}(d_{row,1}, \dots, d_{row,m})$ and that of the $n \times n$ diagonal matrix $\boldsymbol{D}_{col} = \text{diag}(d_{col,1}, \dots, d_{col,n})$, respectively.

For a given integer $1 \le k \le \min\{m, n\}$, we are looking for k-dimensional representatives $\mathbf{r}_1, \dots, \mathbf{r}_m \in \mathbb{R}^k$ of the rows and $\mathbf{q}_1, \dots, \mathbf{q}_n \in \mathbb{R}^k$ of the columns such that they minimize the objective function

$$Q_k = \sum_{i=1}^{m} \sum_{j=1}^{n} c_{ij} \|\mathbf{r}_i - \mathbf{q}_j\|^2 \tag{1.15}$$

subject to

$$\sum_{i=1}^{m} d_{row,i} \mathbf{r}_i \mathbf{r}_i^T = \boldsymbol{I}_k \quad and \quad \sum_{j=1}^{n} d_{col,j} \mathbf{q}_j \mathbf{q}_j^T = \boldsymbol{I}_k. \tag{1.16}$$

When minimized, the objective function Q_k favors k-dimensional placement of the rows and columns such that representatives of columns and rows with large association are forced to be close to each other. As we will see, this is equivalent to the problem of correspondence analysis: see Appendix C.3.

Let us put both the objective function and the constraints in a more favorable form. Let $\boldsymbol{X}$ be the $m \times k$ matrix of rows $\mathbf{r}_1^T, \dots, \mathbf{r}_m^T$, and $\mathbf{x}_1, \dots, \mathbf{x}_k \in \mathbb{R}^n$ denote the columns of $\boldsymbol{X}$, for which fact we use the notation $\boldsymbol{X} = (\mathbf{x}_1, \dots, \mathbf{x}_k)$. Because of the constraint (1.16), the vectors $\boldsymbol{D}_{row}^{-1/2} \mathbf{x}_i$ $(i = 1, \dots, k)$ form an orthonormal system, hence, $\boldsymbol{D}_{row}^{-1/2} \boldsymbol{X}$ is a suborthogonal matrix. Therefore, the first part of the constraint can be formulated as $\boldsymbol{X}^T \boldsymbol{D}_{row} \boldsymbol{X} = \boldsymbol{I}_k$. Likewise, let $\boldsymbol{Y}$ be the $n \times k$ matrix of rows $\mathbf{q}_1^T, \dots, \mathbf{q}_n^T$, and $\mathbf{y}_1, \dots, \mathbf{y}_k \in \mathbb{R}^n$ denote the columns of $\boldsymbol{Y}$, that is, $\boldsymbol{Y} = (\mathbf{y}_1, \dots, \mathbf{y}_k)$. Hence, the second part of the constraint (1.16) can be formulated as $\boldsymbol{Y}^T \boldsymbol{D}_{col} \boldsymbol{Y} = \boldsymbol{I}_k$ and the matrix $\boldsymbol{D}_{col}^{-1/2} \boldsymbol{Y}$ is also suborthogonal.

With this notation, the objective function (1.15) is rewritten as

$$
\begin{aligned}
Q_k &= \sum_{i=1}^{m} d_{row,i} \|\mathbf{r}_i\|^2 + \sum_{j=1}^{n} d_{col,j} \|\mathbf{q}_j\|^2 - \sum_{i=1}^{m} \sum_{j=1}^{n} c_{ij} \mathbf{r}_i^T \mathbf{q}_j \\
&= \sum_{\ell=1}^{k} \mathbf{x}_\ell^T \boldsymbol{D}_{row} \mathbf{x}_\ell + \sum_{\ell=1}^{k} \mathbf{y}_\ell^T \boldsymbol{D}_{col} \mathbf{y}_\ell - \sum_{\ell=1}^{k} \mathbf{x}_\ell^T \boldsymbol{C} \mathbf{y}_\ell \\
&= \operatorname{tr}(\boldsymbol{X}^T \boldsymbol{D}_{row} \boldsymbol{X}) + \operatorname{tr}(\boldsymbol{Y}^T \boldsymbol{D}_{col} \boldsymbol{Y}) - \operatorname{tr}(\boldsymbol{X}^T \boldsymbol{C} \boldsymbol{Y}) \\
&= 2k - \operatorname{tr}(\boldsymbol{X}^T \boldsymbol{C} \boldsymbol{Y}) = 2k - \operatorname{tr}\left[\left(\boldsymbol{D}_{row}^{1/2} \boldsymbol{X} \right)^T \left(\boldsymbol{D}_{row}^{-1/2} \boldsymbol{C} \boldsymbol{D}_{col}^{-1/2} \right) \left(\boldsymbol{D}_{col}^{1/2} \boldsymbol{Y} \right) \right]
\end{aligned}
$$

where the matrix $\boldsymbol{C}_{corr} = \boldsymbol{D}_{row}^{-1/2} \boldsymbol{C} \boldsymbol{D}_{col}^{-1/2}$ is introduced in Appendix C.3 and Bolla *et al.* (2010) as the *normalized contingency table* or *correspondence matrix* belonging to the contingency table $\boldsymbol{C}$.

The correspondence matrix has singular value decomposition, briefly SVD (see Appendix (A.12))

$$
\boldsymbol{C}_{corr} = \sum_{k=0}^{r-1} s_k \mathbf{v}_k \mathbf{u}_k^T, \tag{1.17}
$$

where $r \leq \min\{n, m\}$ is the rank of $\boldsymbol{C}_{corr}$, or equivalently (as there are no identically zero rows or columns), the rank of $\boldsymbol{C}$. Here $1 = s_0 \geq s_1 \geq \cdots \geq s_{r-1} > 0$ are the non-zero singular values of $\boldsymbol{C}_{corr}$. They cannot exceed 1, since they are correlations (see Appendix C.3). Furthermore, 1 is a single singular value if $\boldsymbol{C}_{corr}$ (or equivalently, $\boldsymbol{C}$) is non-decomposable; see Definition A.3.28 of the Appendix. In this case $\mathbf{v}_0 = (\sqrt{d_{row,1}}, \dots, \sqrt{d_{row,m}})^T$ and $\mathbf{u}_0 = (\sqrt{d_{col,1}}, \dots, \sqrt{d_{col,n}})^T$ is the singular vector pair corresponding to $s_0 = 1$.

Note that the singular spectrum of a decomposable contingency table can be composed from the singular spectra of its non-decomposable parts, as well as their singular vector pairs. Therefore, in the future, the non-decomposability of the underlying contingency table will be assumed.

Theorem 1.2.1 (Representation theorem for contingency tables) *Let* $\boldsymbol{C}$ *be a non-decomposable contingency table with correspondence matrix* $\boldsymbol{C}_{corr}$. *Let* $1 = s_0 > s_1 \geq \cdots \geq s_{r-1}$ *be the positive singular values of* $\boldsymbol{C}_{corr}$ *with unit-norm singular vector pairs* $\mathbf{v}_i, \mathbf{u}_i$ $(i = 0, \dots, r-1)$, *and* $k \leq r$ *be a positive integer such that* $s_{k-1} > s_k$. *Then the minimum of (1.15) subject to (1.16) is* $2k - \sum_{i=0}^{k-1} s_i$ *and it is attained with the optimum row representatives* $\mathbf{r}_1^*, \dots, \mathbf{r}_m^*$ *and column representatives* $\mathbf{q}_1^*, \dots, \mathbf{q}_n^*$ *the transposes of which are row vectors of the matrices* $\boldsymbol{X}^* = \boldsymbol{D}_{row}^{-1/2}(\mathbf{v}_0, \mathbf{v}_1, \dots, \mathbf{v}_{k-1})$ *and* $\boldsymbol{Y}^* = \boldsymbol{D}_{col}^{-1/2}(\mathbf{u}_0, \mathbf{u}_1, \dots, \mathbf{u}_{k-1})$, *respectively.*

Proof. In fact, we have to maximize

$$\mathrm{tr}\left[\left(\boldsymbol{D}_{row}^{1/2}\boldsymbol{X}\right)^T \boldsymbol{C}_{corr}\left(\boldsymbol{D}_{col}^{1/2}\boldsymbol{Y}\right)\right]$$

under the constraints that $\boldsymbol{D}_{row}^{1/2}\boldsymbol{X}$ and $\boldsymbol{D}_{col}^{1/2}\boldsymbol{Y}$ are suborthogonal matrices. Hence, Proposition A.12 of the Appendix is applicable.

Definition 1.2.2 *The vectors* $\mathbf{r}_1^*, \ldots, \mathbf{r}_n^*$ *and* $\mathbf{q}_1^*, \ldots, \mathbf{q}_m^*$ *giving the optimum in Theorem 1.2.1 are called optimum k-dimensional representatives of the rows and columns, while the transformed singular vectors* $\boldsymbol{D}_{row}^{-1/2}\mathbf{v}_0, \ldots, \boldsymbol{D}_{row}^{-1/2}\mathbf{v}_{k-1}$ *and* $\boldsymbol{D}_{col}^{-1/2}\mathbf{u}_0, \ldots, \boldsymbol{D}_{col}^{-1/2}\mathbf{u}_{k-1}$ *are called vector components of the contingency table, taking part in the k-dimensional representation of its rows and columns.*

We remark the following.

- Provided 1 is a single singular value (or equivalently, $\boldsymbol{C}$ is non-decomposable), the first columns of the matrices $\boldsymbol{X}^*$ and $\boldsymbol{Y}^*$ are $\boldsymbol{D}_{row}^{-1/2}\mathbf{v}_0$ and $\boldsymbol{D}_{col}^{-1/2}\mathbf{u}_0$, that is, the constantly $\mathbf{1}$ vectors of $\mathbb{R}^m$ and $\mathbb{R}^n$, respectively. Therefore they do not contribute significantly to the separation of the representatives, and the k-dimensional representatives are in a $(k-1)$-dimensional hyperplane of $\mathbb{R}^m$ and $\mathbb{R}^n$, respectively.

- Note that the dimension k does not play an important role here; the vector components can be included successively up to a k such that $s_{k-1} > s_k$. We remark that the singular vectors can arbitrarily be chosen in the isotropic subspaces corresponding to possible multiple singular values, under the orthogonality conditions.

- As for the joint distribution view (when the rows and columns belong to the categories of two categorical variables, see correspondence analysis in Section C.3), this representation has the following optimum properties: the closeness of categories of the same variable reflects the similarity between them, while the closeness of categories of the two different variables reflects their frequent simultaneous occurrence. For example, $\boldsymbol{C}$ being a microarray, the representatives of similar function genes as well as representatives of similar conditions are close to each other; likewise, representatives of genes that are responsible for a given condition are close to the representatives of those conditions.

- One frequently studied example of a rectangular array is the keyword–document matrix. Here the entries are associations between documents and words. Based on network data, the entry in the ith row and jth column is the relative frequency of word j in document i. Latent semantic indexing looks for real scores of the documents and keywords such that the score of any document be proportional to the total scores of the keywords occurring in it, and vice versa, the score of any keyword being proportional to the total scores of the documents containing it. Not surprisingly, the solution is given by the SVD of the contingency table, where the document- and keyword-scores are the coordinates of the left and right singular vectors corresponding to its largest non-trivial singular value which gives the constant of proportionality. This idea is generalized in Frieze *et al.* (1998) in the following way. We can think of the above relation between keywords and documents as the relation with respect to the most important topic (or context, or factor). After this, we are looking for another scoring with respect to the second topic, up to k (where k is a positive integer not exceeding the rank of the table). The solution is given by the singular vector pairs corresponding to the k largest singular values of the table. The problem is also related to the Pagerank, see for example Kleinberg (1997).

- In another view, a contingency table can be considered as part of the weight matrix of a bipartite graph on vertex set $Row \cup Col$. However, it would be hard to always distinguish between these two types of vertices, we rather use the framework of correspondence analysis, and formulate our statements in terms of rows and columns.

1.3 Normalized Laplacian and modularity spectra

Let $G = (V, \boldsymbol{W}, \boldsymbol{S})$ be a weighted graph on the vertex-set V ($|V| = n$), where both the edges and vertices have nonnegative weights. The edge-weights are entries of $\boldsymbol{W}$ as in Section 1.1, whereas the diagonal matrix $\boldsymbol{S} = \mathrm{diag}(s_1, \dots, s_n)$ contains the positive vertex-weights in its main diagonal. Without loss of generality, we can assume that the entries in $\boldsymbol{W}$ and $\boldsymbol{S}$ both sum to 1. For the time being, the vertex-weights have nothing to do with the edge-weights. These individual weights are assigned to the vertices subjectively. For example, in a social network, the edge-weights are similarities between the vertices based on the strengths of their pairwise connections (like frequency of co-starring of actors), while vertex-weights embody the individual strengths of the vertices in the network (like the actors' individual abilities). We will further motivate this idea in Chapter 4.

Now, we look for k-dimensional representatives $\mathbf{r}_1, \dots, \mathbf{r}_n$ of the vertices so that they minimize the objective function (1.1), that is, $Q_k = \sum_{i<j} w_{ij} \|\mathbf{r}_i - \mathbf{r}_j\|^2$ subject to

$$\sum_{i=1}^{n} s_i \mathbf{r}_i \mathbf{r}_i^T = \boldsymbol{I}_k .$$

With the notation and considerations of Section 1.1,

$$\begin{aligned}
\min_{\sum_{i=1}^{n} s_i \mathbf{r}_i \mathbf{r}_i^T = \boldsymbol{I}_k} Q_k &= \min_{\boldsymbol{X}^T \boldsymbol{S} \boldsymbol{X} = \boldsymbol{I}_k} \mathrm{tr}(\boldsymbol{X}^T \boldsymbol{L} \boldsymbol{X}) \\
&= \min_{\boldsymbol{X}^T \boldsymbol{S} \boldsymbol{X} = \boldsymbol{I}_k} \mathrm{tr}[(\boldsymbol{S}^{1/2}\boldsymbol{X})^T (\boldsymbol{S}^{-1/2} \boldsymbol{L} \boldsymbol{S}^{-1/2})(\boldsymbol{S}^{1/2}\boldsymbol{X})] \\
&= \sum_{i=0}^{k-1} \lambda_i(\boldsymbol{L}_S) = \sum_{i=1}^{k-1} \lambda_i(\boldsymbol{L}_S)
\end{aligned}$$

where $\boldsymbol{L}_S = \boldsymbol{S}^{-1/2}\boldsymbol{L}\boldsymbol{S}^{-1/2}$ is the Laplacian normalized by $\boldsymbol{S}$, and because of the constraints, $\boldsymbol{S}^{1/2}\boldsymbol{X}$ is a suborthogonal matrix. Obviously, $\boldsymbol{L}_S$ is also positive semidefinite with eigenvalues $0 = \lambda_0(\boldsymbol{L}_S) \leq \lambda_1(\boldsymbol{L}_S) \leq \cdots \leq \lambda_{n-1}(\boldsymbol{L}_S)$ and corresponding orthonormal eigenvectors $\mathbf{u}_0, \mathbf{u}_1, \ldots, \mathbf{u}_{n-1}$. Furthermore, 0 is a single eigenvalue if and only if G is connected. The optimum k-dimensional representation is obtained by the row vectors of the matrix $\boldsymbol{S}^{-1/2}(\mathbf{u}_0, \mathbf{u}_1, \ldots, \mathbf{u}_{k-1})$.

The special case, when the vertex-weights are the generalized degrees, that is $\boldsymbol{S} = \boldsymbol{D}$, has a distinguished importance.

Definition 1.3.1 *The matrix*

$$\boldsymbol{L}_D = \boldsymbol{D}^{-1/2}\boldsymbol{L}\boldsymbol{D}^{-1/2} = \boldsymbol{I}_n - \boldsymbol{D}^{-1/2}\boldsymbol{W}\boldsymbol{D}^{-1/2}$$

is called the normalized Laplacian of the edge-weighted graph $G = (V, \boldsymbol{W})$.

Remark 1.3.2 *Now, we enumerate some simple statements concerning the normalized Laplacian spectrum.*

(i) Since the matrix $\boldsymbol{D}^{-1/2}\boldsymbol{W}\boldsymbol{D}^{-1/2}$ is the correspondence matrix belonging to the symmetric contingency table $\boldsymbol{W}$, the singular values of this matrix are in the $[0, 1]$ interval: they are special correlations, the largest one being 1 (see Section 1.2 and Appendix C.3). Consequently, the eigenvalues of $\boldsymbol{D}^{-1/2}\boldsymbol{W}\boldsymbol{D}^{-1/2}$ are in the $[-1, 1]$ interval, while those of $\boldsymbol{I}_n - \boldsymbol{D}^{-1/2}\boldsymbol{W}\boldsymbol{D}^{-1/2}$ are in the $[0, 2]$ interval. Let

$$0 = \lambda_0 \leq \lambda_1 \leq \cdots \leq \lambda_{n-1} \leq 2$$

denote the spectrum of the normalized Laplacian $\boldsymbol{L}_D$.

(ii) Trivially, 0 is a single eigenvalue of $\boldsymbol{L}_D$ if and only if G is connected (i.e., $\boldsymbol{W}$ is irreducible), and in this case, the corresponding unit-norm eigenvector is the $\sqrt{\mathbf{d}} = (\sqrt{d_1}, \ldots, \sqrt{d_n})^T$ vector. Furthermore, the normalized Laplacian spectrum of a disconnected graph is the union of those of its connected components.

(iii) Since $\sum_{i=0}^{n-1} \lambda_i = \mathrm{tr}(\boldsymbol{L}_D) = n$, the following estimations for the smallest and largest positive normalized Laplacian eigenvalues of the connected edge-weighted graph $G = (V, \boldsymbol{W})$ on n vertices hold:

$$\lambda_1 = \min_{i \in \{1,\ldots,n-1\}} \lambda_i \leq \frac{1}{n-1}\sum_{i=1}^{n-1} \lambda_i = \frac{n}{n-1} \leq \max_{i \in \{1,\ldots,n-1\}} \lambda_i = \lambda_{n-1}.$$

Note that both of the above inequalities hold with equality at the same time, if and only if G is the complete graph (see the forthcoming Example (A)).

(iv) For a simple graph G, which is not the complete graph, $\lambda_1 \leq 1$ holds. For the proof see Chung (1997).

(v) *Provided G is connected, 2 is an eigenvalue if and only if G is a bipartite graph (i.e., its vertices can be divided into two parts such that there are no edges within these two vertex-subsets, or equivalently, after permuting its rows and columns in the same way,* $\mathbf{W}$ *contains two zero diagonal blocks). The proof for simple graphs is found in Chung (1997), and for general edge-weighted graphs follows from the following considerations. The bipartedness of a connected G is equivalent to the fact that its weight matrix* $\mathbf{W}$ *is irreducible, but decomposable (see Definition A.3.28 of the Appendix). With the arguments of (i), this property extends to the correspondence matrix, which has therefore a multiple singular value 1. Since 1 is a single eigenvalue of* $\mathbf{D}^{-1/2}\mathbf{W}\mathbf{D}^{-1/2}$ *(thanks to G connected), it must have the* -1 *as an eigenvalue, which results in the eigenvalue 2 of* $\mathbf{L}_D$.

Next, we enlist the normalized Laplacian spectra of some simple graphs. They are obtained via similar considerations used in Section 1.1.3. Especially, for regular graphs, the Laplacian and normalized Laplacian eigenvalues are constant multiples of each other.

(A) Since the complete graph K_n is $(n-1)$-regular, its normalized Laplacian eigenvalues are the $\frac{1}{n-1}$ multiples of the Laplacian ones with the same eigenvectors. Therefore,

$$\lambda_0 = 0,\ \lambda_1 = \cdots = \lambda_{n-1} = \frac{n}{n-1}.$$

(B) For the path graph P_n, the normalized Laplacian eigenvalues are

$$\lambda_i = 1 - \cos\frac{i\pi}{n-1}, \qquad i = 0, 1, \ldots, n-1.$$

Note that the largest eigenvalue is 2, since P_n is bipartite. Indeed, let us label the vertices in their natural succession. Then the vertices of odd and even labels constitute the two independent vertex subsets of the bipartite graph. We also remark that for large n, P_n is almost regular, in other words, its degree-matrix is close to $2I_n$. Therefore the normalized Laplacian eigenvalues of P_n are asymptotically $\frac{1}{2}$ multiples of the Laplacian ones; see Example (b) of Section 1.1.3.

(C) For the d-dimensional hypercube Q_d on 2^d vertices, based on the adjacency spectrum (derived in Lovász (1993)), the normalized Laplacian eigenvalues are the numbers $\frac{2i}{d}$ with multiplicity $\binom{d}{i}$, $i = 0, 1, \ldots, d$. Hence, 2 is a single eigenvalue of Q_d, which fact is not surprising, since Q_d is bipartite again.

(D) The normalized Laplacian eigenvalues of the complete bipartite graph K_{n_1,n_2} are

$$\lambda_0 = 0,\ \lambda_1 = \cdots = \lambda_{n_1+n_2-2} = 1,\ \lambda_{n_1+n_2-1} = 2.$$

(E) Especially, the star graph S_d on $d+1$ vertices is the $K_{1,d}$ graph, therefore its eigenvalues are

$$\lambda_0 = 0,\ \lambda_1 = \cdots = \lambda_{d-1} = 1,\ \lambda_d = 2.$$

Normalized Laplacian was used for spectral clustering in several papers (see, e.g., Azran and Ghahramani (2006); Bolla and Tusnády (1990, 1994)), the idea of which will be summarized in Chapter 2. Those results will be based on the observation that the spectral decomposition (briefly, SD) of $\boldsymbol{L}_D$ solves the following quadratic placement problem.

Theorem 1.3.3 (Representation theorem for edge- and vertex-weighted graphs) *Let $G = (V, \boldsymbol{W})$ be a connected edge-weighted graph with normalized Laplacian $\boldsymbol{L}_D$. Let $0 = \lambda_0 < \lambda_1 \le \cdots \le \lambda_{n-1}$ be the eigenvalues of $\boldsymbol{L}_D$ with corresponding unit-norm eigenvectors $\mathbf{u}_0, \mathbf{u}_1, \ldots, \mathbf{u}_{n-1}$. Let $k < n$ be a positive integer such that $\lambda_{k-1} < \lambda_k$. Then the minimum of Q_{k-1} of (1.1) subject to*

$$\sum_{i=1}^{n} d_i \mathbf{r}_i \mathbf{r}_i^T = \boldsymbol{I}_{k-1} \quad \text{and} \quad \sum_{i=1}^{n} d_i \mathbf{r}_i = \mathbf{0}$$

is $\sum_{i=1}^{k-1} \lambda_i$ and it is attained with the optimum $(k-1)$-dimensional representatives $\mathbf{r}_1^*, \ldots, \mathbf{r}_n^*$, the transposes of which are row vectors of $\boldsymbol{X}^* = \boldsymbol{D}^{-1/2}(\mathbf{u}_1, \ldots, \mathbf{u}_{k-1})$.

Proof. Observe that instead of $\boldsymbol{X}$, the *augmented* $n \times k$ matrix $\tilde{\boldsymbol{X}}$ can also be used, which is obtained from $\boldsymbol{X}$ by inserting the column $\mathbf{x}_0 = \mathbf{1}$ of all 1's. In fact, $\mathbf{x}_0 = \boldsymbol{D}^{-1/2}\mathbf{u}_0$, where $\mathbf{u}_0 = \sqrt{\mathbf{d}}$ is the eigenvector corresponding to the eigenvalue 0 of $\boldsymbol{L}_D$. Then

$$\begin{aligned}
\min_{\substack{\sum_{i=1}^n d_i \mathbf{r}_i \mathbf{r}_i^T = \boldsymbol{I}_{k-1} \\ \sum_{i=1}^n d_i \mathbf{r}_i = \mathbf{0}}} Q_{k-1} &= \min_{\substack{\boldsymbol{X}^T \boldsymbol{D} \boldsymbol{X} = \boldsymbol{I}_{k-1} \\ \boldsymbol{X}^T \boldsymbol{D} \mathbf{1} = \mathbf{0}}} \operatorname{tr}[(\boldsymbol{D}^{1/2}\boldsymbol{X})^T \boldsymbol{L}_D (\boldsymbol{D}^{1/2}\boldsymbol{X})] \\
&= \min_{\tilde{\boldsymbol{X}}^T \boldsymbol{D} \tilde{\boldsymbol{X}} = \boldsymbol{I}_k} \operatorname{tr}[(\boldsymbol{D}^{1/2}\tilde{\boldsymbol{X}})^T \boldsymbol{L}_D (\boldsymbol{D}^{1/2}\tilde{\boldsymbol{X}})] = \sum_{i=1}^{k-1} \lambda_i. \qquad (1.18)
\end{aligned}$$

Here we used

$$\begin{aligned}
\operatorname{tr}[(\boldsymbol{D}^{1/2}\tilde{\boldsymbol{X}})^T \boldsymbol{L}_D (\boldsymbol{D}^{1/2}\tilde{\boldsymbol{X}})] &= \sum_{\ell=0}^{k-1} (\boldsymbol{D}^{1/2}\mathbf{x}_\ell)^T \boldsymbol{L}_D (\boldsymbol{D}^{1/2}\mathbf{x}_\ell) \\
&= \sum_{\ell=1}^{k-1} (\boldsymbol{D}^{1/2}\mathbf{x}_\ell)^T \boldsymbol{L}_D (\boldsymbol{D}^{1/2}\mathbf{x}_\ell)
\end{aligned}$$

because of the relation $\boldsymbol{L}_D(\boldsymbol{D}^{1/2}\mathbf{x}_0) = \boldsymbol{L}_D\sqrt{\mathbf{d}} = \mathbf{0}$, since $\sqrt{\mathbf{d}} = \mathbf{u}_0$ is the unit-norm eigenvector of $\boldsymbol{L}_D$ corresponding to the eigenvalue 0. It is important that G is connected and $\boldsymbol{W}$ is normalized such that $\sum_{i=1}^{n} d_i = 1$ which will be assumed in the sequel.

The *modularity matrix* $\boldsymbol{M}$ was defined by Newman and Girvan (2004) and Newman (2004) for simple graphs, and naturally extends to edge-weighted graphs with the normalization $\sum_{i=1}^{n} d_i = 1$ (see Bolla (2011)) as

$$\boldsymbol{M} = \boldsymbol{W} - \mathbf{d}\mathbf{d}^T \tag{1.19}$$

which is the negative of the so-called Q-Laplacian introduced in White and Smyth (2005). It is easy to see that 0 is always an eigenvalue of $\boldsymbol{M}$ with corresponding eigendirection $\mathbf{1}$. However, it is not true that the modularity spectrum of a disconnected graph is the union of modularity spectra of its components, and the above minima are not related immediately to the eigenvalues of this modularity matrix. In case of simple graphs, $\boldsymbol{M}$ is usually indefinite, and it is negative semidefinite for complete or complete multipartite graphs. It is conjectured that the complete and complete multipartite graphs are the only ones for which the largest modularity eigenvalue is 0 (this conjecture is supported by the intensive search of D. Stevanovic).

In Bolla (2011) we introduced the following normalized version of the modularity matrix.

Definition 1.3.4 *Let $G = (V, \boldsymbol{W})$ be an edge-weighted graph with the entries of $\boldsymbol{W}$ summing up to 1. The matrix*

$$\boldsymbol{M}_D = \boldsymbol{D}^{-1/2}\boldsymbol{M}\boldsymbol{D}^{-1/2} = \boldsymbol{D}^{-1/2}\boldsymbol{W}\boldsymbol{D}^{-1/2} - \sqrt{\mathbf{d}}\sqrt{\mathbf{d}}^T \tag{1.20}$$

is called normalized modularity matrix of G.

As we have established, the eigenvalues of $\boldsymbol{D}^{-1/2}\boldsymbol{W}\boldsymbol{D}^{-1/2}$ are in the $[-1, 1]$ interval; the largest eigenvalue is always 1 with corresponding unit-norm eigenvector $\sqrt{\mathbf{d}}$. The only non-zero eigenvalue of the rank 1 term $\sqrt{\mathbf{d}}\sqrt{\mathbf{d}}^T$ is also 1 with the same eigenvector. Therefore, the spectrum of the matrix $\boldsymbol{M}_D$ is the same as the spectrum of $\boldsymbol{D}^{-1/2}\boldsymbol{W}\boldsymbol{D}^{-1/2}$, with the only exception that (due to the subtraction of the term $\sqrt{\mathbf{d}}\sqrt{\mathbf{d}}^T$) the eigenvalue 1 of $\boldsymbol{D}^{-1/2}\boldsymbol{W}\boldsymbol{D}^{-1/2}$ becomes an eigenvalue 0 of $\boldsymbol{M}_D$ with eigenvector $\sqrt{\mathbf{d}}$. Hence, the spectrum of $\boldsymbol{M}_D$ is in $[-1, 1]$ and includes the 0.

These considerations give an exact relation between the normalized Laplacian and modularity matrix: $\boldsymbol{M}_D = \boldsymbol{I} - \boldsymbol{L}_D - \sqrt{\mathbf{d}}\sqrt{\mathbf{d}}^T$. If the eigenvalues of $\boldsymbol{L}_D$ are $0 = \lambda_0 \leq \lambda_1 \cdots \leq \lambda_{n-1} \leq 2$, then the spectrum of $\boldsymbol{M}_D$ consists of the numbers $1 - \lambda_i$ $(i = 1, \ldots, n-1)$ and the zero with corresponding eigenvector $\sqrt{\mathbf{d}}$. Further, the multiplicity of 0 is one more than the multiplicity of the eigenvalue 1 of $\boldsymbol{L}_D$. The multiplicity of 1 is one less than multiplicity of the eigenvalue 0 of $\boldsymbol{L}_D$; hence, 1 cannot be an eigenvalue of $\boldsymbol{M}_D$ if G is connected.

In terms of the normalized modularity matrix, the minimization problem (1.18) can be formulated as a maximization task in the following way.

$$\max_{\boldsymbol{X}^T \boldsymbol{D}\boldsymbol{X}=\boldsymbol{I}_{k-1}} \operatorname{tr}[(\boldsymbol{D}^{1/2}\boldsymbol{X})^T \boldsymbol{M}_D(\boldsymbol{D}^{1/2}\boldsymbol{X})]$$
$$= \max_{\substack{\boldsymbol{X}^T \boldsymbol{D}\boldsymbol{X}=\boldsymbol{I}_{k-1}\\ \boldsymbol{X}^T \boldsymbol{D}\,\mathbf{1}=0}} \operatorname{tr}[(\boldsymbol{D}^{1/2}\boldsymbol{X})^T (\boldsymbol{D}^{-1/2}\boldsymbol{W}\boldsymbol{D}^{-1/2})(\boldsymbol{D}^{1/2}\boldsymbol{X})]$$
$$= k-1- \min_{\substack{\boldsymbol{X}^T \boldsymbol{D}\boldsymbol{X}=\boldsymbol{I}_{k-1}\\ \boldsymbol{X}^T \boldsymbol{D}\,\mathbf{1}=0}} \operatorname{tr}[(\boldsymbol{D}^{1/2}\boldsymbol{X})^T (\boldsymbol{I}_n-\boldsymbol{D}^{-1/2}\boldsymbol{W}\boldsymbol{D}^{-1/2})(\boldsymbol{D}^{1/2}\boldsymbol{X})]$$
$$= k-1- \min_{\substack{\sum_{i=1}^n d_i \mathbf{r}_i \mathbf{r}_i^T=\boldsymbol{I}_{k-1}\\ \sum_{i=1}^n d_i \mathbf{r}_i=0}} Q_{k-1}.$$

The maximum is $k-1-\sum_{i=1}^{k-1}\lambda_i=\sum_{i=1}^{k-1}(1-\lambda_i)$, that is the sum of the $k-1$ largest eigenvalues of $\boldsymbol{M}_D$.

1.4 Representation of joint distributions

In this section we would like to give an abstract description of the issues discussed in the previous sections in terms of two-variate distributions. With the help of joint distributions, representation can be discussed in a more general framework, of which graphs and contingency tables are special finite cases. This section is rather of theoretical importance, however, some parts of it will appear in Chapter 4 when we will take limits of graphs and contingency tables, and consider the continuous limit objects as kernels of integral operators taking conditional expectation with respect to the joint distributions. Here and in the next section we will intensively use the theory of Hilbert spaces and linear operators between them and further, the distribution of random vectors. Therefore we recommend the reader to read first the related parts of Appendix A.2 and B.1.

1.4.1 General setup

Let (ξ, η) be a pair of real-valued random variables – neither of them being constant with probability 1 – defined over the product space $\mathcal{X} \times \mathcal{Y}$ having joint distribution $\mathbb{W}$ with marginals $\mathbb{P}$ and $\mathbb{Q}$, respectively. Assume that the dependence between ξ and η is regular, that is, their joint distribution $\mathbb{W}$ is absolutely continuous with respect to the product measure $\mathbb{P} \times \mathbb{Q}$, and let w denote the Radon–Nikodym derivative of $\mathbb{W}$ with respect to $\mathbb{P} \times \mathbb{Q}$; see Rényi (1959b) for details. In the case of discrete or absolutely continuous distributions we will soon introduce more friendly versions of this derivative.

In the spirit of Breiman and Friedman (1985), let $H = L^2(\xi)$ and $H' = L^2(\eta)$ be the set of random variables which are functions of ξ and η and have zero expectation and finite variance with respect to $\mathbb{P}$ and $\mathbb{Q}$, respectively. Both H and H' are Hilbert spaces with the covariance as inner product; further, they are embedded as subspaces into the L^2 space defined likewise by the (ξ, η) pair over the product space. (Note that we consider Borel-measurable functions which are also measurable with respect to the so-called σ-algebras generated by ξ and η, but we do not want to introduce superfluous notions that will not be used later. For example, in case of discrete, especially categorical variables with finitely many values, a function of such a variable takes on as many values as the original one, with the same probabilities.)

1.4.2 Integral operators between L^2 spaces

Let $K : \mathcal{X} \times \mathcal{Y} \to \mathbb{R}$ be a kernel such that for it

$$\int_{\mathcal{X}} \int_{\mathcal{Y}} K^2(x, y)\, \mathbb{P}(dx)\, \mathbb{Q}(dy) < \infty \tag{1.21}$$

holds. With the kernel K, a linear operator (integral operator) $A : H' \to H$ is defined in the following way: to the random variable $\phi \in H'$ the linear operator A assigns the random variable $\psi \in H$ such that

$$\psi(x) = (A\phi)(x) = \int_{\mathcal{Y}} K(x, y)\phi(y)\, \mathbb{Q}(dy), \qquad x \in \mathcal{X}.$$

By the linearity of A, ψ has zero expectation, and it is easy to check that has finite variance; further

$$\|\psi\| \le \|K\|_2 \|\phi\| < \infty,$$

where $\|\psi\|$ and $\|\phi\|$ denote the standard deviation (squareroot of the variance) of the random variables ψ and ϕ, respectively, while $\|K\|_2$ is the squareroot of the finite integral in (1.21). Further, in view of (A.8), for the operator norm of A, the following holds:

$$\|A\| = \sup_{\|\phi\|=1} \|A\phi\| \le \|K\|_2.$$

It is known that the above L^2 spaces are separable Hilbert spaces, and in view of (1.21), A is a Hilbert–Schmidt, therefore a compact (in other words, completely continuous) linear operator (see Appendix A.2) with SVD of (A.9):

$$A = \sum_{i=1}^{\infty} s_i \langle ., \phi_i \rangle_{H'} \psi_i.$$

Here $\langle ., . \rangle$ denotes the real inner product (covariance) in the corresponding Hilbert space; further, the nonnegative real numbers $s_1 \ge s_2 \ge \cdots \ge 0$ are the singular values with the zero as the only possible point of accumulation, and the corresponding function pairs ψ_i, ϕ_i can be chosen (even in the case of multiple singular values) so that $\{\psi_i\}_{i=1}^{\infty} \subset H$ and $\{\phi_i\}_{i=1}^{\infty} \subset H'$ form complete orthonormal systems. Since A is also a Hilbert–Schmidt operator, $\sum_{i=1}^{\infty} s_i^2 = \|K\|_2^2 < \infty$. Such an SVD is essentially unique (apart from function pairs corresponding to multiple singular values). It is easy to see that the adjoint of A (which is, in fact, the transpose, as we only deal with real valued random variables) has the SVD

$$A^* = \sum_{i=1}^{\infty} s_i \langle ., \psi_i \rangle_H \phi_i$$

and

$$A\phi_i = s_i \psi_i, \quad A^* \psi_i = s_i \phi_i, \quad i = 1, 2, \dots$$

further, s_1 is the spectral norm of both A and A^*.

Now we proceed to the symmetric case: the joint distribution over the product space is symmetric, that is, $w(x, y) = w(y, x)$, $x \in \mathcal{X}$, $y \in \mathcal{Y}$. In this case ξ and η are identically distributed (not independent as their joint distribution is $\mathbb{W}$) and hence, each random variable in H has a counterpart in H' and vice versa, such that they are identically distributed; therefore H and H' are isomorphic in terms of the distributions, too. By the Hilbert–Schmidt theorem A.2.5, the selfadjoint compact linear operator $A : H' \to H$ has SD

$$A = \sum_{i=1}^{\infty} \lambda_i \langle ., \psi_i' \rangle_{H'} \psi_i$$

with real eigenvalues whose only possible point of accumulation is the zero ($\lim_{i\to\infty} \lambda_i = 0$ if the eigenvalues are countably infinitely many), and corresponding orthonormal eigenvectors $\psi_1, \psi_2, \dots$ such that ψ_i and ψ_i' are identically distributed with joint distribution $\mathbb{W}$. Of course, A also has a singular value decomposition, where the singular values are the absolute values of the eigenvalues; if $\lambda_i > 0$, then $s_i = \lambda_i$, $\psi_i = \phi_i$ and they coincide with the unit-norm eigenfunction; if $\lambda_i < 0$, then $s_i = -\lambda_i$, $\psi_i = -\phi_i$ and any of them can be the unit-norm eigenfunction corresponding to λ_i.

1.4.3 When the kernel is the joint distribution itself

The following linear operators taking conditional expectation between the two marginals (with respect to the joint distribution) will play a crucial role in the future, see also Breiman and Friedman (1985). In the general, not necessarily symmetric case, they are defined as

$$P_{\mathcal{X}} : H' \to H, \quad \psi = P_{\mathcal{X}}\phi = \mathbb{E}(\phi \mid \xi), \quad \psi(x) = \int_{\mathcal{Y}} w(x, y)\phi(y)\, \mathbb{Q}(dy)$$

and

$$P_{\mathcal{Y}} : H \to H', \quad \phi = P_{\mathcal{Y}}\psi = \mathbb{E}(\psi \mid \eta), \quad \phi(y) = \int_{\mathcal{X}} w(x, y)\psi(x)\, \mathbb{P}(dx).$$

It is easy to see that $P_{\mathcal{X}}^* = P_{\mathcal{Y}}$ and vice versa, because of the relation

$$\langle P_{\mathcal{X}}\phi, \psi \rangle_H = \langle P_{\mathcal{Y}}\psi, \phi \rangle_{H'} = \mathrm{Cov}_{\mathbb{W}}(\psi, \phi), \tag{1.22}$$

where $\mathrm{Cov}_{\mathbb{W}}$ is the so-called covariance function with respect to the joint distribution $\mathbb{W}$, defined as

$$\mathrm{Cov}_{\mathbb{W}}(\psi, \phi) = \int_{\mathcal{X}\times\mathcal{Y}} \psi(x)\phi(y)\mathbb{W}(dx, dy) = \int_{\mathcal{X}}\int_{\mathcal{Y}} \psi(x)\phi(y)w(x, y)\mathbb{Q}(dy)\mathbb{P}(dx).$$

Assume that

$$\int_{\mathcal{X}}\int_{\mathcal{Y}} w^2(x, y)\mathbb{Q}(dy)\mathbb{P}(dx) < \infty. \tag{1.23}$$

In case of discrete distributions with joint distribution $\{w_{ij}\}$ and marginals $\{p_i\}$ ($p_i = \sum_j w_{ij}$) and $\{q_j\}$ ($q_j = \sum_i w_{ij}$), (1.23) means that

$$\sum_{i\in\mathcal{X}}\sum_{j\in\mathcal{Y}} \left(\frac{w_{ij}}{p_i q_j}\right)^2 p_i q_j = \sum_{i\in\mathcal{X}}\sum_{j\in\mathcal{Y}} \frac{w_{ij}^2}{p_i q_j} < \infty,$$

while in case of absolutely continuous distributions with joint p.d.f. $f(x, y)$ and marginal p.d.f.s $f_1(x)$ ($f_1(x) = \int f(x, y)\, dy$) and $f_2(y)$ ($f_2(y) = \int f(x, y)\, dx$), (1.23) means that

$$\int_{\mathcal{X}}\int_{\mathcal{Y}} \left(\frac{f(x, y)}{f_1(x)f_2(y)}\right)^2 f_1(x)f_2(y)\, dx\, dy = \int_{\mathcal{X}}\int_{\mathcal{Y}} \frac{f^2(x, y)}{f_1(x)f_2(y)}\, dx\, dy < \infty.$$

Under these conditions $P_{\mathcal{X}}$ and $P_{\mathcal{Y}}$ are Hilbert–Schmidt operators, and therefore compact, with SVD

$$P_{\mathcal{X}} = \sum_{i=1}^{\infty} s_i \langle ., \phi_i\rangle_{H'}\psi_i, \quad P_{\mathcal{Y}} = \sum_{i=1}^{\infty} s_i \langle ., \psi_i\rangle_{H}\phi_i \tag{1.24}$$

where for the singular values $1 > s_1 \geq s_2 \geq \cdots \geq 0$ holds, since the operators $P_{\mathcal{X}}$ and $P_{\mathcal{Y}}$ are in fact orthogonal projections from one marginal onto the other, but the projections are restricted to the subspaces H and H', respectively. We remark that denoting by ψ_0 and ϕ_0 the constantly 1 random variables, $\mathbb{E}(\phi_0|\xi) = \psi_0$ and $\mathbb{E}(\psi_0|\eta) = \phi_0$, however, this pair is not considered as a function pair with singular value $s_0 = 1$, since they have no zero expectation. Consequently, we will subtract 1 from the kernel, but with this new kernel, $P_{\mathcal{X}}$ and $P_{\mathcal{Y}}$ will define the same integral operators.

Especially, if $\mathbb{W}$ is symmetric (H and H' are isomorphic in terms of the distributions too), then in view of (1.22), $P_{\mathcal{X}} = P_{\mathcal{Y}}$ is a selfadjoint (symmetric) linear operator, since

$$\langle P_{\mathcal{X}}\phi, \psi\rangle_H = \mathrm{Cov}_{\mathbb{W}}(\phi, \psi) = \mathrm{Cov}_{\mathbb{W}}(\psi, \phi) = \langle P_{\mathcal{Y}}\psi, \phi\rangle_{H'}.$$

The SD of $P_{\mathcal{X}} : H' \to H$ is

$$P_{\mathcal{X}} = \sum_{i=1}^{\infty} \lambda_i \langle ., \psi_i'\rangle_{H'}\psi_i.$$

Here for the eigenvalues, $|\lambda_i| \leq 1$ holds, and the eigenvalue–eigenfunction equation looks like

$$P_{\mathcal{X}}\psi_i' = \lambda_i \psi_i$$

where ψ_i and ψ_i' are identically distributed, whereas their joint distribution is $\mathbb{W}$ $(i = 1, 2, \dots)$.

1.4.4 Maximal correlation and optimal representations

From now on, we will intensively use separation theorems for the singular values and eigenvalues, see Appendix A.2 and A.3. In view of these, the SVD (A.6) gives the solution of the following task of *maximal correlation*, posed by Gebelein (1941) and Rényi (1959a). We are looking for $\psi \in H$, $\phi \in H'$ such that their correlation is maximum with respect to their joint distribution $\mathbb{W}$. Using (1.22) and Proposition A.2.7,

$$\max_{\|\psi\|=\|\phi\|=1} \mathrm{Cov}_{\mathbb{W}}(\psi, \phi) = s_1$$

and it is attained on the non-trivial ψ_1, ϕ_1 pair. In the finite, symmetric case, maximal correlation is related to some conditional probabilities in Bolla and M.-Sáska (2004).

This task is equivalent to the following one:

$$\min_{\|\psi\|=\|\phi\|=1} \|\psi - \phi\|^2 = \min_{\|\psi\|=\|\phi\|=1} (\|\psi\|^2 + \|\phi\|^2 - 2\mathrm{Cov}_{\mathbb{W}}(\psi, \phi)) = 2(1 - s_1). \tag{1.25}$$

Correspondence analysis (see Appendix C.3) is on the one hand, a special case of the problem of maximal correlation being $\mathcal{X}$ and $\mathcal{Y}$ finite sets, but on the other hand, it is a generalization to the extent that we are successively finding maximal correlations under some orthogonality conditions.

The product space is now an $m \times n$ contingency table with row set $\mathcal{X} = \{1, \dots, m\}$ and column set $\mathcal{Y} = \{1, \dots, n\}$, whereas the entries $w_{ij} \geq 0$ $(i = 1, \dots, m;\ j = 1, \dots, n)$ embody the joint distribution over the product space, with row-sums $p_1, \dots, p_m$ and column-sums $q_1, \dots, q_n$ as marginals.

Hence, the effect of $P_{\mathcal{X}} : H' \to H$, $P_{\mathcal{X}}\phi = \psi$ is the following:

$$\psi(i) = \frac{1}{p_i} \sum_{j=1}^{n} w_{ij}\phi(j) = \sum_{j=1}^{n} \frac{w_{ij}}{p_i q_j} \phi(j) q_j, \quad i = 1, \dots, m. \tag{1.26}$$

Therefore, $P_{\mathcal{X}}$ is an integral operator with kernel $K_{ij} = \frac{w_{ij}}{p_i q_j}$ (instead of integration, we have summation with respect to the marginal measure $\mathbb{Q}$).

Consider the SVD

$$P_{\mathcal{X}} = \sum_{k=1}^{r-1} s_k \langle ., \phi_k \rangle_{H'} \psi_k$$

where r is the now finite rank of the contingency table ($r \le \min\{n, m\}$). The singular value $s_0 = 1$ with the trivial $\psi_0 = 1$, $\phi_0 = 1$ factor pair is disregarded as their expectation is 1 with respect to the $\mathbb{P}$- and $\mathbb{Q}$-measures, respectively, therefore, the summation starts from 1. If we used the kernel $K_{ij} - 1$, we could eliminate the trivial factors. Assume that there is no other singular value 1, that is, our contingency table is non-decomposable. Then, by the orthogonality, the subsequent left- and right-hand side singular functions have zero expectation with respect to the $\mathbb{P}$- and $\mathbb{Q}$-measures, and they solve the following successive maximal correlation problem. For $k = 1, \dots, r-1$, in step k we want to find $\max \text{Corr}_{\mathbb{W}}(\psi, \phi)$ subject to

$$\text{Var}_{\mathbb{P}}(\psi) = \text{Var}_{\mathbb{Q}}(\phi) = 1, \quad \text{Cov}_{\mathbb{P}}(\psi, \psi_i) = \text{Cov}_{\mathbb{Q}}(\phi, \phi_i) = 0, \quad i = 0, \dots, k-1.$$

Note that the last condition for $i = 0$ is equivalent to

$$\mathbb{E}_{\mathbb{P}}(\psi) = \mathbb{E}_{\mathbb{Q}}(\phi) = 0.$$

Referring to Appendix C.3, the maximum is s_k and it is attained on the ψ_k, ϕ_k pair.

Now, we are able to define the joint representation of the general Hilbert-spaces H, H' – introduced in Subsection 1.4.1 – with respect to the joint measure $\mathbb{W}$ in the following way.

Definition 1.4.1 *We say that the pair* $(\mathbf{X}, \mathbf{Y})$ *of k-dimensional random vectors with components in H and H′, respectively, form a k-dimensional representation of the product space endowed with the measure* $\mathbb{W}$ *if* $\mathbb{E}_{\mathbb{P}}\mathbf{X}\mathbf{X}^T = \boldsymbol{I}_k$ *and* $\mathbb{E}_{\mathbb{Q}}\mathbf{Y}\mathbf{Y}^T = \boldsymbol{I}_k$ *(i.e., the components of* $\mathbf{X}$ *and* $\mathbf{Y}$ *are uncorrelated with zero expectation and unit variance, respectively). Further, the cost of this representation is defined as*

$$Q_k(\mathbf{X}, \mathbf{Y}) = \mathbb{E}_{\mathbb{W}} \|\mathbf{X} - \mathbf{Y}\|^2.$$

The pair $(\mathbf{X}^*, \mathbf{Y}^*)$ *is an optimal representation if it minimizes the above cost.*

Theorem 1.4.2 (Representation theorem for joint distributions) *Let* $\mathbb{W}$ *be a joint distribution with marginals* $\mathbb{P}$ *and* $\mathbb{Q}$. *Assume that among the singular values of the conditional expectation operator* $P_{\mathcal{X}} : H' \to H$ *(see (1.24)) there are at least k positive ones and denote by* $1 > s_1 \ge s_2 \ge \cdots \ge s_k > 0$ *the largest ones. The minimum cost of a k-dimensional representation is* $2\sum_{i=1}^{k}(1 - s_i)$ *and it is attained with* $\mathbf{X}^* = (\psi_1, \dots, \psi_k)$ *and* $\mathbf{Y}^* = (\phi_1, \dots, \phi_k)$*, where* ψ_i, ϕ_i *is the singular function pair corresponding to the singular value* s_i $(i = 1, \dots, k)$.

Proof.

$$\begin{aligned}\mathbb{E}_{\mathbb{W}}(\mathbf{X}-\mathbf{Y})^T(\mathbf{X}-\mathbf{Y}) &= \mathbb{E}_{\mathbb{W}}(\mathbf{X}^T\mathbf{X}) + \mathbb{E}_{\mathbb{W}}(\mathbf{Y}^T\mathbf{Y}) - \mathbb{E}_{\mathbb{W}}\mathbf{X}^T\mathbf{Y} - \mathbb{E}_{\mathbb{W}}\mathbf{Y}^T\mathbf{X} \\ &= \mathbb{E}_{\mathbb{P}}(\mathrm{tr}[\mathbf{X}\mathbf{X}^T]) + \mathbb{E}_{\mathbb{Q}}(\mathrm{tr}[\mathbf{Y}\mathbf{Y}^T]) - 2\sum_{i=1}^{k}\mathbb{E}_{\mathbb{W}}(X_iY_i) \\ &= 2k - 2\sum_{i=1}^{k}\mathrm{Cov}(X_iY_i).\end{aligned}$$

Applying Proposition A.2.8 of Appendix A.2 for the singular values of the conditional expectation operator, the required statement is obtained.

We remark that in case of a finite $\mathcal{X}$ and $\mathcal{Y}$, the solution corresponds to the SVD of the correspondence matrix: see Appendix C. Though, the correspondence matrix seemingly does not have the same normalization as the kernel, our numerical algorithm for the SVD of a rectangular matrix is capable of finding orthogonal eigenvectors in the usual Euclidean norm, which corresponds to the Lebesgue measure and not to the $\mathbb{P}$- or $\mathbb{Q}$-measures. Observe that in case of an irreducible contingency table, $s_i < 1$ $(i = 1, \ldots, k)$, therefore the minimum cost is strictly positive.

In the symmetric case we can also define a representation. Now the $\mathbf{X}, \mathbf{X}'$ pair is identically distributed, but not independent as they are connected with the symmetric joint measure $\mathbb{W}$.

Definition 1.4.3 *We say that the k-dimensional random vector $\mathbf{X}$ with components in H forms a k-dimensional representation of the product space $H \times H'$ (H and H' are isomorphic) endowed with the symmetric measure $\mathbb{W}$ (and marginal $\mathbb{P}$) if $\mathbb{E}_{\mathbb{P}}\mathbf{X}\mathbf{X}^T = \boldsymbol{I}_k$ (i.e., the components of $\mathbf{X}$ are uncorrelated with zero expectation and unit variance). Further, the cost of this representation is defined as*

$$Q_k(\mathbf{X}) = \mathbb{E}_{\mathbb{W}}\|\mathbf{X} - \mathbf{X}'\|^2,$$

where $\mathbf{X}$ and $\mathbf{X}'$ are identically distributed and the joint distribution of their coordinates X_i and X_i' is $\mathbb{W}$ $(i = 1, \ldots, k)$, while X_i and X'_j are uncorrelated if $i \neq j$. The random vector $\mathbf{X}^*$ is an optimal representation if it minimizes the above cost.

Applying Proposition A.2.10 of Appendix A.2 for the SD of the conditional expectation operator, the following result is obtained.

Theorem 1.4.4 (Representation theorem for symmetric joint distributions) *Let $\mathbb{W}$ be a symmetric joint distribution with marginal $\mathbb{P}$. Assume that among the eigenvalues of the conditional expectation operator $P_{\mathcal{X}}: H' \to H$ (H and H' are isomorphic) there are at least k positive ones and denote by $1 > \lambda_1 \geq \lambda_2 \geq \cdots \geq \lambda_k > 0$ the largest ones. Then the minimum cost of a k-dimensional representation is $2\sum_{i=1}^{k}(1-\lambda_i)$ and it is attained by $\mathbf{X}^* = (\psi_1, \ldots, \psi_k)$ where ψ_i is the eigenfunction corresponding to the eigenvalue λ_i $(i = 1, \ldots k)$.*

In the case of a finite $\mathcal{X}$ (vertex set of an edge-weighted graph), we have a weighted graph with edge-weights w_{ij} ($\sum_{i=1}^{n}\sum_{j=1}^{n} w_{ij} = 1$). The operator $P_{\mathcal{X}}$ deprived of the trivial factor corresponds to its normalized modularity matrix with eigenvalues in the $[-1,1]$ interval (1 cannot be an eigenvalue if the underlying graph is connected), and eigenfunctions which are the transformed eigenvectors. As the numerical algorithm gives an orthonormal set of eigenvectors in Euclidean norm, some back-transformation is needed to get uncorrelated components with unit variance, therefore we use the normalized modularity matrix instead of the kernel $K_{ij} = \frac{w_{ij}}{d_i d_j}$ expected from (1.26), where $d_i = \sum_{j \in \mathcal{X}} w_{ij}$, $i \in \mathcal{X}$, the generalized degrees of the vertices.

We remark that the above formula for the kernel corresponds to the so-called copula transformation of the joint distribution $\mathbb{W}$ into the unit square: see Nelsen (2006). This idea appears when vertex- and edge-weighted graphs are transformed into piecewise constant functions over $[0, 1] \times [0, 1]$; see the definition of graphons by Borgs *et al.* (2008) in Chapter 4. This transformation can be performed in the general non-symmetric and non-finite cases too. Also observe that neither the kernel nor the contingency table or graph is changed under measure-preserving transformations of $\mathcal{X}$; see the theory of exchangeable sequences and arrays Aldous (1981); Austin (2008); Diaconis and Janson (2008); de Finetti (1974); Kallenberg (2001).

We also remark that any or both of the initial random variables ξ, η can as well be a random vector (with real components). For example, if they have p- and q-dimensional Gaussian distribution respectively, then their maximum correlation is the largest canonical correlation between them (see Appendix C.2), and it is realized by appropriate linear combinations of the components of ξ and η, respectively. Moreover, we can find canonical correlations one after the other with corresponding function pairs (under some orthogonality constraints), as many as the rank of the cross-covariance matrix of ξ and η. In fact, the whole decomposition relies on the SVD of a matrix calculated from this cross-covariance matrix and the individual covariance matrices of ξ and η. Hence, in many cases, the maximum correlation problem can be dealt with more generally by means of SVD or SD.

1.5 Treating nonlinearities via reproducing kernel Hilbert spaces

There are fairy tales about some fictitious spaces, where everything is 'smooth' and 'linear'. Such spaces really exist, the hard part is that we should adapt them to our data. The good news is that it is not necessary to actually map our data into them, it suffices to treat only a kernel function, but the bad news is that the kernel must be appropriately selected so that the underlying nonlinearity could be detected.

Reproducing kernel Hilbert spaces were introduced in the middle of the 20th century by Aronszajn (1950); Parzen (1963), and others, but the theory itself is an elegant application of already known theorems of functional analysis, first of all the Riesz–Fréchet theorem and the theory of integral operators, see the works of Fréchet (1907); Riesz (1907, 1909); Riesz and Sz.-Nagy (1952) tracing back to the beginning of the 20th century. Later on, in the last decades of the 20th century and even in our time, reproducing kernel Hilbert spaces have been several times reinvented and applied in modern statistical methods and data mining, for example in Bach and Jordan (2002); Baker (1977); Bengio *et al.* (2004); Shawe-Taylor and Cristianini (2004). But what is the mystery of reproducing kernels and what is the diabolic kernel trick? We would like to reveal this secret and show the technical advantages of this artificially constructed creature. We will start with the motivation for using this concept of data representation.

A popular approach to data clustering (sometimes this is called spectral clustering) is the following. Our data points $\mathbf{x}_1, \ldots, \mathbf{x}_n$ are already in a metric space, called *data space*, but cannot be well classified by the k-means algorithm (see Appendix C.5) for no integer $0 < k < n$. For example, there are obviously two clusters of points in $\mathbb{R}^2$, but they are separated by an annulus and the k-means algorithm with $k = 2$ is not able to find them, see Reisen and Bunke (2010). However, we can map the points into a usually higher dimensional, or more abstract space with a non-linear mapping so that the images are already well clustered in the new, so-called *feature space*. At the end of this section, we will give a mapping of the points into $\mathbb{R}^3$ using a second degree polynomial kernel, because we want to make a separation with second degree curves linear.

In practical higher-dimensional problems, when we do not have the faintest idea about the clusters and no visualization is possible, unfortunately, we cannot give such mappings explicitly. Moreover, the feature space usually has a much higher dimension than the original one, which fact is frequently referred to as the *curse of dimensionality*. However, the point of the kernel method to be introduced is that it is not even necessary to perform the mapping, it suffices to select a kernel – based on the inner product of the original points – that is no longer a linear kernel, but a more complicated, still admissible kernel (the exact meaning is given in Definition 1.5.2), and defines a new inner product within the feature space. Then we process statistical algorithms that need only this inner product and nothing else.

The feature space above is the counterpart of a so-called *reproducing kernel Hilbert space* that we will introduce right now together with the correspondence between it and the feature space. For those who are not familiar with abstract vector spaces, the definition of a Hilbert space is to be found in Appendix A.2. For example, finite dimensional Euclidean spaces (vector spaces with the usual inner product, e.g., $\mathbb{R}^p$) or the $L^2(\mathcal{X})$ space of real-valued, square integrable functions with respect to some finite measure on the compact set $\mathcal{X}$ (where the inner product of two functions is the integral of their product on $\mathcal{X}$) are such.

1.5.1 Notion of the reproducing kernel

A stronger condition imposed on a Hilbert space $\mathcal{H}$ of functions $\mathcal{X} \to \mathbb{R}$ (where $\mathcal{X}$ is an arbitrary set, for the time being) is that the following so-called evaluation mapping be a continuous, or equivalently, a bounded linear functional. The evaluation mapping $L_x : \mathcal{H} \to \mathbb{R}$ works on an $f \in \mathcal{H}$ such that

$$L_x(f) = f(x). \tag{1.27}$$

Definition 1.5.1 *A Hilbert space $\mathcal{H}$ of (real) functions on the set $\mathcal{X}$ is a reproducing kernel Hilbert space, briefly RKHS, if the point evaluation functional L_x of (1.27) exists and is continuous for all $x \in \mathcal{X}$.*

The name reproducing kernel comes from the fact that – by the Riesz–Fréchet representation theorem (see Theorem A.2.1 of Appendix A) – the result of such a continuous mapping can be written as an inner product. This theorem states that a Hilbert space (in our case $\mathcal{H}$) and its dual (in our case the set of $\mathcal{H} \to \mathbb{R}$ continuous linear functionals, e.g. L_x) are isometrically isomorphic. Therefore, to any L_x there uniquely corresponds a $K_x \in \mathcal{H}$ such that

$$L_x(f) = \langle f, K_x \rangle_{\mathcal{H}}, \quad \forall f \in \mathcal{H}. \tag{1.28}$$

Since K_x is itself an $\mathcal{X} \to \mathbb{R}$ function, it can be evaluated at any point $y \in \mathcal{X}$. We define the bivariate function $K : \mathcal{X} \times \mathcal{X} \to \mathbb{R}$ as

$$K(x, y) := K_x(y) \tag{1.29}$$

and call it the *reproducing kernel* for the Hilbert space $\mathcal{H}$. Then using formulas (1.27), (1.28), and (1.29), we get on the one hand,

$$K(x, y) = K_x(y) = L_y(K_x) = \langle K_x, K_y \rangle_{\mathcal{H}},$$

and on the other hand,

$$K(y, x) = K_y(x) = L_x(K_y) = \langle K_y, K_x \rangle_{\mathcal{H}}.$$

By the symmetry of the (real) inner product it follows that the reproducing kernel is symmetric and it is also reproduced as the inner product of special functions in the RKHS:

$$K(x, y) = \langle K_x, K_y \rangle_{\mathcal{H}} = \langle K(x, .), K(., y) \rangle_{\mathcal{H}}, \tag{1.30}$$

hence, K is positive definite (for the precise notion see the forthcoming Definition 1.5.2). This is the diabolic kernel trick. In this way, any point is represented by its similarity to all other points, see Ando (1987); Ramsay and Silverman (2006); Schölkopf and Smola (2002) for more details.

Vice versa, if we are given a positive definite kernel function on $\mathcal{X} \times \mathcal{X}$ at the beginning, then there exists an RKHS such that with appropriate elements of it, the inner product relation (1.30) holds. (In fact, we are not given, we just select an appropriate kernel function.) The mystery of RKHS just lies in this converse statement.

For this purpose, let us first define the most important types of kernel functions and discuss how more and more complicated ones can be derived from the simplest ones.

Definition 1.5.2 *A symmetric two-variate function $K : \mathcal{X} \times \mathcal{X} \to \mathbb{R}$ is called positive definite kernel (equivalently, admissible, valid, or Mercer kernel) if for any $n \in \mathbb{N}$ and*

$x_1, \dots, x_n \in \mathcal{X}$, *the symmetric matrix of entries* $K(x_i, x_j) = K(x_j, x_i)(i, j = 1, \dots n)$ *is positive semidefinite.*

We remark (see Proposition A.3.7 of Appendix A) that a symmetric real matrix is positive semidefinite if and only if it is a Gram matrix, and hence, its entries become inner products, but usually not of the entries in its arguments. However, the simplest kernel function, the so-called *linear kernel*, does this job. It is defined as

$$K_{\text{lin}}(x, y) = \langle x, y \rangle_{\mathcal{X}},$$

if $\mathcal{X}$ is subset of a Euclidean space.

From a valid kernel, one can get other valid kernels with the following operations:

1. If $K_1, K_2 : \mathcal{X} \times \mathcal{X} \to \mathbb{R}$ are positive definite kernels, then the kernel K defined by $K(x, y) = K_1(x, y) + K_2(x, y)$ is also positive definite.

2. If $K_1, K_2 : \mathcal{X} \times \mathcal{X} \to \mathbb{R}$ are positive definite kernels, then the kernel K defined by $K(x, y) = K_1(x, y)K_2(x, y)$ is also positive definite. Especially, if K is a positive definite kernel, then so does cK with any $c > 0$.

The first statement is trivial, the second one follows from Proposition B.1.5 of Appendix B.

Consequently, if h is a polynomial with positive coefficients and $K : \mathcal{X} \times \mathcal{X} \to \mathbb{R}$ is a positive definite kernel, then the kernel $K_h : \mathcal{X} \times \mathcal{X} \to \mathbb{R}$ defined by

$$K_h(x, y) = h(K(x, y)) \tag{1.31}$$

is also positive definite. Since the exponential function can be approximated by polynomials with positive coefficients and the positive definiteness is closed under pointwise convergence (see Berg and Ressel (1984)), the same is true if h is the exponential function: $h(x) = e^x$, or perhaps some transformation of it.

Putting these facts together and using the formula

$$\|x - y\|^2 = \langle x, x \rangle + \langle y, y \rangle - 2\langle x, y \rangle, \tag{1.32}$$

one can easily verify that the following, so-called *Gaussian kernel* is positive definite:

$$K_{\text{Gauss}}(x, y) = e^{-\frac{\|x-y\|^2}{2\sigma^2}}, \tag{1.33}$$

where $\sigma > 0$ is a parameter. Indeed, in view of (1.32), this kernel can be written as the product of two positive definite kernels in the following way:

$$K_{\text{Gauss}}(x, y) = K_1(x, y)K_2(x, y),$$

where

$$K_1(x, y) = e^{-\dfrac{\langle x, x \rangle + \langle y, y \rangle}{2\sigma^2}},$$

and

$$K_2(x, y) = e^{\frac{\langle x, y \rangle}{\sigma^2}}.$$

Here K_2 is positive definite as it is the exponential function of the positive definite kernel $\frac{1}{\sigma^2} K_{\text{lin}}$. To show that K_1 is positive definite, by definition, we have to verify that for any $n \in \mathbb{N}$ and $x_1, \dots, x_n \in \mathcal{X}$, the symmetric matrix of entries

$$K_1(x_i, x_j) = e^{-\frac{\langle x_i, x_i \rangle}{2\sigma^2}} e^{-\frac{\langle x_j, x_j \rangle}{2\sigma^2}}, \quad i, j = 1, \dots n$$

is positive semidefinite. But it is a rank 1 matrix, its only non-zero eigenvalue being equal to its trace, which is positive.

If $\mathcal{X} = \{x_1, \dots, x_n\}$ and $\boldsymbol{S}$ is an $n \times n$ symmetric similarity matrix comprised of the pairwise similarities between the entries of $\mathcal{X}$, then the kernel K defined by the $n \times n$ symmetric, positive definite matrix $e^{\lambda \boldsymbol{S}}$ is called the *diffusion kernel*, where $0 < \lambda < 1$ is the parameter (sometimes called decay factor). For the definition of an analytic function of a symmetric matrix see Equation (A.16) of Appendix A.3. Let us recapitulate that the eigenvalues of the $e^{\lambda \boldsymbol{S}}$ matrix are the numbers $e^{\lambda \lambda_i}$ $(i = 1, \dots, n)$, where λ_i's are real eigenvalues of $\boldsymbol{S}$. Therefore the diffusion kernel is always strictly positive definite, even if $\boldsymbol{S}$ is not positive semidefinite. Above the aforementioned ones, there are a lot of other kernels, see Ando (1987); Reisen and Bunke (2010) for details.

1.5.2 RKHS corresponding to a kernel

Now we are able to formulate the converse statement. Recall that, by the Riesz–Fréchet representation theorem, an RKHS defines a positive definite kernel. In the other direction, for any positive definite kernel we can find a unique RKHS such that the relation formulated in (1.30) holds with appropriate elements of it. The following theorem is due to Aronszajn (1950) who attributed it to E. H. Moore.

Theorem 1.5.3 *For any positive definite kernel $K : \mathcal{X} \times \mathcal{X} \to \mathbb{R}$ there exists a unique, possibly infinite-dimensional Hilbert space $\mathcal{H}$ of functions on $\mathcal{X}$, for which K is a reproducing kernel.*

If we want to emphasize that the RKHS corresponds to the kernel K, we will denote it by $\mathcal{H}_K$. The proof of the Theorem 1.5.3 can be found in Ando (1987); Ramsay and Silverman (2006); Schölkopf and Smola (2002), among others. It is based on the observation that the function space $\text{Span}\{K_x = K(x, .) : x \in \mathcal{X}\}$ uniquely defines a pre-Hilbert space that can be completed into a Hilbert space. This will provide the unique RKHS $\mathcal{H}_K$.

However, we may wish to realize the elements of an RKHS $\mathcal{H}_K$ in a more straightforward Hilbert space $\mathcal{F}$. Assume that there is a (usually not linear) map $\phi : \mathcal{X} \to \mathcal{F}$

such that when $x \in \mathcal{X}$ is mapped into $\phi(x) \in \mathcal{F}$, then

$$K(x, y) = \langle \phi(x), \phi(y) \rangle_{\mathcal{F}}$$

is the desired positive definite kernel. At the same time, in view of (1.30),

$$K(x, y) = \langle K_x, K_y \rangle_{\mathcal{H}_K}$$

where recall that $K_x = K(x, .)$ is an $\mathcal{X} \to \mathbb{R}$ function, hence, cannot be identical to $\phi(x)$, but they can be connected with the following transformation. Let T be a linear operator from $\mathcal{F}$ to the space of functions $\mathcal{X} \to \mathbb{R}$ defined by

$$(Tf)(y) = \langle f, \phi(y) \rangle_{\mathcal{F}}, \quad y \in \mathcal{X}, \ f \in \mathcal{F}.$$

Then

$$(T\phi(x))(y) = \langle \phi(x), \phi(y) \rangle_{\mathcal{F}} = K(x, y) = K_x(y),$$

therefore

$$T\phi(x) = K_x, \quad \forall x \in \mathcal{X} \tag{1.34}$$

and hence, $\mathcal{H}_K$ becomes the range of T. This was the informal proof of the more precise statements about this correspondence elaborated in Ando (1987); Ramsay and Silverman (2006).

1.5.3 Two examples of an RKHS

Here we give the theoretical construction for $\mathcal{H}_k$ and $\mathcal{F}$, together with the functions K_x and the features $\phi(x)$, in two special cases.

(a) Let K be the continuous kernel of a positive definite Hilbert–Schmidt operator which is an integral operator working on the $L^2(\mathcal{X})$ space, where $\mathcal{X}$ is a compact set in $\mathbb{R}$ for simplicity (it could be $\mathbb{R}^p$ as well), see Section 1.4 and Appendix A.2. Now the positive definiteness of K means that

$$\int_{\mathcal{X}} \int_{\mathcal{X}} K(x, y) f(x) f(y) \, dx \, dy \geq 0, \quad \forall f \in L^2(\mathcal{X}),$$

and for the integral operator to be Hilbert–Schmidt, K must be in the $L^2(\mathcal{X} \times \mathcal{X})$ space, that is

$$\int_{\mathcal{X}} \int_{\mathcal{X}} K^2(x, y) \, dx \, dy < \infty$$

holds for it.

It is well known (see, e.g., Riesz and Sz.-Nagy (1952)) that this operator is compact, hence has a discrete spectrum whose only possible point of accumulation is the 0. Because of the symmetry of K, the integral operator is also self-adjoint, and for it, the Hilbert–Schmidt theorem (see Theorem A.2.5

in Appendix A) is applicable. This guarantees that the operator has nonnegative real eigenvalues $\lambda_1 \geq \lambda_2 \geq \cdots \geq 0$ and corresponding eigenfunctions $\psi_1, \psi_2, \ldots$. It also follows that $\sum_{i=1}^{\infty} \lambda_i^2 = \int_{\mathcal{X}} \int_{\mathcal{X}} K^2(x, y)\, dx\, dy < \infty$. By the Mercer theorem (see Theorem A.2.12 of Appendix A), if K is a continuous kernel of a positive definite integral operator on $L^2(\mathcal{X})$, where $\mathcal{X}$ is some compact space, then it can be expanded into the following uniformly convergent series:

$$K(x, y) = \sum_{i=1}^{\infty} \lambda_i \psi_i(x) \psi_i(y), \quad \forall x, y \in \mathcal{X}$$

by the eigenfunctions and the eigenvalues of the integral operator.

The RKHS defined by K is the following:

$$\mathcal{H}_K = \{f : \mathcal{X} \to \mathbb{R} : f(x) = \sum_{i=1}^{\infty} c_i \psi_i(x) \quad \texttt{s.t.} \quad \sum_{i=1}^{\infty} \frac{c_i^2}{\lambda_i} < \infty\}.$$

If $g(x) = \sum_{i=1}^{\infty} d_i \psi_i(x)$ – where $\sum_{i=1}^{\infty} \frac{d_i^2}{\lambda_i} < \infty$ – is another function in $\mathcal{H}_K$, then $f(x) + g(x)$ also corresponds to $\mathcal{H}_K$, due to $(c_i + d_i)^2 \leq 2(c_i^2 + d_i^2)$; the constant multiple of $f(x)$ also corresponds to $\mathcal{H}_K$, therefore $\mathcal{H}_K$ is a subspace of $L^2(\mathcal{X})$. The inner product of f and g is

$$\langle f, g \rangle_{\mathcal{H}_k} = \sum_{i=1}^{\infty} \frac{c_i d_i}{\lambda_i}.$$

Then one can easily verify that

$$K_x = K(x, .) = \sum_{i=1}^{\infty} \lambda_i \psi_i(x) \psi_i$$

is in $\mathcal{H}_K$. Indeed, it operates as

$$K_x(z) = \sum_{i=1}^{\infty} \lambda_i \psi_i(x) \psi_i(z)$$

and

$$\sum_{i=1}^{\infty} \frac{\lambda_i^2 \psi_i^2(x)}{\lambda_i} = K(x, x) < \infty.$$

Therefore,

$$\langle K_x, K_y \rangle_{\mathcal{H}_K} = \sum_{i=1}^{\infty} \frac{\lambda_i \psi_i(x) \lambda_i \psi_i(y)}{\lambda_i} = K(x, y). \tag{1.35}$$

Here the feature space $\mathcal{F}$ is the following counterpart of $\mathcal{H}_K$: it consists of infinite dimensional vectors

$$\phi(x) = (\sqrt{\lambda_1}\psi_1(x), \sqrt{\lambda_2}\psi_2(x), \ldots), \quad x \in \mathcal{X}$$

and the inner product is naturally defined by

$$\langle \phi(x), \phi(y) \rangle_{\mathcal{F}} = \sum_{i=1}^{\infty} \lambda_i \psi_i(x) \psi_i(y),$$

which, in view of (1.35), is indeed equal to $\langle K_x, K_y \rangle_{\mathcal{H}_K}$.

In fact, here there are the functions $\sqrt{\lambda_1}\psi_1, \sqrt{\lambda_2}\psi_2, \ldots$, which form an orthonormal basis in $\mathcal{H}_k$, and because of this transformation, a function $f \in L^2(\mathcal{X})$ is in $\mathcal{H}_k$ if $\|f\|^2_{\mathcal{H}_k} = \sum_{i=1}^{\infty} \frac{c_i^2}{\lambda_i} < \infty$. This condition restricts $\mathcal{H}_k$ to special functions between which the inner product is also adopted to the affine basis transformation. As the eigenfunctions of a Hilbert–Schmidt operator are continuous, an $f \in \mathcal{H}_K$ is also a continuous function. To further characterize the elements of $\mathcal{H}_K$, let us use the Banach–Steinhaus theorem (Theorem A.2.13 of Appendix A) which, in our case, states the following. Consider the collection $\{L_x : x \in X\}$ of continuous linear functionals, where L_x is the evaluation mapping of (1.27). One can easily see that

$$\sup_{x \in \mathcal{X}} \|L_x(f)\| = \sup_{x \in \mathcal{X}} |f(x)| < \infty$$

holds for any $f \in \mathcal{H}_K$, as f is continuous on the compact set $\mathcal{X}$: see Theorem A.1.2. Under these circumstances, the Banach–Steinhaus uniform boundedness principle (see Theorem A.2.13) states that

$$\sup_{x \in \mathcal{X}} \|L_x\| < \infty,$$

that is, with some positive constant B,

$$\sup_{x \in \mathcal{X}} \sup_{\|f\|_{\mathcal{H}_K}=1} |f(x)| \leq B < \infty.$$

Consequently, functions with fixed norm $\|f\|_{\mathcal{H}_K}$ are uniformly bounded, and the uniform bound is proportional to their $\mathcal{H}_K$-norm. Therefore, the global behavior of functions in $\mathcal{H}_K$ affects their local behavior, at least, it bounds the functions on their whole domain. Thus, they are – in a certain sense – smooth functions. This property is due to the fact that these functions are strongly determined by the common kernel.

(b) Now $\mathcal{X}$ is a Hilbert space of finite dimension, say R^p, and its elements will be denoted by boldface $\mathbf{x}$, stressing that they are vectors. If we used K_{lin} on $\mathcal{X} \times \mathcal{X}$, then $K_{\mathbf{x}} = \langle \mathbf{x}, . \rangle_{\mathcal{X}}$, and by the Riesz–Fréchet representation theorem, $\phi(\mathbf{x}) = \mathbf{x}$ would reproduce the kernel, as $K_{\text{lin}}(\mathbf{x}, \mathbf{y}) = \langle \mathbf{x}, \mathbf{y} \rangle_{\mathcal{X}}$ for all $\mathbf{x}, \mathbf{y} \in \mathcal{X}$.

Now the RKHS induced by K_{lin} is identified with the feature space, which is $\mathcal{X} = \mathbb{R}^p$ itself.

In case of more sophisticated kernels, $\mathcal{H}_K$ contains non-linear functions, and therefore, the features $\phi(\mathbf{x})$ can be realized usually in a much higher dimension than that of $\mathcal{X}$. For example, let us consider the so-called *polynomial kernel*

$$K_{\text{poly}}(\mathbf{x}, \mathbf{y}) = (\langle \mathbf{x}, \mathbf{y} \rangle_{\mathcal{X}} + c)^d, \quad c \geq 0, \; d \in \mathbb{N}, \quad (\mathbf{x}, \mathbf{y} \in \mathbb{R}^p)$$

obtained by using a special h in (1.31). It has $\binom{p+d}{d} \approx p^d$ eigenfunctions that span the space of p-variate polynomials with total degree d (this number is actually less if $c = 0$, i.e., the polynomials are of homogeneous degree).

In the following example of Reisen and Bunke (2010); Schölkopf and Smola (2002), $p = 2$, $d = 2$, but instead of a $\binom{4}{2}$ =6-dimensional feature space, a 3-dimensional one will do, since the choice $c = 0$ is possible. Indeed, for $\mathbf{x} = (x_1, x_2) \in \mathcal{X} = \mathbb{R}^2$ let

$$\phi(\mathbf{x}) := (x_1^2, x_2^2, \sqrt{2} x_1 x_2),$$

hence, $\mathcal{F} \subset \mathbb{R}^3$. The idea comes from wanting to separate data points allocated along two concentric circles, and therefore $\mathbb{R}^2 \to \mathbb{R}$ quadratic functions are applied. The separating circle with equation $x_1^2 + x_2^2 = r^2$ (with a radius r between the radii of the two concentric circles) becomes a plane in the new coordinate system. The original $\mathbf{x}$'s in $\mathbb{R}^2$ were separated by an annulus, whereas by projecting $\phi(\mathbf{x})$'s onto an appropriate two-dimensional plane of $\mathbb{R}^3$, a linear separation can be achieved. The two clusters can be separated by the k-means algorithm, as well.

Let us see exactly what kernel is applied here. Using the usual inner product in R^3,

$$\langle \phi(\mathbf{x}), \phi(\mathbf{y}) \rangle_{\mathcal{F}} = x_1^2 y_1^2 + x_2^2 y_2^2 + 2 x_1 x_2 y_1 y_2 = (x_1 y_1 + x_2 y_2)^2 = \langle \mathbf{x}, \mathbf{y} \rangle_{\mathcal{X}}^2,$$

hence, the new kernel is the square of the linear one, which is also positive definite (polynomial kernel with $c = 0$, $d = 2$).

The RKHS $\mathcal{H}_K$ corresponding to the feature space $\mathcal{F}$ now consists of homogeneous degree quadratic functions $\mathbb{R}^2 \to \mathbb{R}$, with the functions $f_1 : (x_1, x_2) \to x_1^2$, $f_2 : (x_1, x_2) \to x_2^2$, and $f_3 : (x_1, x_2) \to \sqrt{2} x_1 x_2$ forming an orthonormal basis in $\mathcal{H}_k$ such that $K(\mathbf{x}, \mathbf{y}) = \sum_{i=1}^3 f_i(\mathbf{x}) f_i(\mathbf{y})$. However, by the correspondence (1.34), the elements of $\mathcal{H}_K$ can be imagined as elements $\phi(\mathbf{x})$ in $\mathbb{R}^3$. Anyway, we do not need to navigate either in the RKHS, nor in the feature space $\mathcal{F}$, but all we need is the new kernel:

$$K(\mathbf{x}, \mathbf{y}) = [K_{\text{lin}}(\mathbf{x}, \mathbf{y})]^2, \quad \forall \mathbf{x}, \mathbf{y} \in \mathcal{X}.$$

In another setup, we may start with the above quadratic kernel K and build up the RKHS $\mathcal{H}_K$ as the span and completion of the following (homogeneous degree quadratic) $\mathcal{X} \to \mathbb{R}$ functions:

$$K_{\mathbf{x}} = \sum_{i=1}^{3} f_i(\mathbf{x}) f_i = x_1^2 f_1 + x_2^2 f_2 + \sqrt{2} x_1 x_2 f_3, \quad \mathbf{x} \in \mathcal{X}.$$

For example, $f_1 = K_{(1,0)}$, $f_2 = K_{(0,1)}$, and $f_3 = \frac{1}{2\sqrt{2}}(K_{(1,1)} - K_{(-1,1)})$.
Then $\mathcal{F} = \{\phi(\mathbf{x}) : \mathbf{x} \in \mathbb{R}^2\} \subset \mathbb{R}^3$, and

$$\langle K_{\mathbf{x}}, K_{\mathbf{y}} \rangle_{\mathcal{H}_K} = \langle \phi(\mathbf{x}), \phi(\mathbf{y}) \rangle_{\mathcal{F}} = K(\mathbf{x}, \mathbf{y}),$$

in accord with the theory, producing some kind of representation for special non-linear $\mathbb{R}^2 \to \mathbb{R}$ functions.

Observe that in both examples $\phi(x)$ is a vector with coordinates which are the basis vectors of the RKHS evaluated at x. In the first exercise $\phi(x)$ is an infinite, whereas in the second one, a finite dimensional vector. Note that $\mathcal{H}_K$ is an affine and sparsified version of the Hilbert space of $\mathcal{X} \to \mathbb{R}$ functions, between which the inner product is adopted to the requirement that it would reproduce the kernel.

1.5.4 Kernel – based on a sample – and the empirical feature map

In practical applications, we usually have a finite sample $\mathcal{X} = \{\mathbf{x}_1, \dots, \mathbf{x}_n)$. Based on it, the *empirical feature map* $\hat{\phi} : \mathcal{X} \to \mathbb{R}^n$ can be constructed in the following way (see e.g., Schölkopf and Smola (2002)):

$$\hat{\phi}(\mathbf{x}) = \boldsymbol{K}^{-1/2} \phi_n(\mathbf{x}), \tag{1.36}$$

with $\phi_n(\mathbf{x}) = (K(\mathbf{x}, \mathbf{x}_1), \dots, K(\mathbf{x}, \mathbf{x}_n))^T$, the counterpart of $K(\mathbf{x}, .)$ based on the n-element set $\mathcal{X}$, and the $n \times n$ symmetric real matrix $\boldsymbol{K} = (K_{ij})$ of entries $K_{ij} = K(\mathbf{x}_i, \mathbf{x}_j)$, $i, j = 1, \dots, n$. Assume that $\boldsymbol{K}$ is positive definite, otherwise (if positive semidefinite with at least one zero eigenvalue) we will use generalized inverse when calculating $\boldsymbol{K}^{-1/2}$. Let us apply (1.36) for $\mathbf{x}_i$'s. Since

$$\phi_n(\mathbf{x}_i) = \boldsymbol{K} \mathbf{e}_i,$$

where $\mathbf{e}_i$ is the ith unit vector in $\mathbb{R}^n$ (it has 0 coordinates, except the ith one which is equal to 1), the relation

$$\hat{\phi}(\mathbf{x}_i) = \boldsymbol{K}^{-1/2} \phi_n(\mathbf{x}_i) = \boldsymbol{K}^{1/2} \mathbf{e}_i$$

holds. Further,

$$\langle \hat{\phi}(\mathbf{x}_i), \hat{\phi}(\mathbf{x}_j) \rangle = (\boldsymbol{K}^{1/2} \mathbf{e}_i)^T (\boldsymbol{K}^{1/2} \mathbf{e}_j) = \mathbf{e}_i^T \boldsymbol{K} \mathbf{e}_j = K_{ij}, \quad i, j = 1, \dots, n.$$

Although we cannot see well in the artificially constructed spaces, this whole abstraction was not in vain. Observe that for the data points $\mathbf{x}_i$'s we need not even calculate $\hat{\phi}(\mathbf{x}_i)$'s; the spectral clustering of these images can be done based on their pairwise distances:

$$\begin{aligned}\|\hat{\phi}(\mathbf{x}_i) - \hat{\phi}(\mathbf{x}_j)\|^2 &= \langle \hat{\phi}(\mathbf{x}_i), \hat{\phi}(\mathbf{x}_j)\rangle + \langle \hat{\phi}(\mathbf{x}_i), \hat{\phi}(\mathbf{x}_i)\rangle - 2\langle \hat{\phi}(\mathbf{x}_j), \hat{\phi}(\mathbf{x}_j)\rangle \\ &= K(\mathbf{x}_i, \mathbf{x}_i) + K(\mathbf{x}_j, \mathbf{x}_j) - 2K(\mathbf{x}_i, \mathbf{x}_j)\end{aligned}$$

$(i, j = 1, \ldots, n)$. Thus, to evaluate the pairwise distances between any pairs of the n features, merely the kernel values are needed. Sometimes the kernel is some transformation of a similarity matrix of n objects, even if we do not have them as finite-dimensional points. In other cases, we have finite dimensional measurements on the objects, but merely the $n \times n$ empirical covariance matrix is stored. If our data are multivariate Gaussian, this matrix suffices for further processing, in other cases, we can calculate a polynomial or Gaussian kernel based on it, with the explanation that it may be the true similarity between our non-Gaussian data which are, in fact, in an abstract space. For example, if $K(\mathbf{x}_i, \mathbf{x}_j) = \langle \mathbf{x}_i, \mathbf{x}_j\rangle^2$, then the $n \times n$ similarity matrix of the features, applying a Gaussian kernel afterwards, gives matrix entries

$$K_{\text{Gauss}}(\hat{\phi}(\mathbf{x}_i), \hat{\phi}(\mathbf{x}_j)) = e^{-\frac{\|\hat{\phi}(\mathbf{x}_i) - \hat{\phi}(\mathbf{x}_j)\|^2}{2\sigma^2}} = e^{-\frac{\langle \mathbf{x}_i, \mathbf{x}_i\rangle^2 + \langle \mathbf{x}_j, \mathbf{x}_j\rangle^2 - 2\langle \mathbf{x}_i, \mathbf{x}_j\rangle^2}{2\sigma^2}}$$

which can be further processed through Laplacian-based clustering: see Chapter 2.

In this way, linear methods are applicable in an implicitly constructed space, instead of having to use non-linear methods in the original one. Here we only use the kernel which is calculated from the inner products of the data points through several transformations. The philosophy behind the above techniques is that sometimes sophisticated, composite kernels are more capable of revealing the structure of our data or to classify them, especially if they are not from a Gaussian distribution or consist of different types of measurements (e.g., location, brightness, color, texture). Just as in geometry, where the Euclidean distance is not necessarily the best choice, it is not always the linear kernel which is most useful in data representation.

But what kind of a kernel should be used? This is the important question. Many authors, for example von Luxburg (2006); von Luxburg *et al.* (2008), recommend the Gaussian kernel. For the data points $\mathbf{x}_1, \ldots, \mathbf{x}_n$ to be classified they construct the Gaussian kernel and the $n \times n$ symmetric, positive definite kernel matrix is considered as weight matrix $\boldsymbol{W}$ of a graph (in Ng *et al.* (2001) the authors use the zero diagonal). Then they perform spectral clustering based on the Laplacian or normalized Laplacian matrix corresponding to $\boldsymbol{W}$: see Chapter 2. In this way, applying the k-means algorithm for the so obtained k-dimensional (in fact, $(k-1)$-dimensional) representatives, they obtain satisfactory clusters. This is because the data points of $\mathcal{X}$ are mapped into a feature space $\mathcal{F}$ such that the only implicitly known images $\phi(\mathbf{x}_i)$ $(i = 1, \ldots, n)$ define a graph similarity, starting with which, the usual representation based spectral clustering works well. Hence, the graph construction is just an intermediate step for the subsequent metric clustering. Even if we are given a graph in advance, we may calculate the k-dimensional representatives of the vertices (with a relatively small k, based on the Laplacian eigenvectors), and then we classify them using kernel methods (e.g., substituting them into the Gaussian kernel).

The advantage of the Gaussian kernel is that it is also translation-invariant. The kernel $K : \mathbb{R}^p \times \mathbb{R}^p \to \mathbb{R}$ is *translation-invariant* if

$$K(\mathbf{x}, \mathbf{y}) = k(\mathbf{x} - \mathbf{y})$$

with some $\mathbb{R}^p \to \mathbb{R}$ function k. In this case, the feature space has infinite dimension and the RKHS determined by a translation-invariant K is described by Fourier theory, see Girosi *et al.* (1995). Since the Fourier transforms of functions in $\mathcal{H}_K$ decay rapidly, the induced RKHS consists of smooth functions. Now, let K be a p-dimensional Gaussian kernel with parameter $\sigma > 0$, defined in (1.33). Then

$$K(\mathbf{x}, \mathbf{y}) = k(\mathbf{x} - \mathbf{y}) = e^{-\frac{\|\mathbf{x}-\mathbf{y}\|^2}{2\sigma^2}} \quad (\mathbf{x}, \mathbf{y} \in \mathbb{R}^p),$$

$K_{\mathbf{x}}$'s are the so-called radial basis functions, and the functions of the induced RKHS are convolutions of functions of $L^2(\mathbb{R}^p)$ with $e^{-\frac{\|\mathbf{x}\|^2}{\sigma^2}}$. The RKHS decreases from $L^2(\mathbb{R}^p)$ to $\emptyset$ as σ increases from 0 to ∞. In this way, Gaussian kernels may be used as smoothing functions, for example, in the ACE (Alternating Conditional Expectation) algorithm elaborated for the generalized non-parametric regression problem, see Breiman and Friedman (1985) for details.

If the underlying space $\mathcal{X}$ is a probability space of random variables with finite variance, and the product space is endowed with a joint distribution (see Section 1.4), we may look for the so-called $\mathcal{F}$-correlation of two random variables which is the largest possible correlation between their ϕ-maps in the feature space. In Bach and Jordan (2002) it is proved that if $\mathcal{F}$ is the feature space corresponding to the RKHS defined by the Gaussian kernel on $\mathbb{R}$ (with any positive parameter σ), then the $\mathcal{F}$-correlation of two random variables is zero if and only if they are independent. Therefore, by mapping our data into the feature space, usual linear methods – like principal component analysis or canonical correlation analysis (see Sections C.1 and C.2 of the Appendix) – become non-linear ones, and we are able to find independent components instead of uncorrelated ones, which fact has significance if our data come from a non-Gaussian distribution. This is the base of the so-called kernel independent component analysis; see Bach and Jordan (2002) and Schölkopf *et al.* (1998) for details. Note that with a finite-dimensional feature space, $\mathcal{F}$-correlation cannot characterize independence.

Finally, let us remark that because of the smoothness of the functions in an RKHS, not only the ϕ-maps of the data points $\mathbf{x}_i$'s are available, but also $\phi(\mathbf{x})$ for an $\mathbf{x}$ in the small neighborhood of a data point. There is the Nyström formula of similar flavor than (1.36) to do so, see Baker (1977); Bengio *et al.* (2004). It can also be helped if the kernel is just some similarity function between pairs of data (not in a metric space), and though it is symmetric, not positive definite. In this case we can approximate it with a positive definite one. The technique applied is similar to that of the multidimensional scaling (see Appendix C.6) and other low rank approximations of Appendix C. In addition, we can better work with a low rank matrix; especially, if n is 'large' we need to find only some leading eigenvalues and eigenvectors of the $n \times n$ matrix $\boldsymbol{K}$. Using the Gaussian kernel, the entries of $\boldsymbol{K}$ corresponding to pairs of data points that are far away, are negligibly 'small' and made zero, which fact gives rise to use of algorithms developed for SD of sparse matrices, see for example Achlioptas and McSherry (2007).

Summarizing, RKHS techniques can be useful if we want to recover non-linear separation in our data. What we can do in general is that we calculate a linear kernel based on the sample and use admissible transformations to define newer and newer positive definite kernels, for example, a polynomial kernel of degree d if we guess that our data points can be separated with some curve of degree d. If only the relative position of the data points is of importance, we can build a Gaussian kernel based on their pairwise distances. The diffusion kernel is advisable to use in situations, when only a distance matrix of the objects is given and it is not Euclidean, cf. Definition C.6.2. In this case, either we use multidimensional scaling to embed the objects into a Euclidean space (and then use linear, polynomial, or Gaussian kernels), or else we select a diffusion kernel based on the not necessarily positive definite similarity matrix obtained from the given distance matrix (e.g., the entries are transformed by some monotonous decreasing function). Possibly, the similarities are correlations between the variables (if we want to classify the variables in a multidimensional dataset).

Even if we know the type of the kernel to be used, we must adapt its parameters to the data. With the new kernel and the pairwise distances, either multidimensional scaling or Laplacian-based spectral clustering can be performed here.

We remark that for image segmentation purposes, Shi and Malik (2000) use two or more Gaussian kernels: one contains the Euclidean distances of the pixels, and the others those of their brightness, color, texture, etc. Eventually, they take the product of the two or more positive definite kernels. If we multiply kernels, it means that entries in the same position of the kernel matrices are multiplied together. During the whole calculation for n data points, we only use the $n \times n$ symmetric, positive definite, usually sparse kernel matrix. In Yan *et al.* (2007) the authors also recommend kernelization, but they do not specify the kernel to be used.

References

Achlioptas D and McSherry F 2007 Fast computation of low-rank matrix approximations. *J. ACM* **54** (2), Article 9.

Aldous DJ 1981 Representations for partially exchangeable arrays of random variables. *J. Multivariate Anal.* **11** (4), 581–598.

Anderson WN and Morley TD 1985 Eigenvalues of the Laplacian of a graph. *Linear Multilinear Algebra* **18**, 141–145.

Ando T 1987 *RKHS and Quadratic Inequalities*, Lecture Notes, Hokkaido University, Sapporo, Japan.

Aronszajn N 1950 Theory of reproducing kernels. *Trans. Am. Math. Soc.* **68**, 337–404.

Austin T 2008 On exchangeable random variables and the statistics of large graphs and hypergraphs. *Prob. Surv.* **5**, 80–145.

Azran A and Ghahramani Z 2006 Spectral methods for automatic multiscale data clustering. In *Proc. IEEE Computer Society Conference on Computer Vision and Pattern Recognition (CVPR 2006), New York NY* (eds Fitzgibbon A, Taylor CJ and Lecun Y), IEEE Computer Society, Los Alamitos, California, pp. 190–197.

Bach FR and Jordan MI 2002 Kernel independent component analysis. *J. Mach. Learn. Res.* **3**, 1–48.

Baker C 1977 *The Numerical Treatment of Integral Equations*, Clarendon Press.

Ballester C, Calvó-Armengol A and Zenou Y 2006 Who's who in networks. Wanted: The key player. *Econometrica* **74** (5), 1403–1417.

Bengio Y, Delalleau O, Le Roux N *et al.* 2004 Learning eigenfunctions links spectral embedding and kernel PCA. *Neural Comput.* **16**, 2197–2219.

Berg Ch, Christensen J and Ressel P 1984 *Harmonic Analysis on Semigroups*, Springer.

Biggs NL 1974 *Algebraic Graph Theory*, Cambridge University Press, Cambridge.

Bolla M 1989 Spectra, Euclidean representation, and vertex-colourings of hypergraphs, Technical Report 78/1989, Mathematical Institute of the Hungarian Academy of Sciences.

Bolla M and Tusnády G 1990 Spectra and colourings of weighted graphs, Technical Report 52/1990, Mathematical Institute of the Hungarian Academy of Sciences.

Bolla M 1993 Spectra and Euclidean representation of hypergraphs. *Discret. Math.* **117**, 19–39.

Bolla M 2011 Penalized versions of the Newman–Girvan modularity and their relation to multi-way cuts and k-means clustering. *Phys. Rev. E* **84**, 016108.

Bolla M and Tusnády G 1994 Spectra and optimal partitions of weighted graphs. *Discret. Math.* **128**, 1–20.

Bolla M and M.-Sáska G 2004 Optimization problems for weighted graphs and related correlation estimates. *Discrete Mathematics* **282**, 23–33.

Bolla M, Friedl K and Krámli A 2010 Singular value decomposition of large random matrices (for two-way classification of microarrays). *J. Multivariate Anal.* **101**, 434–446.

Borgs C, Chayes JT, Lovász L T.-Sós V and Vesztergombi K 2008 Convergent sequences of dense graphs I: Subgraph frequences, metric properties and testing. *Advances in Math.* **219**, 1801–1851.

Breiman L and Friedman JH 1985 Estimating optimal transformations for multiple regression and correlation. *J. Am. Stat. Assoc.* **80**, 580–619.

Chung F 1997 *Spectral Graph Theory*, CBMS Regional Conference Series in Mathematics **92**, American Mathematical Society, Providence RI.

Cvetković DM, Doob M and Sachs H 1979 *Spectra of Graphs*, Academic Press, New York.

Cvetković DM, Rowlinson P and Simić S 1997 Eigenspaces of Graphs, *Encyclopedia of Mathematics and Its Applications* Vol. 66, Lavoisier Libraire, S.A.S. France.

Dhillon IS 2001 Co-clustering documents and words using bipartite spectral graph partitioning, in *Proc. ACM Int. Conf. Knowledge Disc. Data Mining (KDD 2001)* (eds Provost FJ and Srikant R), Association for Computer Machinery, New York, pp. 269–274.

Diaconis P and Janson S 2008 Graph limits and exchangeable random graphs. *Rend. Mat. Appl.* (VII. Ser.) **28**, 33–61.

Fiedler M 1973 Algebraic connectivity of graphs. *Czech. Math. J.* **23** (98), 298–305.

de Finetti B 1974 *Probability, Induction and Statistics*, John Wiley & Sons, Ltd.

Fréchet M 1907 Sur les ensembles de fonctions et les opérations linéaires. *C. R. Acad. Sci. Paris* **144**, 1414–1416.

Frieze A, Kannan R and Vempala S 1998 Fast Monte-Carlo algorithms for finding low-rank approximations. In *Proc. 39th Annual IEEE Symposium on Foundations of Computer Science (FOCS 1998), Palo Alto, California*, IEEE Computer Society, Los Alamitos, pp. 370–386.

Gebelein H 1941 Das statistische Problem der Korrelation als Variations und Eigenwertproblem und sein Zusammenhang mit der Ausgleichsrechnung. *Z. Angew. Math. Mech.* **21**, 364–379.

Girosi F, Jones M and Poggio T 1995 Regularization theory and neural networks architectures. *Neural Comput.* **7**, 219–269.

Hoffman AL 1969 The change in the least eigenvalue of the adjacency matrix of a graph under imbedding. *SIAM J. Appl. Math.* **17**, 664–671.

Jackson MO and Zenou Y (forthcoming) Games on Networks. To appear in *Handbook of Game Theory* **4** (eds Young P and Zamir S). Elsevier Science.

Kallenberg O 2001 *Foundations of Modern Probability*, 2nd edn., Springer.

Kelmans AK 1967 Properties of the charasteristic polynomial of a graph. *Cibernetics in the Science of Communication* **4**, Energija, Moskva–Leningrad, 27–41 (in Russian).

Kleinberg J 1997 *Authoritative Sources in Hyperlinked Environment.* IBM Research Report RJ 10076 (91892).

Lovász L 1993 *Combinatorial Problems and Exercises.* Akadémiai Kiadó–North Holland, Budapest–Amsterdam.

Liotta G ed. 2004 *Graph drawing.* Lecture Notes in Computer Science **2912**, Springer.

Maas C 1987 Transportation in graphs and the admittance spectrum. *Discret. Appl. Math.* **16**, 31–49.

Merris R 1987 Characteristic vertices of trees. *Linear Multilinear Algebra* **22**, 115–131.

Mohar B 1988 Isoperimetric inequalities, growth and the spectrum of graphs. *Linear Algebra Appl.* **103**, 119–131.

Mohar B 1991 The Laplacian spectrum of graphs, in *Graph Theory, Combinatorics, and Applications* Vol 2 (eds Alavi Y, Chartrand G, Oellermann OR and Schwenk AJ), John Wiley & Sons, Ltd, pp. 871–898.

Nelsen RB 2006 *An Introduction to Copulas*, Springer.

Newman MEJ 2004 Detecting community structure in networks. *Eur. Phys. J. B* **38**, 321–330.

Newman MEJ and Girvan M 2004 Finding and evaluating community structure in networks. *Phys. Rev. E* **69**, 026113.

Ng AY Jordan MI and Weiss Y 2001 On spectral clustering: analysis and an algorithm, in *Proceedings of the 14th Neural Information Processing Systems Conference, NIPS 2001* (eds Dietterich TG, Becker S and Ghahramani Z), MIT Press, Cambridge, USA, pp. 849–856.

Parzen E 1963 Probability density functionals and reproducing kernel Hilbert spaces, in *Proceedings of the Symposium on Time Series Analysis* (ed. Rosenblatt M), Brown University, Providence, RI, USA, pp. 155–169.

Pisanski T and Shawe-Taylor J 2000 Characterizing graph drawing with eigenvectors. *J. Chem. Inf. Comput. Sci.* **40**, 567–571.

Ramsay JO and Silverman BW 2006 *Functional Data Analysis*, Springer.

Reisen K and Bunke H 2010 *Graph Classification and Clustering Based on Vector Space Embedding*, World Scientific.

Rényi A 1959a On measures of dependence. *Acta Math. Acad. Sci. Hung.* **10**, 441–451.

Rényi A 1959b New version of the probabilistic generalization of the large sieve. *Acta Math. Acad. Sci. Hung.* **10**, 218–226.

Riesz F 1907 Sur une espéce de géométrie analytique des systémes de fonctions sommables. *C. R. Acad. Sci. Paris* **144**, 1409–1411.

Riesz F 1909 Sur les opérations fonctionnelles linéaires. *C. R. Acad. Sci. Paris* **149**, 974–977.

Riesz F and Sz.-Nagy B 1952 *Leçons d'Analyse Fonctionnelle*, Academic Publishing House, Budapest.

Shawe-Taylor J and Cristianini N 2004 *Kernel Methods for Pattern Analysis*, Cambridge Univ. Press, Cambridge.

Shi J and Malik J 2000 Normalized cuts and image segmentation. *IEEE Trans. Pattern Anal. Mach. Intell.* **22** (8), 888–905.

Schölkopf B Smola AJ and Müller K-R 1998 Nonlinear component analysis as a kernel eigenvalue problem. *Neural Comput.* **10**, 1299–1319.

Schölkopf B and Smola AJ 2002 *Learning with Kernels*, MIT Press, Cambridge, USA.

von Luxburg U 2006 A tutorial on spectral clustering. *Stat. Comput.* **17**, 395–416.

von Luxburg U, Belkin M and Bousquet O 2008 Consistency of spectral clustering. *Ann. Stat.* **36**, 555–586.

White S and Smyth P 2005 A spectral clustering approach to find communities in graphs, in *Proc. SIAM International Conference on Data Mining* (eds Kargupta H, Srivastava J, Kamath Ch and Goodman A), SIAM, Newport Beach, pp. 76–84.

Yan S, Xu D, Zhang B, *et al.* 2007 Graph embedding and extensions: a general framework for dimensionality reduction. *IEEE Trans. Pattern Anal. Mach. Intell.* **29** (1), 40–48.

2

Multiway cuts and spectra

Minimum ratio- or normalized multiway cut problems are discussed together with modularity cuts. Since the optima of the corresponding objective functions are taken on partition vectors corresponding to the hidden clusters, they are related to Laplacian, normalized Laplacian, or modularity spectra, whereas the precision of the estimates depends on the distance between the subspaces spanned by the corresponding eigenvectors and partition vectors. By an analysis of variance argument, this distance is the sum of the inner variances of the underlying clusters, the objective function of the k-means algorithm.

2.1 Estimating multiway cuts via spectral relaxation

Clusters (in other words, modules or communities) of graphs are typical (strongly or loosely connected) subsets of vertices that can be identified, for example, with social groups or interacting enzymes in social or metabolic networks, respectively; they form special partition classes of the vertices. To measure the performance of a clustering, different kinds of multiway cuts are introduced and estimated by means of Laplacian spectra. The key motif of these estimations is that minima and maxima of the quadratic placement problems of Section 1.1 are attained on some appropriate eigenspaces of the Laplacian, while optimal multiway cuts are special values of the same quadratic objective function realized by piecewise constant vectors. Hence, the optimization problem, formulated in terms of the Laplacian eigenvectors, is the continuous relaxation of the underlying maximum or minimum multiway cut problem. The objective functions defined on the partitions of the vertices are sometimes called modularities, see Newman (2010); Newman and Girvan (2004a).

Spectral Clustering and Biclustering: Learning Large Graphs and Contingency Tables, First Edition.
Marianna Bolla.

2.1.1 Maximum, minimum, and ratio cuts of edge-weighted graphs

In the sequel, we will use the general framework of an edge-weighted graph introduced in Chapter 1. Let $G = (V, \boldsymbol{W})$ be an edge-weighted graph on n vertices with edge-weight matrix $\boldsymbol{W}$ and generalized vertex-degrees $d_1, \dots, d_n$. For a fixed integer $1 \le k \le n$, let $P_k = (V_1, \dots, V_k)$ be a k-partition of the vertices, where the disjoint, non-empty vertex subsets $V_1, \dots, V_k$ will be referred to as clusters or modules.

The number of k-partitions is k^n, or the Stirling's partition number $\left\{ {n \atop k} \right\}$ if there are no empty clusters, see Lovász (1993). Let $\mathcal{P}_k$ denote the set of all k-partitions. Optimization over $\mathcal{P}_k$ is usually NP-complete, except in some special classes of graphs.

To illustrate the relaxation technique, first we perform an easy estimation in the simplest case of $k = 2$. To this end, we introduce some definitions.

Definition 2.1.1 *The weighted cut between the non-empty vertex-subsets $U, T \subset V$ of the edge-weighted graph $G = (V, \boldsymbol{W})$ is*

$$w(U, T) = \sum_{i \in U} \sum_{j \in T} w_{ij}.$$

Note that $w(U, T)$ is the sum of the weights of edges connecting vertices of U and T. For now, U and T are not necessary disjoint subsets, though, in the sequel we will mainly use this notion for disjoint cluster pairs of a partition of V.

The so-called *maximum cut problem* looks for the maximum of the above weighted cut over 2-partitions of the vertices.

Definition 2.1.2 *The maximum cut of the edge-weighted graph $G = (V, \boldsymbol{W})$ is*

$$\mathrm{maxcut}(G) = \max_{U \subset V} w(U, \overline{U}), \tag{2.1}$$

where $\overline{U}$ is the complement of U in V.

The following statement is due to Mohar and Poljak (1990), though, in the proof we will use the machinery of Chapter 1.

Proposition 2.1.3 *Let $0 = \lambda_0 < \lambda_1 \le \cdots \le \lambda_{n-1}$ denote the eigenvalues of the Laplacian $\boldsymbol{L}$ of the connected edge-weighted graph $G = (V, \boldsymbol{W})$ on n vertices. Then*

$$\mathrm{maxcut}(G) \le \frac{n}{4} \lambda_{n-1}.$$

Proof. Proposition A.3.11 of the Appendix guarantees that

$$\lambda_{n-1} = \max_{\|\mathbf{x}\|=1} \mathbf{x}^T \boldsymbol{L} \mathbf{x}. \tag{2.2}$$

Due to Equation (1.3), it follows that

$$\mathbf{x}^T \boldsymbol{L}\mathbf{x} = Q_1 = \sum_{i<j} w_{ij}(r_i - r_j)^2 \tag{2.3}$$

where the coordinates of the vector $\mathbf{x} = (r_1, \dots, r_n)^T$ are considered as one-dimensional representatives of the vertices.

Let U^* be the subset of V giving the maximum in (2.1). With it, let us define the following 1-dimensional representation of the vertices:

$$r_i = \begin{cases} n - |U^*| \text{ if } & i \in U^* \\ -|U^*| \quad \text{if } & i \in \overline{U^*}. \end{cases} \tag{2.4}$$

Let us normalize the vector $\mathbf{x}$ of coordinates r_i's. Because of

$$\|\mathbf{x}\|^2 = \sum_{i=1}^{n} r_i^2 = |U^*|(n - |U^*|)^2 + (n - |U^*|)|U^*|^2 = |U^*|(n - |U^*|)n,$$

the so obtained unit-norm vector $\tilde{\mathbf{x}}$ will have coordinates $\tilde{r}_i = \frac{r_i}{\sqrt{|U^*|(n-|U^*|)n}}$. With this special vector, we will go below the maximum in (2.2), that is

$$\lambda_{n-1} \geq \tilde{\mathbf{x}}^T \boldsymbol{L}\tilde{\mathbf{x}} = \sum_{i<j} w_{ij}(\tilde{r}_i - \tilde{r}_j)^2 = w(U^*, \overline{U^*})\frac{(n - |U^*| + |U^*|)^2}{|U^*|(n - |U^*|)n},$$

where we also used (2.3). Therefore, utilizing that $w(U^*, \overline{U^*}) = \text{maxcut}(G)$,

$$\text{maxcut}(G) \leq \frac{|U^*|(n - |U^*|)}{n}\lambda_{n-1} \leq \frac{n}{4}\lambda_{n-1}.$$

We remark the following.

- Observe that the role of U^* and $\overline{U^*}$ is symmetric in the proof. The last estimation suggests that the upper bound for the maximum cut can be sharp for balanced partition of V, that is, when U^* and $\overline{U^*}$ nearly have the same cardinality. If one looks for the maximum in (2.1) under the condition that the clusters of the 2-partition are of equal sizes, we get the maximum bisection problem which makes sense, of course, for even n.

- The proof also gives us a hint how to find the optimal U^*: the eigenvector $\mathbf{u}_{n-1}$ should be close to a piecewise constant vector over an appropriate 2-partition of the vertices. Later on, in this section, we will prove that it can be achieved by the 2-partition which is the output of the k-means algorithm with 2 clusters applied for the coordinates of the vector $\mathbf{u}_{n-1}$. With this spectral relaxation, the maximum cut can be approached in polynomial time for finding the largest eigenvalue and corresponding eigenvector of the Laplacian matrix. However, the exact solution of the maximum cut problem is NP-complete, even for simple graphs, called the maximum bipartite subgraph problem, though for some special classes of graphs, it is polynomially solvable (for example, for planar graphs).

The minimum cut of an edge-weighted graph is defined analogously, and for simple graphs, it is the edge-connectivity of Fiedler (1973). The solution is often given by an uneven 2-partition, for example, if there is an almost isolated vertex connected to few other vertices, it may form a cluster itself. To prevent this situation and rather find real-life loosely connected clusters, we require some balancing for the cluster sizes. For this purpose, we define the minimum cut and ratio cut for not only 2 clusters, but for any k-partition of the vertices and minimize it over the set of k-partitions. Roughly speaking, the minimum k-way cut minimizes the sum of the weights of intersecting edges (between the clusters), whereas the ratio cut (see e.g., Alpert and Yao (1995); Bolla (1989, 1993); Hagen and Kahng (1992)) in addition, penalizes partitions with very unequal cluster sizes.

Definition 2.1.4 *Let $G = (V, \mathbf{W})$ be an edge-weighted graph and $P_k = (V_1, \ldots, V_k)$ a k-partition of its vertices. The k-way cut of G corresponding to the k-partition P_k is*

$$\mathrm{cut}(P_k, G) = \sum_{a=1}^{k-1} \sum_{b=a+1}^{k} w(V_a, V_b)$$

and the minimum k-way cut of G is

$$\mathrm{mincut}_k(G) = \min_{P_k \in \mathcal{P}_k} \mathrm{cut}(P_k, G). \tag{2.5}$$

Definition 2.1.5 *Let $G = (V, \mathbf{W})$ be an edge-weighted graph and $P_k = (V_1, \ldots, V_k)$ a k-partition of its vertices. The k-way ratio cut of G corresponding to the k-partition P_k is*

$$g(P_k, G) = \sum_{a=1}^{k-1} \sum_{b=a+1}^{k} \left(\frac{1}{|V_a|} + \frac{1}{|V_b|} \right) w(V_a, V_b) = \sum_{a=1}^{k} \frac{w(V_a, \overline{V}_a)}{|V_a|}$$

and the minimum k-way ratio cut of G is

$$g_k(G) = \min_{P_k \in \mathcal{P}_k} g(P_k, G).$$

The equivalent form in the formula of $g(P_k, G)$ is obtained by an easy calculation.

Proposition 2.1.6 *Let $0 = \lambda_0 < \lambda_1 \leq \cdots \leq \lambda_{n-1}$ denote the eigenvalues of the Laplacian $\mathbf{L}$ of the connected edge-weighted graph $G = (V, \mathbf{W})$ on n vertices. Then*

$$\mathrm{mincut}_2(G) \geq \frac{n-1}{n} \lambda_1. \tag{2.6}$$

Proof. We will use a similar relaxation technique as in the Proof of Proposition 2.1.3. Here Proposition A.3.13 of the Appendix guarantees that

$$\lambda_1 = \min_{\substack{\|\mathbf{x}\|=1 \\ \mathbf{x}^T\mathbf{1}=0}} \mathbf{x}^T \boldsymbol{L}\mathbf{x}. \tag{2.7}$$

Let us recall that due to Equation (1.3), the relation

$$\mathbf{x}^T \boldsymbol{L}\mathbf{x} = Q_1 = \sum_{i<j} w_{ij}(r_i - r_j)^2$$

holds with the coordinates of the vector $\mathbf{x} = (r_1, \ldots, r_n)^T$ which are considered as one-dimensional representatives of the vertices.

Let U^* be the subset of V giving the minimum in (2.5) for $k = 2$. With it, we define the same representation (2.4) of the vertices as in the Proof of Proposition 2.1.3 and we use the same normalization, so that the vector $\tilde{\mathbf{x}} = (\tilde{r}_1, \ldots, \tilde{r}_n)$ satisfies $\|\tilde{\mathbf{x}}\| = 1$ and $\tilde{\mathbf{x}}^T\mathbf{1} = 0$, thanks to $\sum_{i=1}^n \tilde{r}_i = 0$. With this special vector, we will go above the minimum in (2.7), that is

$$\lambda_1 \leq \tilde{\mathbf{x}}^T \boldsymbol{L}\tilde{\mathbf{x}} = \sum_{i<j} w_{ij}(\tilde{r}_i - \tilde{r}_j)^2 = w(U^*, \overline{U^*})\frac{(n - |U^*| + |U^*|)^2}{|U^*|(n - |U^*|)n}.$$

Therefore, utilizing that $w(U^*, \overline{U^*}) = \text{mincut}_2(G)$,

$$\text{mincut}_2(G) \geq \frac{|U^*|(n - |U^*|)}{n}\lambda_1 \geq \frac{n-1}{n}\lambda_1.$$

A number of comments are in order.

- For a simple graph G, Fiedler (1973) called the quantity $\text{mincut}_2(G)$ the edge-connectivity of G, because it is equal to the minimum number of edges that should be removed to make G disconnected. He used the notation $e(G)$ for the *edge-connectivity* of the simple graph G, and $v(G)$ for its *vertex-connectivity* (minimum number of vertices which should be removed to make G disconnected). In his papers, Fiedler (1972, 1973) proved that $v(G) \leq e(G)$, and for any graph G on n vertices, that differs from the complete graph K_n, the relation

$$\lambda_1 \leq v(G) \leq e(G) \tag{2.8}$$

  holds, which gives a sharper estimate for λ_1 by $e(G)$ than Inequality (2.6). However, for the complete graph K_n, equality holds in (2.6). Indeed, on the one hand, Example (a) of Section 1.1.3 shows that $\lambda_1(K_n) = n$, and on the other hand, $e(K_n) = n - 1$, since the minimum cut is realized by a 2-partition consisting of a single vertex and all the other vertices.

- Fiedler (1973) also provided two lower estimates for λ_1 by $e(G)$:

$$\lambda_1 \geq 2e(G)(1 - \cos\frac{\pi}{n}) \tag{2.9}$$

and

$$\lambda_1 \geq C_1 e(G) - C_2 d_{\max}, \tag{2.10}$$

where $C_1 = 2(\cos\frac{\pi}{n} - \cos\frac{2\pi}{n})$, $C_2 = 2\cos\frac{\pi}{n}(1 - \cos\frac{\pi}{n})$, and $d_{\max} = \max_i d_i$ is the maximum vertex degree. Compared to (2.8), this estimation makes sense in the $n \geq 3$ case. The bound of (2.10) is better than that of (2.9) if and only if $e(G) \geq \frac{1}{2} d_{\max}$. The two estimates are equal and sharp for the path graph P_n with $e(P_n) = 1$ and $\lambda_1(P_n) = 2(1 - \cos\frac{\pi}{n})$, see Example (b) of Section 1.1.3. The path graph can be split into two clusters by removing any of its edges; however, we would not state that it has two underlying clusters. The ratio cut of P_n is minimized by removing the middle edge (for even n) or one of the middle edges (for odd n), thus, providing balanced clusters. Note that P_n is a tree, and hence, a sparse graph, that is not expected to have a remarkable cluster structure, unlike the dense graphs, incarnating a statistical sample.

- Because of this bilateral relation between λ_1 and $e(G)$, the smallest positive Laplacian eigenvalue of a connected graph is able to detect the strength of its connectivity; therefore, Fiedler called λ_1 the *algebraic connectivity* of G, and denoted it by $a(G)$. This relation between $e(G)$ and $a(G)$ was also recovered by Donath and Hoffman (1973) and Hoffman (1972). Graphs with 'large' algebraic connectivity play an important role in communication networks, since the information goes through them very quickly. The so-called concentrators and expanders are graphs with high connectivity properties, see Alon (1986); Alon and Milman (1987); Babai and Szegedy (1992); Chung (1997) and the parallel sorting of Ajtai *et al.* (1983).

- The proof of Proposition 2.1.6 gives us the following hint how to find the optimal U^*: the eigenvector $\mathbf{u}_1$ should be close to a piecewise constant vector over an appropriate 2-partition of the vertices. Note that because of its orthogonality to the vector $\mathbf{1}$, the vector $\mathbf{u}_1$ consists of both positive and negative coordinates, and Juhász and Mályusz (1980) separated the two clusters according to the signs. In the sequel, we will use the k-means algorithm for this purpose, in a more general setup. Note that the vector $\mathbf{u}_1$ is frequently called the *Fiedler-vector*.

The problem of finding the minima of the above multiway cuts over k-partitions is NP-complete. However, spectral techniques working in polynomial time are at our disposal. How accurately these minima can be approximated by means of spectral clustering depends on how close the partition vectors can get to the eigenvectors corresponding to the k smallest Laplacian or normalized Laplacian eigenvalues. The measure of the closeness of the involved subspaces is the k-variance defined in Appendix C.5 and minimized by the k-means algorithm. More precisely, we will apply the k-means algorithm for the optimal representatives of the vertices introduced in Chapter 1. For this purpose, instead of partitions, we will use partition vectors.

The k-partition P_k is uniquely determined by the $n \times k$ balanced partition matrix $\mathbf{Z}_k = (\mathbf{z}_1, \dots, \mathbf{z}_k)$, where the a-th *balanced k-partition vector* $\mathbf{z}_a = (z_{1a}, \dots, z_{na})^T$

is the following:

$$z_{ia} = \begin{cases} \frac{1}{\sqrt{|V_a|}} & \text{if} \quad i \in V_a \\ 0 & \text{otherwise.} \end{cases} \tag{2.11}$$

The matrix $\mathbf{Z}_k$ is trivially suborthogonal, and the set of balanced k-partition matrices is denoted by $\mathcal{Z}_k^B$. With the special representation in which the representatives $\tilde{\mathbf{r}}_1, \dots, \tilde{\mathbf{r}}_n \in \mathbb{R}^k$ are row vectors of $\mathcal{Z}_k^B$, the ratio cut of $G = (V, \mathbf{W})$ corresponding to the k-partition P_k (see Definition 2.1.5) can be rewritten as

$$g(P_k, G) = \sum_{i=1}^{n-1} \sum_{j=i+1}^{n} w_{ij} \|\tilde{\mathbf{r}}_i - \tilde{\mathbf{r}}_j\|^2 = \sum_{a=1}^{k} \mathbf{z}_a^T \mathbf{L} \mathbf{z}_a = \text{tr}(\mathbf{Z}_k^T \mathbf{L} \mathbf{Z}_k). \tag{2.12}$$

We want to minimize it over balanced k-partition matrices $\mathbf{Z}_k \in \mathcal{Z}_k^B$.

Assume that G is connected. Let $0 = \lambda_0 < \lambda_1 \leq \dots \leq \lambda_{n-1}$ denote the eigenvalues of its Laplacian matrix $\mathbf{L}$ with corresponding unit-norm, pairwise orthogonal eigenvectors $\mathbf{u}_0, \mathbf{u}_1, \dots, \mathbf{u}_{n-1}$. Namely, $\mathbf{u}_0 = \frac{1}{\sqrt{n}}\mathbf{1}$.

Theorem 2.1.7 *For the minimum k-way ratio cut of the edge-weighted graph $G = (V, \mathbf{W})$ the lower estimate*

$$g_k(G) \geq \sum_{i=1}^{k-1} \lambda_i$$

holds.

Proof. The discrete problem is relaxed to a continuous one. Let $\mathbf{r}_1, \dots, \mathbf{r}_n$ denote k-dimensional representatives of the vertices. Let $\mathbf{X}$ be the $n \times k$ matrix with these representatives as row-vectors. The Representation theorem for edge-weighted graphs (see Theorem 1.1.2) states that

$$\min_{\mathbf{X}^T\mathbf{X}=\mathbf{I}_k} \sum_{i=1}^{n-1} \sum_{j=i+1}^{n} w_{ij} \|\mathbf{r}_i - \mathbf{r}_j\|^2 = \min_{\mathbf{X}^T\mathbf{X}=\mathbf{I}_k} \text{tr}(\mathbf{X}^T \mathbf{L} \mathbf{X}) = \sum_{i=0}^{k-1} \lambda_i,$$

and equality is attained with $\mathbf{X} = (\mathbf{u}_0, \dots, \mathbf{u}_{k-1})$.

As a balanced k-partition matrix is a special suborthogonal matrix,

$$g_k(G) = \min_{\mathbf{Z}_k \in \mathcal{Z}_k^B} \text{tr}(\mathbf{Z}_k^T \mathbf{L} \mathbf{Z}_k) \geq \sum_{i=0}^{k-1} \lambda_i \tag{2.13}$$

and equality can be attained only in the $k = 1$ trivial case, otherwise the eigenvectors $\mathbf{u}_i$ $(i = 1, \dots, k-1)$ cannot be partition vectors, since their coordinates sum to 0 because of the orthogonality to the $\mathbf{u}_0$ vector.

In the case of $k = 2$, in view of Theorem 2.1.7, $g_2(G)$ is bounded from below by λ_1, akin to the edge-connectivity of Fiedler (1973). The proof also suggests that the quality of the above estimation depends on how close the k bottom eigenvectors are to partition vectors. Concerning this issue and the optimum choice of k, we have the following results. The clustering P_k of the vertices divides the edges of the underlying graph into within- and between-cluster edges. (If the vertices of the clusters are colored with k different colors, then we can speak in terms of mono- and bicolored edges.) Accordingly, the weight matrix is decomposed as $\boldsymbol{W} = \boldsymbol{W}_w + \boldsymbol{W}_b$. Since for the corresponding degree-matrices $\boldsymbol{D} = \boldsymbol{D}_w + \boldsymbol{D}_b$ holds, their Laplacians are also added together: $\boldsymbol{L} = \boldsymbol{L}_w + \boldsymbol{L}_b$. Let the k-partition P_k be such that the smallest positive eigenvalue (say, ρ) of $\boldsymbol{L}_w$ is greater than the largest eigenvalue (say, ε) of $\boldsymbol{L}_b$. Here ρ and ε depend on $P_k = (V_1, \dots, V_k)$, but we denote this dependence only if necessary. Let us examine ρ and ε.

Since the matrix $\boldsymbol{L}_b$ is block-diagonal, its spectrum consists of the spectra of its blocks' Laplacians, therefore ρ is the smallest algebraic connectivity of the clusters and can be estimated from below by constant times the minimum of the edge-connectivities of the edge-weighted subgraphs induced by the clusters $V_1, \dots, V_k$, see (2.9) and (2.10). Therefore, ρ is relatively large, if the clusters themselves have a large spectral gap which is defined as their smallest positive Laplacian eigenvalue (if not, the clusters can be split into further clusters, and k is increased).

By the Gersgorin disc theorem A.3.2, ε can be estimated from above by $2\max_{i\in\{1,\dots,n\}} \sum_{j:\, c(j) \neq c(i)} w_{ij}$, where $c(i)$ is the cluster membership of vertex i in the clustering (coloring) $V_1, \dots, V_k$. It is small if from each vertex, there are few, small-weight edges emanating to clusters different of the vertex's own cluster. A sharper estimation can be given by using the following result of Mohar and Poljak (1990): the maximum Laplacian eigenvalue of an edge-weighted graph with generalized degrees d_i's can be estimated from above by

$$\lambda_{\max} \leq \max_{w_{ij} \neq 0} (d_i + d_j).$$

In view of this,

$$\varepsilon \leq \max_{w_{ij} \neq 0} \left(\sum_{\ell:\, c(\ell) \neq c(i)} w_{i\ell} + \sum_{\ell:\, c(\ell) \neq c(j)} w_{j\ell} \right).$$

The Weyl's perturbation theorem (see Theorem A.3.18) applied for the kth and $(k+1)$th eigenvalues (in increasing order) of $\boldsymbol{L}$ and $\boldsymbol{L}_w$ (taking $\boldsymbol{L}_b$ as perturbation) yields that on the one hand,

$$\lambda_{k-1}(\boldsymbol{L}_w) + \lambda_0(\boldsymbol{L}_b) \leq \lambda_{k-1}(\boldsymbol{L}) \leq \lambda_{k-1}(\boldsymbol{L}_w) + \lambda_{n-1}(\boldsymbol{L}_b), \tag{2.14}$$

that is, $0 + 0 \leq \lambda_{k-1} \leq 0 + \varepsilon$. On the other hand,

$$\lambda_k(\boldsymbol{L}_w) + \lambda_0(\boldsymbol{L}_b) \leq \lambda_k(\boldsymbol{L}) \leq \lambda_k(\boldsymbol{L}_w) + \lambda_{n-1}(\boldsymbol{L}_b), \tag{2.15}$$

that is, $\rho + 0 \leq \lambda_k \leq \rho + \varepsilon$. Summarizing the estimates of (2.14) and (2.15), for any k-partition P_k such that $\varepsilon(P_k) \leq \rho(P_k)$, we get that

$$0 < \lambda_{k-1} \leq \varepsilon(P_k) \leq \rho(P_k) \leq \lambda_k \leq \rho(P_k) + \varepsilon(P_k).$$

Therefore,

$$\frac{\lambda_{k-1}}{\lambda_k} \leq \frac{\varepsilon(P_k)}{\rho(P_k)}, \tag{2.16}$$

where the inequality also holds for a k-partition P_k for which $\varepsilon(P_k)$ happens to exceed $\rho(P_k)$. Consequently,

$$\frac{\lambda_{k-1}}{\lambda_k} \leq \min_{P_k \in \mathcal{P}_k} \frac{\varepsilon(P_k)}{\rho(P_k)}. \tag{2.17}$$

In Section C.5 of the Appendix we defined the notion of k-variance of points in a finite dimensional Euclidean space. Briefly, let us denote by $S_k^2(P_k; \boldsymbol{X}^*)$ the k-variance of the optimal vertex representatives comprising row vectors of $\boldsymbol{X}^* = (\mathbf{u}_0, \ldots, \mathbf{u}_{k-1})$ in the clustering P_k of the vertices. For it, an estimation as in (2.16) will be proved.

Proposition 2.1.8 *Let $G = (V, \boldsymbol{W})$ be edge-weighted graph and P_k be k-partition of its vertices such that $\varepsilon(P_k) < \rho(P_k)$, where $\varepsilon(P_k)$ is the largest, while $\rho(P_k)$ is the smallest positive eigenvalue of the Laplacian $\mathbf{L}_b$ and $\mathbf{L}_w$, defined above, respectively. Let $\boldsymbol{X}^*$ be optimum k-dimensional representation of the vertices. Then*

$$S_k^2(P_k; \boldsymbol{X}^*) \leq k \frac{\varepsilon(P_k)}{\rho(P_k)}.$$

Trivially, $S_k^2(P_k; \boldsymbol{X}^*) \leq k$ (since the k variance of the rows of $\boldsymbol{X}^*$ is the sum of k-variances of its column's coordinates, which are at most their constantly 1 variances). Therefore the estimate of Proposition 2.1.8 also holds for a k-partition P_k with $\varepsilon(P_k) \geq \rho(P_k)$. These considerations result in the following important statement.

Proposition 2.1.9

$$S_k^2(\boldsymbol{X}^*) \leq k \min_{P_k \in \mathcal{P}_k} \frac{\varepsilon(P_k)}{\rho(P_k)}.$$

Thus, both the k-variance and the gap in the spectrum between λ_{k-1} and λ_k can be estimated from above by the minimum $\frac{\varepsilon(P_k)}{\rho(P_k)}$ ratio, but they cannot be directly estimated by each other as in case of the normalized Laplacian eigenvalues, see Section 2.2. This ratio is small in the presence of clusters with high intra- and low inter-cluster densities.

To prove Proposition 2.1.8, we use a lemma of Bolla and Tusnády (1994).

Lemma 2.1.10 *Let the $n \times n$ symmetric, positive semidefinite matrix $\boldsymbol{B}$ be of rank $n - k$ with k-dimensional kernel space F, and let ρ denote the minimum of its positive eigenvalues. Let $\boldsymbol{P}$ be $n \times n$ positive semidefinite matrix (perturbation) such* *that* $\|\boldsymbol{P}\| = \varepsilon$. *Then the* $n \times n$ *positive semidefinite matrix* $\boldsymbol{A} = \boldsymbol{B} + \boldsymbol{P}$ *has at least* k *eigenvalues which are at most* ε. *Further, denoting by* $\mathbf{y}_1, \dots, \mathbf{y}_k$ *the unit-norm eigenvectors corresponding to the* k *smallest eigenvalues of* $\boldsymbol{A}$, *and decomposing them as*

$$\mathbf{y}_i = \mathbf{v}_i + \mathbf{r}_i, \qquad \mathbf{v}_i \in F, \quad \mathbf{r}_i \perp F,$$

for the norms of the orthogonal components, the relations

$$\|\mathbf{r}_i\|^2 \le \frac{\varepsilon}{\rho} \quad (i = 1, \dots, k)$$

hold true.

Proof. The first part of the statement follows from Theorem A.3.19 of the Appendix (the Weyl's perturbation theorem for symmetric matrices), since the k smallest eigenvalues of the matrix $\boldsymbol{B}$ are zeros (with corresponding eigenspace F), and the perturbation $\boldsymbol{P}$ has spectral norm ε. As for the second part, let $\mathbf{v} \in F, \ \|\mathbf{v}\| = 1$ be an arbitrary vector. Then, on the one hand,

$$\mathbf{v}^T \boldsymbol{A} \mathbf{v} = \mathbf{v}^T \boldsymbol{P} \mathbf{v} \le \|\boldsymbol{P}\| \le \varepsilon. \tag{2.18}$$

On the other hand, as any $\mathbf{y} \in R^n$ can uniquely be decomposed as $\mathbf{y} = \mathbf{v} + \mathbf{r}$, where $\mathbf{v} \in F$ and $\mathbf{r} \perp F$,

$$\mathbf{y}^T \boldsymbol{A} \mathbf{y} = \mathbf{y}^T \boldsymbol{B} \mathbf{y} + \mathbf{y}^T \boldsymbol{P} \mathbf{y} = \mathbf{r}^T \boldsymbol{B} \mathbf{r} + \mathbf{y}^T \boldsymbol{P} \mathbf{y} \ge \varrho \|\mathbf{r}\|^2. \tag{2.19}$$

Let $\mathbf{y}_1, \dots, \mathbf{y}_k$ be an orthonormal set of eigenvectors corresponding to the k smallest eigenvalues (being at most ε) of $\boldsymbol{A}$. Then, according to the Inequalities (2.18) and (2.19), the relations

$$\varepsilon \ge \mathbf{y}_i^T \boldsymbol{A} \mathbf{y}_i \ge \rho \|\mathbf{r}_i\|^2 \quad (i = 1, \dots, k)$$

hold, which finishes the proof.

Note that the statement of Lemma 2.1.10 makes sense in the $\varepsilon < \rho$ case, since trivially, $\|\mathbf{r}_i\| \le 1$ is always true.

Proof. [**Proposition** 2.1.8] Let $F \subset R^n$ be the k-dimensional kernel of $\boldsymbol{L}_w$. Observe that it consists of piecewise constant vectors over the k-partition $P_k = (V_1, \dots, V_k)$ inducing the decomposition $\boldsymbol{L} = \boldsymbol{L}_w + \boldsymbol{L}_b$ of the Laplacian of $G = (V, \boldsymbol{W})$. Indeed, $\boldsymbol{L}_w = \boldsymbol{L}_1 \oplus + \cdots + \oplus \boldsymbol{L}_k$ is the Kronecker-sum of the Laplacians $\boldsymbol{L}_i$'s of G_i's, where G_i is the induced edge-weighted subgraph of G over the vertex set V_i. Therefore, the spectrum of $\boldsymbol{L}_w$ consists of the spectra of $\boldsymbol{L}_i$'s. Since $\boldsymbol{L}_i$ has the eigenvalue 0 with corresponding eigenvector $\frac{1}{\sqrt{|V_i|}}$, the matrix $\boldsymbol{L}_w$ will have the 0 eigenvalue with multiplicity k, and the corresponding eigenspace is spanned by partition vectors described in (2.11). Their linear combinations are piecewise constant vectors over P_k.

Let $\mathbf{y}_1, \dots \mathbf{y}_k$ denote an orthonormal set of eigenvectors corresponding to the k smallest eigenvalues of the Laplacian $\boldsymbol{L}$. Since these eigenvalues are at most ε, Lemma 2.1.10 can be applied for the squared distances between $\mathbf{y}_i$'s and F:

$$\text{dist}^2(\mathbf{y}_i, F) \le \frac{\varepsilon}{\varrho}, \quad (i = 1, \dots, k).$$

By summing it for $i = 1, \dots, k$, the proof is ready, because, by an easy analysis of variance (ANOVA) argument,

$$S_k^2(P_k; \boldsymbol{X}^*) = \sum_{i=1}^{k} \text{dist}^2(\mathbf{y}_i, F).$$

Indeed, $\text{dist}^2(\mathbf{y}_i, F)$ is the minimum squared distance between $\mathbf{y}_i$ and F. In view of the Steiner's theorem, the minimum is attained with the piecewise constant vector of coordinates having at most k different values $c_{1i}, \dots, c_{ki}$ over the sets $V_1, \dots, V_k$ of the underlying k-partition. Namely, if $i \in V_a$, then the ith coordinate of the distance-minimizing piecewise constant vector is $c_{ai} = \frac{1}{|V_a|} \sum_{j \in V_a} y_{ji}$, yielding

$$\text{dist}^2(\mathbf{y}_i, F) = \sum_{a=1}^{k} \sum_{j \in V_a} (y_{ji} - c_{ai})^2,$$

where y_{ji} is the jth coordinate of the vector $\mathbf{y}_i$. By summing it for $i = 1, \dots, k$ and rearranging the finite summation, $\sum_{i=1}^{k} \text{dist}^2(\mathbf{y}_i, F)$ equals $S_k^2(P_k; \boldsymbol{X}^*)$ with cluster centers $\mathbf{c}_a = (c_{a1}, \dots, c_{ak})$, $a = 1, \dots, k$.

2.1.2 Multiway cuts of hypergraphs

For hypergraphs, different kinds of multiway cuts can be defined. By Equation (1.9), to any hypergraph $H = (V, E)$ an edge-weighted graph $G = (V, \mathbf{W})$ can be assigned such that the Laplacian of H is the Laplacian of G. Accordingly, minimum multiway cuts and ratio cuts are defined for H such that they can be directly related to the same quantities of G. Since the formula of (1.9) takes into consideration the cardinalities of the hyperedges, the minimum multiway and ratio cuts of a hypergraph depend on these cardinalities. Therefore, a cardinality-free version of them, as well as a version which takes into consideration how many vertices of the individual clusters a cut-hyperedge contains, will also be introduced. We will discuss relations between these cuts and the Laplacian spectrum of H.

Definition 2.1.11 *Let $H = (V, E)$ be a hypergraph and $P_k = (V_1, \dots, V_k)$ a k-partition of its vertices. The k-way cut of H corresponding to the k-partition P_k is*

$$\text{cut}(P_k, H) = \sum_{e \in E} \frac{1}{|e|} \sum_{a=1}^{k-1} \sum_{b=a+1}^{k} |e \cap V_a| \cdot |e \cap V_b|$$

and the minimum k-way cut of H is

$$\text{mincut}_k(H) = \min_{P_k \in \mathcal{P}_k} \text{cut}(P_k, H). \tag{2.20}$$

Definition 2.1.12 *Let $H = (V, E)$ be a hypergraph and $P_k = (V_1, \ldots, V_k)$ a k-partition of its vertices. The k-way ratio cut of H corresponding to the k-partition P_k is*

$$g(P_k, H) = \sum_{e \in E} \frac{1}{|e|} \sum_{a=1}^{k-1} \sum_{b=a+1}^{k} \left(\frac{1}{|V_a|} + \frac{1}{|V_b|} \right) |e \cap V_a| \cdot |e \cap V_b|$$

and the minimum k-way ratio cut of H is

$$g_k(H) = \min_{P_k \in \mathcal{P}_k} g(P_k, H).$$

Definition 2.1.13 *Let $H = (V, E)$ be a hypergraph and $P_k = (V_1, \ldots, V_k)$ a k-partition of its vertices. The k-sector of H corresponding to the k-partition P_k is the following set of its hyperedges:*

$$\text{sector}(P_k, H) = \{e \in E \ : \ \text{there exist}\, i \neq j\, \text{s. t.}\, e \cap V_i \neq \emptyset \ \text{and} \ e \cap V_j \neq \emptyset\}.$$

The cardinality of the minimum k-sector of H is

$$\theta_k(H) = \min_{P_k \in \mathcal{P}_k} |\text{sector}(P_k, H)|. \tag{2.21}$$

Note that the k-partition P_k defines the coloring c of the vertices with k different colors: $c(v) = i$ if $v \in V_i$. In terms of this coloring, $\text{sector}(P_k, H)$ consists of the multicolored edges of H (i.e., hyperedges of H having at least two vertices of different colors).

Remark 2.1.14 *The above quantities are trivially monotonous in the sense that*

$$\text{mincut}_2(H) \le \text{mincut}_3(H) \le \ldots \le \text{mincut}_n(H),$$
$$g_2(H) \le g_3(H) \le \ldots \le g_n(H),$$
$$\theta_2(H) \le \theta_3(H) \le \ldots \le \theta_n(H) = |E|.$$

Let $\tilde{L}$ be the Laplacian of the hypergraph H defined in Definition 1.1.4. For the bottom of its spectrum, in Bolla (1993), the following two-sided estimation was proved.

Theorem 2.1.15 *Let $0 = \tilde{\lambda}_0 < \tilde{\lambda}_1 \le \ldots \le \tilde{\lambda}_{n-1}$ be the Laplacian eigenvalues of the connected hypergraph H, and k be a fixed integer ($2 \le k \le n$). Then*

$$c_n \theta_k(H) \le \sum_{i=1}^{k-1} \tilde{\lambda}_i \le g_k(H)$$

holds with $c_n = \frac{6}{n(n^2-1)}$.

In the case of $k = 2$, a more precise estimation for λ_1 can be given.

Proposition 2.1.16 *For the smallest positive Laplacian eigenvalue of the connected hypergraph H the following lower estimate holds:*

$$\tilde{\lambda}_1 \geq \begin{cases} 2(1-\cos\frac{\pi}{n})\text{mincut}_2(H) & \text{if} \quad 0 \leq \text{mincut}_2(H) \leq \frac{1}{2}d_{v\max} \\ C_1\text{mincut}_2(H) - C_2 d_{v\max} & \text{if} \quad \frac{1}{2}d_{v\max} < \text{mincut}_2(H) \end{cases}$$

where $d_{v\max} = \max_i d_{vi}$ *is the maximum vertex valence of* H*, whereas the constants* C_1 *and* C_2 *are the same as in Equation (2.10).*

The proof, to be found in Bolla (1993), is analogous to that of the proof of Equation (2.10) of Fiedler (1973).

The upper bound in Theorem 2.1.15 shows that the existence of k relatively small eigenvalues is a necessary condition for the existence of a good classification of the vertices of H (with a small minimal k-way partition cut). Thus, the spectrum gives us some idea about the choice of the number k of the clusters for which good coloring may exist. But the spectrum itself does not divulge anything about the optimal k-partition, moreover it does not give a sufficient condition for the existence of a good clustering. The lower bound in Theorem 2.1.15 depends on the constant c_n, and there are graphs for which the lower bound is attained in order of magnitude, for example for lattices and spiders (see Examples (d) and (e) in Section 1.1.3), which cannot be classified into k clusters in a meaningful way.

Now we want to recognize optimal k-partitions by means of the classification of k-dimensional representatives of the vertices in an optimal k-dimensional Euclidean representation of the hypergraph. The classification is performed by the k-means algorithm. We will be confined to the case when a 'very' well-separated k-partition of the above k-dimensional points exists.

Definition 2.1.17 *A k-partition* $P_k = (V_1, \ldots, V_k)$ *is called well-separated k-partition of the vertex-set V in the k-dimensional Euclidean representation* $\mathbf{r}_1, \ldots, \mathbf{r}_n$ *of the vertices if for the coloring c, corresponding ing to* P_k*, the relation*

$$\frac{\min_{c(v_i) \neq c(v_j)} \|\mathbf{x}_i - \mathbf{x}_j\|}{\max_{c(v_i) = c(v_j)} \|\mathbf{x}_i - \mathbf{x}_j\|} \geq 1$$

holds.

Theorem 2.1.18 *Assume that for some integer* $1 < k < n$ *there exists a well-separated k-partition of the optimal representatives of the vertices of* $H = (V, E)$*,* *for the clusters of which the diameters are at most* ε, *where* $\varepsilon < \frac{1}{2\sqrt{n}}$. *Then*

$$g_k(H) \le q^2 \sum_{i=1}^{k-1} \tilde{\lambda}_i$$

where $q = 1 + \frac{2\varepsilon}{1-\varepsilon\sqrt{n}}$.

For the proof see Bolla (1993). Comparing the results of Theorems 2.1.15 and 2.1.18, under the constraints of the latter one, we obtain that

$$\sum_{i=1}^{k-1} \tilde{\lambda}_i \le g_k(H) \le q^2 \sum_{i=1}^{k-1} \tilde{\lambda}_i, \quad \text{where} \quad 1 < q < 2.$$

This means that, provided ε is less than $\frac{1}{2\sqrt{n}}$, then q is at most 2, and hence, $g_k(H)$ and $\sum_{i=1}^{k-1} \tilde{\lambda}_i$ differ at most by a factor of 4.

2.2 Normalized cuts

Here we will use the normalized Laplacian matrix to find so-called minimum normalized cuts of edge-weighted graphs. Normalized cuts also favor balanced partitions, but the balancing is in terms of the cluster-volumes defined by the generalized degrees.

Definition 2.2.1 *Let* $G = (V, \mathbf{W})$ *be an edge-weighted graph with generalized degrees* $d_1, \dots, d_n$ *and assume that* $\sum_{i=1}^n d_i = 1$. *For the vertex-subset* $U \subset V$ *let* $\mathrm{Vol}(U) = \sum_{i \in U} d_i$ *denote the volume of* U. *The* k*-way normalized cut of* G *corresponding to the* k*-partition* $P_k = (V_1, \dots, V_k)$ *of* V *is defined by*

$$\begin{aligned} f(P_k, G) &= \sum_{a=1}^{k-1} \sum_{b=a+1}^{k} \left(\frac{1}{\mathrm{Vol}(V_a)} + \frac{1}{\mathrm{Vol}(V_b)} \right) w(V_a, V_b) \\ &= \sum_{a=1}^{k} \frac{w(V_a, \overline{V}_a)}{\mathrm{Vol}(V_a)} = k - \sum_{a=1}^{k} \frac{w(V_a, V_a)}{\mathrm{Vol}(V_a)}. \end{aligned} \tag{2.22}$$

The minimum k*-way normalized cut of* G *is*

$$f_k(G) = \min_{P_k \in \mathcal{P}_k} f(P_k, G). \tag{2.23}$$

The equivalence of the seemingly different expressions in (2.22) can be easily verified, using that $\sum_{a=1}^k \mathrm{Vol}(V_a) = \sum_{i=1}^n d_i = 1$. It is easy to see that $f_k(G)$ punishes k-partitions with 'many' inter-cluster edges of 'large' weights and with 'strongly' differing volumes. The quantity $f_2(G)$ was introduced in Mohar (1989) for simple graphs and in Meila and Shi (2001) for edge-weighted graphs; further, for a general k

in Azran and Ghahramani (2006) and Bolla and M.-Sáska (2002), though they called it k-density of G.

Now, $f_k(G)$ will be related to the k smallest normalized Laplacian eigenvalues. Recall that the normalized Laplacian $\boldsymbol{L}_D$ (see Definition 1.3.1) is unaffected by scaling the edge-weights, its spectrum is in the $[0, 2]$ interval and 0 is a single eigenvalue whenever G is connected.

Theorem 2.2.2 *Assume that $G = (V, \boldsymbol{W})$ is connected and let $0 = \lambda_0 < \lambda_1 \leq \cdots \leq \lambda_{n-1} \leq 2$ denote the eigenvalues of its normalized Laplacian matrix. Then*

$$\sum_{i=1}^{k-1} \lambda_i \leq f_k(G) \tag{2.24}$$

and in the case when the optimal k-dimensional representatives of the vertices (see Theorem 1.1.2) can be classified into k well-separated clusters $V_1, \dots, V_k$ in such a way that the maximum cluster diameter ε satisfies the relation $\varepsilon \leq \min\{1/\sqrt{2k}, \sqrt{2}\min_i \sqrt{p_i}\}$, where $p_i = \text{Vol}(V_i)$, $i = 1, \dots, k$, then

$$f_k(G) \leq c^2 \sum_{i=1}^{k-1} \lambda_i,$$

where $c = 1 + \varepsilon c'/(\sqrt{2} - \varepsilon c')$ and $c' = 1/\min_i \sqrt{p_i}$.

To prepare the proof, analogously to the balanced partition vectors defined in (2.11), we will introduce the notion of the normalized partition vectors. The k-partition P_k is uniquely determined by the $n \times k$ normalized partition matrix $\boldsymbol{Z}_k = (\mathbf{z}_1, \dots, \mathbf{z}_k)$, where the a-th *normalized k-partition vector* $\mathbf{z}_a = (z_{1a}, \dots, z_{na})^T$ is the following:

$$z_{ia} = \begin{cases} \frac{1}{\sqrt{\text{Vol}(V_a)}} & \text{if} \quad i \in V_a \\ 0 & \text{otherwise.} \end{cases} \tag{2.25}$$

The matrix $\boldsymbol{D}^{1/2}\boldsymbol{Z}_k$ is obviously suborthogonal, where $\boldsymbol{D}$ is the diagonal degree-matrix. The set of normalized k-partition matrices is denoted by $\mathcal{Z}_k^N$. The normalized cut of $G = (V, \boldsymbol{W})$ corresponding to the k-partition P_k can be rewritten as

$$f(P_k, G) = \sum_{a=1}^{k} \mathbf{z}_a^T \boldsymbol{L}\mathbf{z}_a = \text{tr}(\boldsymbol{Z}_k^T \boldsymbol{L}\boldsymbol{Z}_k) = \text{tr}[(\boldsymbol{D}^{1/2}\boldsymbol{Z}_k)^T \boldsymbol{L}_D(\boldsymbol{D}^{1/2}\boldsymbol{Z}_k)] \tag{2.26}$$

and $f_k(G)$ is its minimum over $\mathcal{Z}_k^N$.

Proof.

- *Lower bound.* The discrete problem is again relaxed to a continuous one. Let $\tilde{\mathbf{r}}_1, \dots, \tilde{\mathbf{r}}_n$ be k-dimensional representatives of the vertices subject to

$\sum_{i=1}^{n} d_i \tilde{\mathbf{r}}_i \tilde{\mathbf{r}}_i^T = \boldsymbol{I}_k$. Let $\tilde{\boldsymbol{X}}$ denote the $n \times k$ matrix with these representatives as row-vectors. Theorem 1.3.3 implies that with the augmented $n \times k$ matrix $\tilde{\boldsymbol{X}}$ (introduced in the proof of this theorem),

$$\begin{aligned}
&\min_{\tilde{\boldsymbol{X}}^T \boldsymbol{D}\tilde{\boldsymbol{X}}=\boldsymbol{I}_k} \sum_{i=1}^{n-1} \sum_{j=i+1}^{n} w_{ij} \|\tilde{\mathbf{r}}_i - \tilde{\mathbf{r}}_j\|^2 \\
&= \min_{\tilde{\boldsymbol{X}}^T \boldsymbol{D}\tilde{\boldsymbol{X}}=\boldsymbol{I}_k} \operatorname{tr}[(\boldsymbol{D}^{1/2}\tilde{\boldsymbol{X}})^T \boldsymbol{L}_D (\boldsymbol{D}^{1/2}\tilde{\boldsymbol{X}})] = \sum_{i=1}^{k-1} \lambda_i = \sum_{i=0}^{k-1} \lambda_i
\end{aligned} \tag{2.27}$$

holds, and equality is attained with $\boldsymbol{X}^* = (\boldsymbol{D}^{-1/2}\mathbf{u}_0, \dots, \boldsymbol{D}^{-1/2}\mathbf{u}_{k-1})$, where $\mathbf{u}_0, \dots \mathbf{u}_{k-1}$ are unit-norm, pairwise orthogonal eigenvectors corresponding to the eigenvalues $\lambda_0, \dots, \lambda_{k-1}$ of $\boldsymbol{L}_D$. We also saw in the Proof of Theorem 1.3.3 that $\boldsymbol{D}^{-1/2}\mathbf{u}_0 = \mathbf{1}$.

Since the normalized k-partition matrix $\boldsymbol{Z}_k$ satisfies $\boldsymbol{Z}_k^T \boldsymbol{D} \boldsymbol{Z}_k = \boldsymbol{I}_k$, the equivalent form (2.26) for $f(P_k, G)$ implies that

$$f_k(G) = \min_{\boldsymbol{Z}_k \in \mathcal{Z}_k^N} \operatorname{tr}(\boldsymbol{Z}_k^T \boldsymbol{L} \boldsymbol{Z}_k) \geq \sum_{i=0}^{k-1} \lambda_i \tag{2.28}$$

and equality can be attained only in the $k = 1$ trivial case, otherwise the vectors $\boldsymbol{D}^{-1/2}\mathbf{u}_i$ cannot be normalized partition vectors, since any $\mathbf{u}_i$ $(i = 1, \dots, k-1)$ has both positive and negative coordinates because of the orthogonality to the $\mathbf{u}_0 = \sqrt{\mathbf{d}}$ vector.

- *Upper bound.* To effectuate the upper estimation, let $P_k = (V_1, \dots, V_k)$ be a k-partition obtained by k-means classification of the optimal k-dimensional Euclidean vertex representatives, $\mathbf{r}_1^*, \dots, \mathbf{r}_n^*$ (row vectors of $\boldsymbol{X}^*$ all having 1 as first coordinate). In fact, the clusters $V_1, \dots, V_k$ are obtained by minimizing the weighted k-variance (C.6) of these representatives. According to our assumption,

$$\varepsilon = \max_{c(i)=c(j)} \|\mathbf{r}_i^* - \mathbf{r}_j^*\| \leq \min\left\{\frac{1}{\sqrt{2k}}, \sqrt{2}\min_i \sqrt{p_i}\right\},$$

where $c(i)$ denotes the cluster membership of vertex i. The representatives satisfy the condition $\sum_{j=1}^{n} d_j \mathbf{r}_j^* \mathbf{r}_j^{*T} = \boldsymbol{X}^* \boldsymbol{D} \boldsymbol{X}^{*T} = \boldsymbol{I}_k$.
Let $\bar{\mathbf{r}}^{(i)}$ denote the weighted center of the ith cluster:

$$\bar{\mathbf{r}}^{(i)} = \frac{1}{p_i} \sum_{j \in V_i} d_j \mathbf{r}_j^*, \quad i = 1, \dots, k.$$

Further, let $\mathbf{y}_i$ denote the k-dimensional vector with coordinates

$$y_{ij} = \begin{cases} \frac{1}{\sqrt{p_i}} & \text{if} \quad j \in V_i \\ 0 & \text{otherwise} \end{cases}$$

and $\boldsymbol{Y} = (\mathbf{y}_1, \ldots, \mathbf{y}_k)$. In fact, with $\boldsymbol{P} = \mathrm{diag}(p_1, \ldots, p_k)$, the relation $\boldsymbol{Y} = \boldsymbol{P}^{-1/2}$ holds. Let $\boldsymbol{R}$ be a $k \times k$ orthogonal matrix. With the notation $\mathbf{y}'_i = \boldsymbol{R}\mathbf{y}_i$ and $\boldsymbol{Y}' = \boldsymbol{R}\boldsymbol{Y}$ we are looking for a system $\boldsymbol{Y}'$ such that $\mathbf{y}'_i$ is 'close' to the cluster center $\bar{\mathbf{r}}^{(i)}$ for $i = 1, \ldots, k$. To this end, we use the multivariate analysis of variance (MANOVA) technique (see Section C.4 of the Appendix). We adopt the decomposition (C.4) to the situation when the variances are calculated with respect to the degree distribution, that is, we use the weights $d_1, \ldots, d_n$ in the formula. In this way, the $k \times k$ empirical covariance matrix of $\mathbf{r}^*_j$'s is decomposed into within-cluster and between-cluster covariances in the following way (the weighted mean of the coordinates of $\mathbf{r}^*_j$s is zero except the first one that is identically 1, but it will not contribute to the variances):

$$\sum_{j=1}^{n} d_j \mathbf{r}^*_j \mathbf{r}^{*T}_j = \sum_{i=1}^{k} \sum_{j \in V_i} d_j \left(\mathbf{r}^*_j - \bar{\mathbf{r}}^{(i)}\right) \left(\mathbf{r}^*_j - \bar{\mathbf{r}}^{(i)}\right)^T + \sum_{i=1}^{k} p_i \bar{\mathbf{r}}^{(i)} \bar{\mathbf{r}}^{(i)T}$$

or briefly,

$$\boldsymbol{I}_k = \sum_{i=1}^{k} \boldsymbol{A}_i + \boldsymbol{B} = \boldsymbol{A} + \boldsymbol{B}$$

where $\boldsymbol{A}_i = \sum_{j \in V_i} d_j (\mathbf{r}^*_j - \bar{\mathbf{r}}^{(i)})(\mathbf{r}^*_j - \bar{\mathbf{r}}^{(i)})^T$, $i = 1, \ldots, k$. Here $\mathrm{tr}(\boldsymbol{A}_i)$ is the k-variance of representatives in cluster i, therefore $\mathrm{tr}(\boldsymbol{A}_i) \le \sum_{c(j)=i} d_j \varepsilon^2 = p_i \varepsilon^2$, and $\mathrm{tr}(\boldsymbol{A}) = \sum_{i=1}^{k} \mathrm{tr}(\boldsymbol{A}_i) \le \varepsilon^2$. Since $\boldsymbol{A}$ is symmetric, positive semidefinite, its maximum eigenvalue is at most ε^2. Hence, $\boldsymbol{A}$ will be viewed as a perturbation on $\boldsymbol{B}$. The matrix $\boldsymbol{B} = \boldsymbol{I}_k - \boldsymbol{A}$ is also positive semidefinite and by the Weyl's perturbation theorem for symmetric matrices (see Theorem A.3.19) it follows that denoting by $\beta_1, \ldots \beta_k$ its eigenvalues, for them the relation

$$0 \le 1 - \beta_i \le \varepsilon^2, \qquad i = 1, \ldots, k$$

holds. With the notation $\bar{\boldsymbol{X}} = (\bar{\mathbf{r}}^{(1)}, \ldots, \bar{\mathbf{r}}^{(k)})$ our matrix $\boldsymbol{B}$ is equal to $\bar{\boldsymbol{X}} \boldsymbol{P} \bar{\boldsymbol{X}}^T$.

Now, let us find an $\boldsymbol{R}$ such that, with $(\mathbf{y}'_1, \ldots, \mathbf{y}'_k) = \boldsymbol{R}\boldsymbol{Y}$, the sum $\sum_{i=1}^{k} p_i \|\bar{\mathbf{r}}^{(i)} - \mathbf{y}'_i\|^2$ be the least possible.

$$\begin{aligned}
\sum_{i=1}^{k} p_i \|\bar{\mathbf{r}}^{(i)} - \mathbf{y}'_i\|^2 &= \mathrm{tr}[(\bar{\boldsymbol{X}} - \boldsymbol{R}\boldsymbol{Y})\boldsymbol{P}(\bar{\boldsymbol{X}} - \boldsymbol{R}\boldsymbol{Y})^T] \\
&= \mathrm{tr}(\bar{\boldsymbol{X}} \boldsymbol{P} \bar{\boldsymbol{X}}^T) + \mathrm{tr}(\boldsymbol{R}\boldsymbol{Y}\boldsymbol{P}\boldsymbol{Y}^T\boldsymbol{R}^T) - 2\mathrm{tr}(\bar{\boldsymbol{X}}\boldsymbol{P}\boldsymbol{Y}^T\boldsymbol{R}) \qquad (2.29) \\
&\ge \sum_{i=1}^{k} \beta_i + k - 2\sum_{i=1}^{k} s_i
\end{aligned}$$

where $s_1, \ldots, s_k$ are the singular values of the matrix $\bar{\boldsymbol{X}}\boldsymbol{P}\boldsymbol{Y}^T$. Indeed, the first term is $\mathrm{tr}\boldsymbol{B}$, the second is $\mathrm{tr}\boldsymbol{I}_k$, while to the third one Proposition A.3.24 of the

Appendix is applicable as follows. With our notation, $\mathrm{tr}(\bar{X}PY^T R)$ is maximal (with respect to R) if the matrix $\bar{X}PY^T R$ is symmetric and the maximum is equal to the sum of the singular values of $\bar{X}PY^T$. By choosing such an R, equality can be attained. Taking into consideration that

$$(\bar{X}PY^T)(\bar{X}PY^T)^T = \bar{X}PY^T YP\bar{X}^T = \bar{X}PP^{-1}P\bar{X}^T$$
$$= \bar{X}P\bar{X}^T = B,$$

the eigenvalues of B can be enumerated in such a way that $\beta_i = s_i^2, i = 1, \dots, k$. But we saw that s_i^2 is of order $1 - \varepsilon^2$, therefore, via Taylor's expansion, $1 - s_i \approx \frac{\varepsilon^2}{2} + \frac{\varepsilon^4}{4}$ is a good approximation. Hence, with the choice of R giving equality in (2.29) we have that

$$\sum_{i=1}^{k} p_i \left\| \bar{\mathbf{r}}^{(i)} - \mathbf{y}_i' \right\|^2 = \sum_{i=1}^{k} s_i^2 - k + 2k - 2\sum_{i=1}^{k} s_i$$
$$\sum_{i=1}^{k} \left(s_i^2 - 1 \right) + 2\sum_{i=1}^{k} (1 - s_i) \approx 2k\varepsilon^4$$

that is less than ε^2 provided that $\varepsilon \le 1/\sqrt{2k}$ holds. Consequently, $p_i \|\bar{\mathbf{r}}^{(i)} - \mathbf{y}_i'\|^2 \le \varepsilon^2$ and $\|\bar{\mathbf{r}}^{(i)} - \mathbf{y}'\| \le \varepsilon c'$.
Let the $\mathbf{y}_i'$ nearest to $\bar{\mathbf{r}}^{(i)}$ be denoted by $\mathbf{y}(\mathbf{r}_j^*)$ for every j in V_i (thus $\mathbf{y}(\mathbf{r}_j^*) = \mathbf{y}_i'$, $\forall\, j \in V_i$). Let δ denote the minimum distance between the different $\mathbf{y}_i'$'s, that is

$$\delta = \min_{a \neq b} \left\| \mathbf{y}_a' - \mathbf{y}_b' \right\| = \min_{a \neq b} \|\mathbf{y}_a - \mathbf{y}_b\| = \min_{a \neq b} \sqrt{\frac{1}{p_a} + \frac{1}{p_b}} \ge \sqrt{2}.$$

Then the estimation

$$f_k(G) \le f_k(P_k, G) = \sum_{i=1}^{n-1} \sum_{j=i+1}^{n} w_{ij} \left\| \mathbf{y}\left(\mathbf{r}_i^*\right) - \mathbf{y}\left(\mathbf{r}_j^*\right) \right\|^2$$
$$\le \sum_{i=1}^{n-1} \sum_{j=i+1}^{n} w_{ij} \left(c \left\| \mathbf{r}_i^* - \mathbf{r}_j^* \right\| \right)^2 = c^2 \sum_{i=1}^{k-1} \lambda_i$$

holds with the constant

$$c = \frac{\delta}{\delta - \varepsilon c'} = 1 + \frac{\varepsilon c'}{\delta - \varepsilon c'} \le 1 + \frac{\varepsilon c'}{\sqrt{2} - \varepsilon c'}$$

where we used Equation (2.27) and the optimality of the representation $\mathbf{r}_1^*, \dots, \mathbf{r}_n^*$. This finishes the proof.

Note that the constant c of the upper estimation is greater than 1, and it is the closer to 1, the smaller ε is. The latter requirement is satisfied if there exists a 'very' well-separated k-partition of the k-dimensional Euclidean representatives. From Theorem 2.2.2 we can also conclude that the gap in the spectrum is a necessary but not a sufficient condition of a good classification. In addition, the Euclidean representatives should be well classified in the appropriate dimension.

The following theorem directly estimates the weighted 2-variance of the optimal representatives of the vertices by the ratio of the two smallest positive normalized Laplacian eigenvalues.

Theorem 2.2.3 *Let $G = (V, \mathbf{W})$ be a connected edge-weighted graph with generalized degrees $d_1, \dots, d_n$ and assume that $\sum_{i=1}^n d_i = 1$. Let $0 = \lambda_0 < \lambda_1 \le \cdots \le \lambda_{n-1}$ denote the eigenvalues of the normalized Laplacian matrix of G. Then for the weighted 2-variance of the optimal vertex representatives, comprising row vectors of the matrix $\boldsymbol{X}_2^*$, the following upper estimate holds:*

$$\tilde{S}_2^2(\boldsymbol{X}_2^*) \le \frac{\lambda_1}{\lambda_2}.$$

In main lines, we will follow the proof of Bolla and Tusnády (1994). Before plunging into it, let us recall that $\boldsymbol{X}_2^* = (\boldsymbol{D}^{-1/2}\mathbf{u}_0, \boldsymbol{D}^{-1/2}\mathbf{u}_1)$, where $\mathbf{u}_0$ and $\mathbf{u}_1$ are unit-norm, orthogonal eigenvectors corresponding to the eigenvalues λ_0 and λ_1 of $\boldsymbol{L}_D$, respectively. Because $\boldsymbol{D}^{-1/2}\mathbf{u}_0$ is the $\mathbf{1}$ vector, the first column of $\boldsymbol{X}_2^*$ can as well be omitted in the representation and we only use the coordinates of the vector $\mathbf{x}^* := \boldsymbol{D}^{-1/2}\mathbf{u}_1 = (x_1^*, \dots, x_n^*)$. Therefore,

$$s^2 = \tilde{S}_2^2\left(\boldsymbol{X}_2^*\right) = \min_{\substack{c(i)\in\{1,2\},\, i=1,\dots,n \\ m_1, m_2}} \sum_{i=1}^n d_i \left(x_i^* - m_{c(i)}\right)^2, \tag{2.30}$$

where the minimization over the 2-partitions of V is uniquely defined by the cluster memberships $c(i)$'s of the vertices ($c(i) = 1$ or $c(i) = 2$ depending on whether vertex i corresponds to the first or second cluster), and the cluster centers are $m_1, m_2 \in \mathbb{R}$.

Proof. As $\mathbf{u}_1$ is the unit-norm vector and orthogonal to the $\mathbf{u}_0 = \sqrt{\mathbf{d}}$ vector, for the coordinates of $\mathbf{x}^*$ the following relations hold:

$$\sum_{i=1}^n d_i x_i^* = 0 \quad \text{and} \quad \sum_{i=1}^n d_i x_i^{*2} = 1. \tag{2.31}$$

Now we will find a vector $\mathbf{y} = (y_1, \dots, y_n)$ such that for it, the conditions

$$\sum_{i=1}^n d_i y_i = 0 \tag{2.32}$$

and

$$\sum_{i=1}^n d_i x_i^* y_i = 0 \tag{2.33}$$

are met. We are looking for $\mathbf{y}$ in the following form:

$$y_i := \left|x_i^* - a\right| - b, \quad (i = 1, \dots, n) \tag{2.34}$$

where a and b are appropriate real numbers.

We will show that there exist real numbers a and b such that the y_i's defined by them satisfy conditions (2.32) and (2.33). Indeed, when we already have a, the above conditions together with $\sum_{i=1}^{n} d_i = 1$ yield

$$b = \sum_{i=1}^{n} d_i \left|x_i^* - a\right| \tag{2.35}$$

for b. With this choice of b, the fulfillment of (2.33) means that

$$\sum_{i=1}^{n} d_i x_i^* \left|x_i^* - a\right| = 0.$$

Since the left-hand side of the above equation is a continuous function of a, and it is equal to 1 if $a \le \min_i x_i^*$, and -1 if $a \ge \max_i x_i^*$, by the Bolzano—Weierstrass theorem, the function $\sum_{i=1}^{n} d_i x_i^* |x_i^* - a|$ must have a root (in a) somewhere between $\min_i x_i^*$ and $\max_i x_i^*$. Choosing such an a and the corresponding b via (2.35), the coordinates of $\mathbf{y}$ are then uniquely determined by (2.34). Let us define the cluster centers by

$$m_1 = a - b \quad \text{and} \quad m_2 = a + b.$$

It is easy to see that

$$y_i = \left|x_i^* - a\right| - b = \begin{cases} m_1 - x_i^* & \text{if} \quad x_i^* < a \\ x_i^* - m_2 & \text{if} \quad x_i^* \ge a, \end{cases}$$

therefore

$$|y_i| = \min\left\{\left|x_i^* - m_1\right|, \left|x_i^* - m_2\right|\right\} \tag{2.36}$$

holds for $i = 1, \dots, n$. Denote

$$\sigma^2(\mathbf{y}) = \sum_{i=1}^{n} d_i y_i^2$$

the variance of the coordinates of $\mathbf{y}$. Since, due to (2.36), the weighted 2-variance of the coordinates of $\mathbf{x}^*$ is one of the terms behind the minimum in (2.30), $\sigma^2(\mathbf{y}) \ge s^2$. In the case of $\sigma(\mathbf{y}) = 0$, the 2-variance s^2 is also equal to 0, and the statement of the theorem is automatically true (but this cannot occur if 0 is a single eigenvalue of $\boldsymbol{L}_D$). Therefore, $\sigma(\mathbf{y}) > 0$ can be assumed. Define the vector $\mathbf{z} \in \mathbb{R}^n$ of the following coordinates:

$$z_i = \frac{y_i}{\sigma(y)}, \quad i = 1, \dots, n$$

and let $\mathbf{x}_i = (x_i^*, z_i)$ be a 2-dimensional vector, a possible representative of vertex i. Further, let

$$X = (\mathbf{x}^*, \mathbf{z}) \quad \text{and} \quad X^* = (\mathbf{x}^*, D^{-1/2}\mathbf{u}_3)$$

be $n \times 2$ matrices, where X contains the representatives $\mathbf{x}_i$'s, while X^* contains the optimal 2-dimensional representatives in its rows. (In fact, they are 3-dimensional representatives, but we disregard the first, constantly 1, coordinates.) Then, on the one hand,

$$\max_{x_i^* \neq x_j^*} \frac{|z_i - z_j|}{|x_i^* - x_j^*|} \leq \frac{1}{\sigma(\mathbf{y})},$$

since due to the definition of y_i, the relation

$$|y_i - y_j| \leq |x_i^* - x_j^*| \quad (i \neq j)$$

holds, that is, $y_i's$ (as functions of x_i^*'s) satisfy the Lipschitz condition (with constant 1).

On the other hand, by Equation (1.3) and Theorem 1.3.3,

$$\frac{\lambda_1 + \lambda_2}{\lambda_1} = \frac{\operatorname{tr}(X^{*T} L X^*)}{\mathbf{x}^{*T} L \mathbf{x}^*} \leq \frac{\operatorname{tr}(X^T L X)}{\mathbf{x}^{*T} L \mathbf{x}^*} = \frac{\sum_{i=1}^{n-1} \sum_{j=i+1}^{n} w_{ij} \|\mathbf{x}_i - \mathbf{x}_j\|^2}{\sum_{i=1}^{n-1} \sum_{j=i+1}^{n} w_{ij} (x_i^* - x_j^*)^2}$$

$$= \frac{\sum_{i=1}^{n-1} \sum_{j=i+1}^{n} w_{ij} \left[\left(x_i^* - x_j^*\right)^2 + (z_i - z_j)^2 \right]}{\sum_{i=1}^{n-1} \sum_{j=i+1}^{n} w_{ij} (x_i^* - x_j^*)^2}$$

$$\leq 1 + \max_{x_i^* \neq x_j^*} \frac{(z_i - z_j)^2}{\left(x_i^* - x_j^*\right)^2} \leq 1 + \frac{1}{\sigma^2(\mathbf{y})} \leq 1 + \frac{1}{s^2},$$

which – by subtracting 1 from both the left- and right-hand sides and taking the reciprocals – finishes the proof.

Theorem 2.2.3 indicates the following two-clustering property of the two smallest positive normalized Laplacian eigenvalues: the greater the gap between them, the better the optimal 2-dimensional (if fact, one-dimensional) representatives of the vertices can be classified into two clusters. This fact, via Theorem 2.2.2 implies that the gap between the eigenvalues λ_1 and λ_2 of L_D is sufficient for the graph to have a small 2-way normalized cut. For $k > 2$, the situation is more complicated, as will be discussed in the next section and Chapter 3.

2.3 The isoperimetric number and sparse cuts

For the two-cluster case, the normalized cut of Section 2.2 is the symmetric version of the isoperimetric number (sometimes called Cheeger constant) introduced in the context of Riemannian manifolds (see e.g., Cheeger (1970)) and much earlier, in mathematical physics (see Pólya and Szegő (1951)). There is a wide literature of this topic together with expander graphs, see for example Alon and Milman (1985); Buser (1984); Diaconis and Stroock (1991); Hoory, Linial and Widgerson (2006); Lubotzky (1994); Mohar (1988, 1989) and Chung (1997), for a summary. We just discuss the most important relations of this topic to sparse, balanced cuts and clustering.

Definition 2.3.1 *Let $G = (V, \mathbf{W})$ be an edge-weighted graph with generalized degrees $d_1, \dots, d_n$ and assume that $\sum_{i=1}^{n} d_i = 1$. The isoperimetric number of G is*

$$h(G) = \min_{\substack{U \subset V \\ \mathrm{Vol}(U) \le \frac{1}{2}}} \frac{w(U, \overline{U})}{\mathrm{Vol}(U)}. \tag{2.37}$$

Since $\mathrm{Vol}(U)$ is the sum of the weights of edges emanating from U, while $w(U, \overline{U})$ is the sum of the weights of those connecting U and $\overline{U}$; the relation $0 \le h(G) \le 1$ is trivial. Further, $h(G) = 0$ if and only if G is disconnected; therefore, only the isoperimetric number of a connected graph is of interest. The isoperimetric number will later be considered as conditional probability, but first we investigate its relation to the smallest positive normalized Laplacian eigenvalue. Note that for simple graphs, $h(G)$ is not identical to the combinatorial isoperimetric number $i(G)$ which uses the cardinality of the subsets instead of their volumes in the denominator of (2.37), and hence, can exceed 1, see Mohar (1989) for details. More precisely, the combinatorial isoperimetric number of the simple graph G is defined by

$$i(G) = \min_{\substack{S \subset V \\ 0 < |S| \le \frac{n}{2}}} \frac{e(S, \overline{S})}{|S|}.$$

It is sometimes called edge-expansion, and mainly used for regular graphs, see Hoory, Linial and Widgerson (2006). The authors of the aforementioned paper also note that a graph is a 'good' expander if it is simultaneously sparse and highly connected.

Intuitively, $h(G)$ is 'small' if 'few low-weight' edges connect together two disjoint vertex-subsets (forming a partition of the vertices) with 'not significantly' differing volumes; therefore, a 'small' $h(G)$ is an indication for a sparse cut of G. On the contrary, a 'large' $h(G)$ means that any vertex-subset of G has a large boundary compared to its volume, where the boundary of $U \subset V$ is the weighted cut between U and its complement in V. This is called good edge-expanding property of G, but we do not want to give the exact definition of an expander graph which depends on many parameters and discussed in details (distinguishing between edge- and vertex-expansion) in many other places, see for example Alon (1986); Alon and Milman (1987); Hoory, Linial and Widgerson (2006); Lubotzky (1994).

Now, a bilateral relation between $h(G)$ and the normalized Laplacian eigenvalue λ_1 is stated for edge-weighted graphs in the following theorem. Similar statements are proved in Alon and Milman (1985); Chung (1997); Mohar (1989) for simple graphs and in Sinclair and Jerrum (1989) for edge-weighted graphs, but without the upcoming improved upper bound.

Theorem 2.3.2 **(Cheeger inequality)** *Let $G = (V, \boldsymbol{W})$ be a connected edge-weighted graph with isoperimetric number $h(G)$, and let λ_1 denote the smallest positive eigenvalue of its normalized Laplacian $\boldsymbol{L}_D$. Then*

$$\frac{\lambda_1}{2} \leq h(G) \leq \min\{1, \sqrt{2\lambda_1}\}$$

always holds true. Furthermore, provided $\lambda_1 \leq 1$, the upper estimate improves to

$$h(G) \leq \sqrt{\lambda_1(2-\lambda_1)}.$$

Note that $\lambda_1 \leq 1$ is not a peculiar requirement as, by Remark 1.3.2 (iv), except the complete graph, every simple graph satisfies this requirement.

Proof.

- *Lower bound.* It follows from the lower estimate (2.24) of the normalized cut (see Theorem 2.2.2). Indeed, in the $k = 2$ case, this gives $\lambda_1 \leq f_2(G)$. By the definition of the normalized cut,

$$f_2(G) = \min_{(U,\overline{U})} \frac{w(U,\overline{U})}{\mathrm{Vol}(U)\mathrm{Vol}(\overline{U})} \leq 2 \min_{\mathrm{Vol}(U)\leq\frac{1}{2}} \frac{w(U,\overline{U})}{\mathrm{Vol}(U)} = 2h(G),$$

  where we used that $\mathrm{Vol}(U) + \mathrm{Vol}(\overline{U}) = 1$, and because of the symmetry, assuming $\mathrm{Vol}(U) \leq \frac{1}{2}$ and $\mathrm{Vol}(\overline{U}) \geq \frac{1}{2}$ is not a restriction. These facts together imply that $\lambda_1 \leq 2h(G)$ which provides the required lower bound.

- *Upper bound.* We will follow the proof of Bolla and M.-Sáska (2004). With the notation of Section 2.2, let $0 = \lambda_0 < \lambda_1 \leq \cdots \leq \lambda_{n-1} \leq 2$ be the spectrum of $\boldsymbol{L}_D$ and $\mathbf{u}_1$ be unit-norm eigenvector corresponding to λ_1. Let $\mathbf{x}^* = \boldsymbol{D}^{-1/2}\mathbf{u}_1$ contain the optimal representatives of the vertices in its coordinates (we omit the trivial dimension) for which $\sum_{i=1}^n d_i x_i^* = 0$ and $\sum_{i=1}^n d_i x_i^{*2} = 1$ holds; see also (2.31) in the proof of Theorem 2.2.3. Without loss of generality, $\mathbf{x}^*$ is directed such that

$$\sum_{i:\,x_i^*<0} d_i \geq \sum_{i:\,x_i^*\geq 0} d_i.$$

  From now on, the superscript of $\mathbf{x}^*$ is discarded for notational convenience. We rearrange the coordinates of $\mathbf{x}$ in increasing order:

$$x_1 \leq \cdots \leq x_{r-1} < 0 \leq x_r \leq \cdots \leq x_n.$$

  Actually, we took advantage of the fact that there are both negative and positive numbers among the coordinates of $\mathbf{x}$, because of the relation $\sum_{i=1}^n d_i x_i = 0$. Say, the number of strictly negative coordinates is $r-1$, $r \geq 2$. The vertex set $V = \{1, \ldots, n\}$ is rearranged, accordingly. Put $V_- := \{1, \ldots, r-1\}$ and $V_+ := \{r, \ldots, n\}$.

By the above assumption, for the coordinates of **x** we have that

$$\sum_{i=1}^{r-1} d_i \geq \sum_{i=r}^{n} d_i. \tag{2.38}$$

Set $\mathbf{y} := \mathbf{x}_+$, that is the coordinates of the vector **y** are

$$y_i = \begin{cases} x_i & \text{if} \quad x_i \geq 0 \\ 0 & \text{if} \quad x_i < 0. \end{cases}$$

We will choose special two-partitions of the rearranged vertex-set induced by the subsets $U_k = \{k, \ldots, n\}$ and put

$$c_k = w(U_k, \overline{U_k}) \quad (k = 2, \ldots, n). \tag{2.39}$$

Obviously,

$$h(G) \leq c = \min_{2 \leq k \leq n} \frac{c_k}{\min\{\mathrm{Vol}(U_k), \mathrm{Vol}(\overline{U_k})\}}. \tag{2.40}$$

We remark that in view of (2.38), the relation

$$\min\{\mathrm{Vol}(U_k), \mathrm{Vol}(\overline{U_k})\} = \mathrm{Vol}(U_k) = \sum_{i=k}^{n} d_i \quad \text{for} \quad k = r, \ldots, n \tag{2.41}$$

is valid.
As $\boldsymbol{D}^{1/2}\mathbf{x}$ is an eigenvector of $\boldsymbol{L}_D = \boldsymbol{I}_n - \boldsymbol{D}^{-1/2}\boldsymbol{W}\boldsymbol{D}^{-1/2}$ with eigenvalue λ_1,

$$\lambda_1 \boldsymbol{D}\mathbf{x} = \boldsymbol{D}\mathbf{x} - \boldsymbol{W}\mathbf{x},$$

or equivalently, for the coordinates,

$$\lambda_1 d_i x_i = d_i x_i - \sum_{j=1}^{n} w_{ij} x_j = \sum_{j=1}^{n} w_{ij}(x_i - x_j), \quad i = 1 \ldots, n \tag{2.42}$$

holds.
Multiplying both sides of (2.42) by x_i and summing for indices $i \in V_+$, we get that

$$\lambda_1 \sum_{i \in V_+} d_i x_i^2 = \sum_{i \in V_+} x_i \sum_{j=1}^{n} w_{ij}(x_i - x_j),$$

or equivalently,

$$\lambda_1 = \frac{\sum_{i \in V_+} x_i \sum_{j=1}^{n} w_{ij}(x_i - x_j)}{\sum_{i \in V_+} d_i x_i^2} =: \frac{A}{\sum_{i=1}^{n} d_i y_i^2}. \tag{2.43}$$

Now, we will estimate the numerator (A) from below as follows:

$$
\begin{aligned}
A &= \sum_{i \in V_+} \sum_{j \in V_+} w_{ij} x_i (x_i - x_j) + \sum_{i \in V_+} \sum_{j \in V_-} w_{ij} x_i (x_i - x_j) \\
&= \sum_{\substack{i \in V_+,\, j \in V_+ \\ i > j}} \left[w_{ij} x_i (x_i - x_j) + w_{ji} x_j (x_j - x_i) \right] \\
&\quad + \sum_{i \in V_+} \sum_{j \in V_-} w_{ij} x_i^2 - \sum_{i \in V_+} \sum_{j \in V_-} w_{ij} x_i x_j \\
&\overset{(1)}{=} \sum_{\substack{i \in V_+,\, j \in V_+ \\ i > j}} w_{ij} (x_i - x_j)^2 + \sum_{i \in V_+} \sum_{j \in V_-} w_{ij} y_i^2 - \sum_{i \in V_+} \sum_{j \in V_-} w_{ij} x_i x_j \\
&\overset{(2)}{\geq} \sum_{\substack{i \in V_+,\, j \in V_+ \\ i > j}} w_{ij} (y_i - y_j)^2 + \sum_{i \in V_+} \sum_{j \in V_-} w_{ij} (y_i - y_j)^2 \\
&\overset{(3)}{=} \sum_{i \in V_+} \sum_{j < i} w_{ij} (y_i - y_j)^2 = \frac{1}{2} \sum_{i=1}^{n} \sum_{j=1}^{n} w_{ij} (y_i - y_j)^2.
\end{aligned}
$$

In the steps (1) and (2) we used the fact that y_i is equal to x_i on V_+ and 0 on V_-. We decreased the expression between the two steps by $-\sum_{i \in V_+} \sum_{j \in V_-} w_{ij} x_i x_j$ which is a nonnegative quantity due to the different signs of x_i and x_j for indices $i \in V_+$ and $j \in V_-$. In the step (3) we utilized that for such indices $i > j$ automatically holds true. We also used the symmetry of $\mathbf{W}$ several times. Now, let us go back to (2.43). Using the lower estimate for A we get that

$$
\lambda_1 \geq \frac{\frac{1}{2} \sum_{i=1}^{n} \sum_{j=1}^{n} w_{ij} (y_i - y_j)^2}{\sum_{i=1}^{n} d_i y_i^2} =: Q. \tag{2.44}
$$

The quantity Q defined above will be important later when we improve the estimate. Q will be further decreased as follows.

$$
\begin{aligned}
Q &= \frac{\frac{1}{2} \left[\sum_{i=1}^{n} \sum_{j=1}^{n} w_{ij} (y_i - y_j)^2 \right] \cdot \left[\sum_{i=1}^{n} \sum_{j=1}^{n} w_{ij} (y_i + y_j)^2 \right]}{\sum_{i=1}^{n} d_i y_i^2 \cdot \left[\sum_{i=1}^{n} \sum_{j=1}^{n} w_{ij} (y_i + y_j)^2 \right]} \\
&= \frac{1}{2} \frac{\left[\sum_{i=1}^{n} \sum_{j=1}^{n} w_{ij} |y_i - y_j|^2 \right] \cdot \left[\sum_{i=1}^{n} \sum_{j=1}^{n} w_{ij} |y_i + y_j|^2 \right]}{\sum_{i=1}^{n} d_i y_i^2 \cdot \left[\sum_{i=1}^{n} \sum_{j=1}^{n} w_{ij} (y_i + y_j)^2 \right]} \\
&\geq \frac{1}{2} \frac{\left[\sum_{i=1}^{n} \sum_{j=1}^{n} w_{ij} |y_i - y_j| \cdot |y_i + y_j| \right]^2}{\sum_{i=1}^{n} d_i y_i^2 \cdot \left[\sum_{i=1}^{n} \sum_{j=1}^{n} w_{ij} (y_i + y_j)^2 \right]}
\end{aligned}
$$

$$= \frac{1}{2} \frac{\left[\sum_{i=1}^{n}\sum_{j=1}^{n} w_{ij}\left|y_i^2 - y_j^2\right|\right]^2}{\sum_{i=1}^{n} d_i y_i^2 \cdot \left[\sum_{i=1}^{n}\sum_{j=1}^{n} w_{ij}(y_i + y_j)^2\right]^2}$$

$$= \frac{1}{2} \frac{\left[2\sum_{i>j} w_{ij}\left|y_i^2 - y_j^2\right|\right]^2}{\sum_{i=1}^{n} d_i y_i^2 \cdot \left[\sum_{i=1}^{n}\sum_{j=1}^{n} w_{ij}(y_i + y_j)^2\right]}$$

$$= 2 \frac{\left[\sum_{i>j} w_{ij}\left(y_i^2 - y_j^2\right)\right]^2}{\sum_{i=1}^{n} d_i y_i^2 \cdot \left[\sum_{i=1}^{n}\sum_{j=1}^{n} w_{ij}(y_i + y_j)^2\right]} =: 2\frac{A_1^2}{B}. \tag{2.45}$$

In the third line we used the Cauchy–Schwarz inequality for the expectation of the random variables $|Y - Y'|$ and $|Y + Y'|$ with the symmetric joint distribution given by $\boldsymbol{W}$, where Y and Y' are identically distributed according to the marginal degree distribution and taking on values y_i's (see Section 1.4 for details).

To estimate A_1 from below, we will use the fact that $y_i \geq y_j$ for $i > j$ and write the terms $y_i^2 - y_j^2$ as a telescopic sum:

$$y_i^2 - y_j^2 = \left(y_i^2 - y_{i-1}^2\right) + \cdots + \left(y_{j+1}^2 - y_j^2\right) \quad \text{for} \quad i > j.$$

By this,

$$\begin{aligned} A_1 &= \sum_{i>j} w_{ij}\left(y_i^2 - y_j^2\right) = \sum_{k=2}^{n} \left(y_k^2 - y_{k-1}^2\right) \sum_{i \geq k > j} w_{ij} \overset{(4)}{=} \sum_{k=2}^{n} \left(y_k^2 - y_{k-1}^2\right) c_k \\ &= \sum_{k=r}^{n} \left(y_k^2 - y_{k-1}^2\right) c_k \overset{(5)}{\geq} \sum_{k=r}^{n} \left(y_k^2 - y_{k-1}^2\right) c \sum_{i=k}^{n} d_i \geq \sum_{k=r}^{n} \left(y_k^2 - y_{k-1}^2\right) h(G) \sum_{i=k}^{n} d_i \\ &= h(G) \sum_{k=r}^{n} \left(y_k^2 - y_{k-1}^2\right) \sum_{i=k}^{n} d_i \overset{(6)}{=} h(G) \sum_{k=r}^{n} y_k^2 d_k \end{aligned}$$

where in (4) we used the definition of c_k, in (5) the relations (2.40) and (2.41) were exploited, while in (6) a partial summation was performed.
The denominator B is estimated from above as follows:

$$\begin{aligned} B &= \sum_{i=1}^{n} d_i y_i^2 \left[\sum_{i=1}^{n}\sum_{j=1}^{n} w_{ij}(y_i + y_j)^2\right] \leq \sum_{i=1}^{n} d_i y_i^2 \left[\sum_{i=1}^{n}\sum_{j=1}^{n} w_{ij}\left(2y_i^2 + 2y_j^2\right)\right] \\ &= \sum_{i=1}^{n} d_i y_i^2 \left[4\sum_{i=1}^{n} y_i^2 d_i\right] = 4\left(\sum_{i=1}^{n} y_i^2 d_i\right)^2. \end{aligned}$$

There remains to collect the terms together:

$$\lambda_1 \geq \frac{2A_1^2}{B} \geq \frac{2h^2(G)\left(\sum_{k=1}^n y_k^2 d_k\right)^2}{4\left(\sum_{i=1}^n y_i^2 d_i\right)^2} = \frac{h^2(G)}{2},$$

and hence, the upper estimate $h(G) \leq \sqrt{2\lambda_1}$ follows.
We can improve this upper bound by using the exact value of B and going back to (2.44) that implies

$$\sum_{i=1}^n \sum_{j=1}^n w_{ij}(y_i - y_j)^2 = 2Q \sum_{i=1}^n d_i y_i^2.$$

An equivalent form of B is

$$\begin{aligned} B &= \sum_{i=1}^n d_i y_i^2 \cdot \left[\sum_{i=1}^n \sum_{j=1}^n w_{ij}(y_i + y_j)^2 \right] \\ &= \sum_{i=1}^n d_i y_i^2 \left[\sum_{i=1}^n \sum_{j=1}^n w_{ij}\left(2y_i^2 + 2y_j^2 - (y_i - y_j)^2\right) \right] \\ &= \sum_{i=1}^n d_i y_i^2 \left[4\sum_{i=1}^n y_i^2 d_i - \sum_{i=1}^n \sum_{j=1}^n w_{ij}(y_i - y_j)^2 \right] = 2\left(\sum_{i=1}^n d_i y_i^2\right)^2 (2 - Q). \end{aligned}$$

Starting the estimation of Q at (2.45) and continuing with the B above, yields

$$Q \geq 2\frac{A_1^2}{B} \geq 2\frac{h^2(G)\left(\sum_{k=1}^n y_k^2 d_k\right)^2}{2\left(\sum_{i=1}^n d_i y_i^2\right)^2 (2 - Q)} = \frac{h^2(G)}{2 - Q}.$$

In view of (2.44), Q is nonnegative, implying

$$Q \geq \frac{h^2(G)}{2 - Q} \quad \text{or equivalently,} \quad 1 - \sqrt{1 - h^2(G)} \leq Q \leq 1 + \sqrt{1 - h^2(G)}.$$

Summarizing, we derive that

$$\lambda_1 \geq Q \geq 1 - \sqrt{1 - h^2(G)} \quad \text{or equivalently,} \quad \sqrt{1 - h^2(G)} \geq 1 - \lambda_1.$$

For $\lambda_1 > 1$ this is a trivial statement. For $\lambda_1 < 1$ it implies that $h(G) \leq \sqrt{\lambda_1(2 - \lambda_1)} < 1$, while for $\lambda_1 = 1$ we get the trivial bound $h(G) \leq 1$. This finishes the proof.

In the framework of joint distributions (see Section 1.4) $h(G)$ can be viewed as a conditional probability and related to the symmetric maximal correlation in the following way. The weight matrix $\mathbf{W}$ (with sum of its entries 1) defines a discrete symmetric joint distribution $\mathbb{W}$ (see Appendix B) with equal marginals $\mathbb{D} = \{d_1, \dots, d_n\}$. Let H

denote the Hilbert space of $V \to \mathbb{R}$ random variables taking on at most n different values with probabilities $d_1, \ldots, d_n$, and having zero expectation and finite variance. Let us take two identically distributed (i.d.) copies $\psi, \psi' \in H$ with joint distribution $\mathbb{W}$. Then, obviously,

$$h(G) = \min_{\substack{B \subset \mathbb{R} \text{ Borel-set} \\ \psi, \psi' \in H \text{ i.d.} \\ \mathbb{P}_{\mathbb{D}}(\psi \in B) \leq 1/2}} \mathbb{P}_{\mathbb{W}}(\psi' \in \overline{B} | \psi \in B).$$

The symmetric maximal correlation defined by the symmetric joint distribution $\mathbb{W}$ is the following:

$$r_1 = \max_{\psi, \psi' \in H \text{ i.d.}} \mathrm{Corr}_{\mathbb{W}}(\psi, \psi') = \max_{\substack{\psi, \psi' \in H \text{ i.d.} \\ \mathrm{Var}_{\mathbb{D}} \psi = 1}} \mathrm{Cov}_{\mathbb{W}}(\psi, \psi').$$

In view of (1.25) and Theorem 1.4.4 of Section 1.4, $r_1 = 1 - \lambda_1$, provided $\lambda_1 \leq 1$.

With this notation, the result of Theorem 2.3.2 can be written in the equivalent form as follows.

Proposition 2.3.3 *Let $\mathbb{W}$ be the symmetric joint distribution of two discrete random variables taking on at most n different values, where the joint probabilities of $\mathbb{W}$ are the entries of the weight matrix $\boldsymbol{W}$. If the symmetric maximal correlation r_1 is nonnegative, then with it, the estimation*

$$\frac{1 - r_1}{2} \leq \min_{\substack{B \subset \mathbb{R} \text{ Borel-set} \\ \psi, \psi' H \text{ i.d.} \\ \mathbb{P}_{\mathbb{D}}(\psi \in B) \leq 1/2}} \mathbb{P}_{\mathbb{W}}(\psi' \in \overline{B} | \psi \in B) \leq \sqrt{1 - r_1^2}$$

holds.

Proof. Since $\lambda_1 = 1 - r_1$, the lower bound trivially follows. $r_1 \geq 0$ implies that $\lambda_1 \leq 1$, so the improved upper bound of Theorem 2.3.2 becomes $\sqrt{(1 - r_1)(1 + r_1)}$ which finishes the proof.

Consequently, the symmetric maximal correlation somehow regulates the minimum conditional probability that, provided a random variable takes values in a category set (with probability less than 1/2), then another copy of it (their joint distribution is given by $\boldsymbol{W}$) will take values in the complementary category set. The larger r_1, the smaller this minimum conditional probability is. In particular, if r_1 is the largest absolute value eigenvalue of $\boldsymbol{I} - \boldsymbol{L}_D$ (apart from the trivial 1), then r_1 is the usual maximal correlation.

The other important application of the isoperimetric inequality is related in many aspects to random walks. We just touch upon to this topic here, for interested readers we refer to Aldous (1989); Babai and Szegedy (1992); Chung (1997); Diaconis and Stroock (1991); Lovász (1993); Lovász and Simonovits (1993); Lovász and Winkler (1995); Telcs (2006); von Luxburg (2006).

In fact, time-reversible Markov chains can be viewed as random walks on undirected, possibly edge-weighted graphs ($\boldsymbol{W}$ is symmetric). The walk can be described by a discrete time stochastic process $\xi_0, \xi_1, \ldots, \xi_t, \ldots$ with finite state space $\{1, \ldots, n\}$. The transition probabilities

$$\mathbb{P}(\xi_{t+1} = j \mid \xi_t = i) = \frac{w_{ij}}{d_i}$$

do not depend on t and are entries of the *transition probability matrix* $\boldsymbol{D}^{-1}\boldsymbol{W}$. The transition probability matrix is not symmetric, but its spectrum is the same as that of the symmetric matrix $\boldsymbol{D}^{-1/2}\boldsymbol{W}\boldsymbol{D}^{-1/2}$, since the eigenvalue–eigenvector equation

$$\boldsymbol{D}^{-1/2}\boldsymbol{W}\boldsymbol{D}^{-1/2}\mathbf{u} = \lambda\mathbf{u}$$

is equivalent to

$$\boldsymbol{D}^{-1}\boldsymbol{W}(\boldsymbol{D}^{-1/2}\mathbf{u}) = \lambda(\boldsymbol{D}^{-1/2}\mathbf{u}).$$

Therefore, the transition probability matrix has real eigenvalues in the $[-1, 1]$ interval, they are the numbers $1 - \lambda_i$, where λ_i is the ith largest eigenvalue of $\boldsymbol{L}_D$, and corresponding eigenvectors which are the vector components of the optimal representation of Theorem 1.3.3. Further, its largest eigenvalue is always 1 with corresponding eigendirection $\mathbf{1}$, and the multiplicity of 1 as an eigenvalue is equal to the number of the connected components of G. The random walk is ergodic if it has a unique stationary distribution. The necessary and sufficient condition of ergodicity is the irreducibility ($\lambda_1 > 0$) and aperiodicity ($\lambda_{n-1} < 2$). Therefore a random walk on a connected and non-bipartite graph exhibits a unique stationary distribution which is just $\{d_1, \ldots, d_n\}$. The so-called *mixing rate* shows how rapidly the random walk converges to this stationary distribution. By the Cheeger inequality (Theorem 2.3.2) it follows that a relatively large λ_1 induces rapid mixing and short *cover time* which is the expected amount of time to reach every vertex (starting from a given distribution): see Lovász (1993). The electric network analogue, including *conductance* of random walks on graphs is discussed thoroughly, for example by Telcs (2006); Tetali (1991); Thomassen (1990).

We are rather interested in the case when λ_1 is near zero and the random walk cannot go through the graph quickly because of bottlenecks in it. Such a bottleneck can be the weighted cut between two disjoint and mutually exhaustive vertex-subsets which give the minimum in the definition of $h(G)$. More generally, if there are $k - 1$ near zero eigenvalues of $\boldsymbol{L}_D$, then we may expect k clusters such that the random walk stays with high probability within the clusters and goes through between the cluster pairs ($\binom{k}{2}$ bottlenecks) with smaller probability. This assumption is formulated by the normalized cuts (introduced in Definition 2.2.1) and sparse cuts, to be introduced.

Note that because of the relation $f_2(G) \leq 2h(G)$ (we saw this when we proved the lower bound in the Cheeger inequality), Theorem 2.3.2 and Theorem 2.2.2 provide us with the following estimation of the 2-way normalized cut of G with the help of its smallest positive normalized Laplacian eigenvalue, in the $\lambda_1 \leq 1$ typical case:

$$\lambda_1 \leq f_2(G) \leq 2\sqrt{\lambda_1(2-\lambda_1)}.$$

There are several natural generalizations of the isoperimetric number for $k > 2$ and of the Cheeger inequality for the upper end of the normalized Laplacian spectrum. We will discuss some recent results that directly relate so-called k-way sparse cuts to the eigenvalue λ_{k-1}.

In Trevisan (2009), the so-called dual Cheeger inequality is proved which estimates the measure $\beta(G)$ of bipartiteness of G by means of the upper spectral gap of $\boldsymbol{L}_D$, that is, the difference between λ_{n-1} and 2. To be consistent with the previous notation, $0 = \lambda_0 \leq \cdots \leq \lambda_{n-1} \leq 2$ will stand for the normalized Laplacian spectrum.

Definition 2.3.4 *The bipartiteness ratio of the simple, d-regular graph G on the n-element vertex set V is*

$$\beta(G) = \min_{\substack{S \subset V \\ (L,R):2-\text{partition of } S}} \frac{2e(L) + 2e(R) + e(S,\overline{S})}{d|S|}$$

where $e(L)$ and $e(R)$ stands for the number of edges between vertices of L and R, respectively, and $e(S,\overline{S})$ denotes the number of cut-edges.

Note that the left and right non-empty, disjoint subsets of S ($L \cup R = S$) do not necessarily exhaust V. Since there are no loops, $e(L,L) = 2e(L)$ and $e(R,R) = 2e(R)$.

Proposition 2.3.5 (Dual Cheeger inequality)

$$\frac{1}{2}(2-\lambda_{n-1}) \leq \beta(G) \leq \sqrt{2(2-\lambda_{n-1})}$$

where λ_{n-1} is the largest eigenvalue of the normalized Laplacian $\boldsymbol{L}_D$.

Note that the original paper uses the eigenvalues of the matrix $\boldsymbol{D}^{-1/2}\boldsymbol{W}\boldsymbol{D}^{-1/2}$, especially, $|1-\lambda_{n-1}|$ in the formulation of the dual Cheeger inequality. Observe that by Remark 1.3.2 (iii), $\lambda_{n-1} \geq \frac{n}{n-1} \geq 1$, therefore $1-\lambda_{n-1} < 0$ always holds.

Observe that if G is bipartite, then $\lambda_{n-1} = 2$ and $\beta(G) = 0$. The proposition implies that $\beta(G)$ is small if λ_{n-1} is close to 2, that is, a large λ_{n-1} is an indication for G having a closely bipartite-induced subgraph that is relatively large and sparsely connected to the other part of G.

We also remark that the notion of the bipartiteness ratio can naturally be extended to an edge-weighted graph $G = (V, \boldsymbol{W})$ in the following way:

$$\beta(G) = \min_{\substack{S \subset V \\ (L,R)\ 2\text{-partition of } S}} \frac{w(L,L) + w(R,R) + w(S,\overline{S})}{\text{Vol}(S)}.$$

Probably, a similar estimation with it for the upper spectral gap of an edge-weighted graph exists. Analogously to Theorem 2.2.3, in Bolla (2011a) the upper spectral gap was used to estimate the 2-variance of the 1-dimensional vertex representatives, based on the coordinates of the vector $\boldsymbol{D}^{-1/2}\mathbf{u}_{n-1}$. The relation between bipartite subgraphs and the smallest adjacency eigenvalue (corresponding to the largest normalized Laplacian one if the graph is regular) is also treated in Alon and Sudakov (2000).

Now we discuss some recent results of Lee *et al.* (2012); Louis *et al.* (2011, 2012) on possible extensions of the Cheeger inequality to multiway cuts, called *higher-order Cheeger inequalities*. For the edge-weighted graph $G = (V, \boldsymbol{W})$, the *expansion* of the vertex-subset $S \subseteq V$ is defined by

$$\phi(S, G) = \frac{w(S, \overline{S})}{\mathrm{Vol}(S)}. \tag{2.46}$$

Note that in Lee *et al.* (2012) it is formulated for simple, d-regular graphs with $e(S, \overline{S})$ in the numerator and $d|S|$ in the denominator. In Louis *et al.* (2012), the authors define this expansion for edge-weighted graphs, but use $\min\{\mathrm{Vol}(S), \mathrm{Vol}(\overline{S})\}$ in the denominator. However, this does not make any difference in the upcoming definition.

Definition 2.3.6 *For a given integer $1 < k < n$, the k-way expansion constant of the edge weighted graph $G = (V, \boldsymbol{W})$ on n vertices is*

$$\rho_k(G) = \min_{\substack{S_1,\dots,S_k \subset V \\ S_i \neq \emptyset,\, i,\dots,k \\ S_i \cap S_j = \emptyset,\, i \neq j}} \max_{i \in \{1,\dots,k\}} \phi(S_i, G).$$

Note that here the collection of pairwise disjoint subsets $S_1, \dots, S_k$ is not necessarily a k-partition, since they do not always exhaust V.

It is easy to see that $\rho_2(G) = h(G)$ and therefore, the Cheeger inequality bounds it from below and from above in terms of λ_1. In Lee *et al.* (2012) a similar relation is proved for $\rho_k(G)$, for a general k, in terms of the kth smallest normalized Laplacian eigenvalue λ_{k-1}. We cite this result with our notation.

Theorem 2.3.7

$$\frac{\lambda_{k-1}}{2} \leq \rho_k(G) \leq \mathcal{O}(k^2)\sqrt{\lambda_{k-1}}.$$

With fewer sets than the order of the eigenvalue, the upper estimate improves to

$$\rho_k(G) \leq \mathcal{O}(\sqrt{\lambda_{2k-1} \log k}). \tag{2.47}$$

The proof of Theorem 2.3.7 is algorithmic. It uses the optimum k-dimensional representatives of the vertices (see Theorem 1.3.3) and applies geometric considerations to them. Namely, it is shown that the total mass of them is localized on k disjoint regions of $\mathbb{R}^k$ (in fact, in a $(k-1)$-dimensional hyperplane of it). Observe that this notion is closely related to our weighted k-variance (C.6) of the Appendix. Otherwise, to find these sparsest cuts is NP-complete.

Both in Lee *et al.* (2012) and Louis *et al.* (2012) a sparsest small set, producing some kind of sparse cut of G and comprising at most the $1/k$ fraction of the total volume, is defined as follows with our notation.

Definition 2.3.8 *For a given integer $1 < k < n$, the small-set sparsity of the edge weighted graph $G = (V, \mathbf{W})$ on n vertices is defined by*

$$\phi_k(G) = \min_{\substack{S \subset V \\ \mathrm{Vol}(S) \le \mathrm{Vol}(V)/k}} \phi(S, G).$$

Obviously, $\phi_k(G) \le \rho_k(G)$, for every positive integer $k < n$. Therefore, Inequality (2.47) implies that

$$\phi_{k/2}(G) \le \mathcal{O}(\sqrt{\lambda_{k-1}} \log k)$$

which improves a statement of Louis *et al.* (2011) that

$$\phi_{\sqrt{k}}(G) \le C(\sqrt{\lambda_{k-1}} \log k),$$

where C is a fixed constant.

We remark that the small set problem is closely related to Unique Games. In this context, Arora *et al.* (2010) showed that $\phi_k(G) < C\sqrt{\lambda_{(k-1)^{100}} \log_k n}$, where C is some absolute constant.

To find the sparsest k-partition, in Louis *et al.* (2011) an iterative algorithm is defined which finds the sparsest 2-way cut in each step and removes the cut-edges. Meanwhile, the graph becomes disconnected, and the algorithm operates on the components of it. The authors also prove that for any edge-weighted graph $G = (V, \mathbf{W})$ and any integer $1 \le k \le |V|$, there exist ck disjoint subsets $S_1, \dots, S_{ck}$ of V such that

$$\max_i \phi(S_i, G) \le C\sqrt{\lambda_{k-1} \log k}$$

where $c < 1$ and C are absolute constants. Moreover, these sets can be identified in polynomial time. In Lee *et al.* (2012), the authors find a so-called non-expanding k-partition $V_1, \dots, V_k$ of V such that

$$\phi(V_i, G) \lesssim k^4 \sqrt{\lambda_{k-1}}, \quad i = 1, \dots, k.$$

We remark that between the k-way expansion constant and the normalized cut the following relation holds:

$$\rho_k(G) \le \min_{(V_1,\dots,V_k) = P_k} \max_{i \in \{1,\dots,k\}} \phi(V_i, G) \le \min_{(V_1,\dots,V_k) = P_k} \sum_{i=1}^{k} \phi(V_i, G) = f_k(G)$$

where we took into consideration the equivalent forms (2.22) of the normalized cut. If the minimum in $\rho_k(G)$ is attained at a k-partition, then

$$f_k(G) = \min_{(V_1,\dots,V_k)=P_k} \sum_{i=1}^{k} \phi(V_i, G) \le k \min_{(V_1,\dots,V_k)=P_k} \max_{i\in\{1,\dots,k\}} \phi(V_i, G) = k\rho_k(G).$$

In this case, Theorem 2.3.7 also implies $f_k(G) \le \mathcal{O}(k^3)\sqrt{\lambda_{k-1}}$.

Consequently, there are the k-way sparsest cuts which are closely related to λ_{k-1}. However, a small λ_{k-1} is not always an indication of a small normalized cut. In fact, λ_{k-1} can be small if there exists an $S_1, \dots, S_k$ sparsely connected system which does not necessarily exhaust V. The other parts of V may have, for example, closely bipartite subgraphs, etc. When in Section 3.3 we define so-called regular cuts, we will illustrate that there are the small and large normalized Laplacian eigenvalues which together recover the graph's structure, and regular cuts may contain sparse and dense cuts as well.

2.4 The Newman–Girvan modularity

The Newman–Girvan modularity introduced in Newman and Girvan (2004a) directly focuses on modules of higher intra-community connections than expected based on the model of independent attachment of the vertices with probabilities proportional to their degrees. To maximize this modularity, hierarchical clustering methods based on the edge betweenness measure of Clauset *et al.* (2004); Newman (2004b,c); Newman and Girvan (2004a), and vector partitioning algorithms based on the eigenvectors of the modularity matrix of Newman (2006) are introduced. In Duch and Arenas (2005) an extremal optimization algorithm is presented.

We will extend the linear algebraic machinery developed for Laplacian-based spectral clustering to the modularity based community detection. To this end, two penalized versions of the Newman–Girvan modularity are introduced in the general framework of an edge-weighted graph, see Newman (2004d), and their relation to projections onto the subspace of partition vectors and to k-variance of the clusters formed by the vertex representatives is investigated. These considerations give useful information on the choice of k and on the nature of the community structure.

Definition 2.4.1 *The Newman-Girvan modularity corresponding to the k-partition* $P_k = (V_1, \dots, V_k)$ *of the vertex-set of the edge-weighted graph* $G = (V, \mathbf{W})$*, where the entries of* $\mathbf{W}$ *sum to 1, is*

$$M(P_k, G) = \sum_{a=1}^{k} \sum_{i,j\in V_a} (w_{ij} - d_i d_j) = \sum_{a=1}^{k} [w(V_a, V_a) - \mathrm{Vol}^2(V_a)].$$

For given integer $1 \le k \le n$*, the k-module Newman-Girvan modularity of the edge-weighted graph G is*

$$M_k(G) = \max_{P_k \in \mathcal{P}_k} M(P_k, G).$$

For a simple graph, $w(V_a, V_a) = 2e(V_a)$ is twice the number of edges with both endpoints in V_a, and $\mathrm{Vol}(V_a)$ is the number of edges emanating from V_a, and we have to normalize by $2e$, where e is the total number of edges. The entries $d_i d_j$ of the null-model matrix $\mathbf{d}\mathbf{d}^T$ correspond to the hypothesis of independence. In other words, under the null-hypothesis, vertices i and j are connected to each other independently, with probability $d_i d_j$ proportional (actually, because the sum of the weights is 1, equal) to their generalized degrees ($i, j = 1, \dots, n$). Hence, for given k, maximizing $M(P_k, G)$ is equivalent to looking for k modules of the vertices with intra-community connections higher than expected under the null-hypothesis. As $\sum_{a=1}^{k} \sum_{b=1}^{k} \sum_{i \in V_a} \sum_{j \in V_b} (w_{ij} - d_i d_j) = 0$, the above task is equivalent to minimizing

$$\sum_{a \neq b} \sum_{i \in V_a,\, j \in V_b} (w_{ij} - d_i d_j) = \sum_{a \neq b} [w(V_a, V_b) - \mathrm{Vol}(V_a)\mathrm{Vol}(V_b)], \qquad (2.48)$$

that is, to looking for k clusters of the vertices with inter-cluster connections lower than expected under the hypothesis of independence. In the minimum cut problem the cumulated inter-cluster connections themselves were minimized. Therefore, the spectral method introduced in White and Smyth (2005) for maximizing the Newman-Girvan modularity is closely related to that of Bolla and Tusnády (1990); Meila and Shi (2001); Ng *et al.* (2001) for minimizing the normalized cut.

We want to penalize partitions with clusters of extremely different sizes. To measure the size of cluster V_a, either the number of its vertices $|V_a|$ or its volume $\mathrm{Vol}(V_a)$ is used. Fortunato (2010) remarks that the Newman–Girvan modularity seems to attain its maximum for clusters of near equal sizes, though there is no explanation for it. Of course, communities of real-life networks have more practical relevance if they do not differ too much in sizes. In Newman (2004b) and Reichardt and Bornholdt (2007), the authors also define a good modularity structure as one having near equal sizes of modules. However, they do not make use of this idea in their objective function. Actually, Reichardt and Bornholdt (2007) prove that the Newman–Girvan modularity is a special ground state energy, and in Chapter 4, the convergence of the ground state energies is used to prove the testability of some balanced multiway cut densities. However, these conditional extrema cannot be immediately related to spectra. As a compromise, we modify the modularity itself so that it would penalize clusters of significantly different sizes. Of course, real-life communities are sometimes very different in sizes or volumes. Our method is capable of finding fundamental clusters, and further analysis is needed to separate small communities from the large ones. Another possibility is to distinguish a core of the graph that is free of low-degree vertices for which, usually near zero eigenvalues are responsible.

As in the case of $k > 2$, there are more inter-cluster sums than intra-cluster ones; it is in (2.48) where we penalize clusters of two different sizes or volumes by introducing a factor $\frac{1}{|V_a|} + \frac{1}{|V_b|}$ or $\frac{1}{\mathrm{Vol}(V_a)} + \frac{1}{\mathrm{Vol}(V_b)}$ for the $a \neq b$ pair that shifts the argmin towards balanced pairs. For the above reasons, analogously to the ratio and normalized cuts, the following notions were introduced in Bolla (2011b).

Definition 2.4.2 *The balanced Newman-Girvan modularity corresponding to the k-partition $P_k = (V_1, \dots, V_k)$ of the vertex-set of $G = (V, \boldsymbol{W})$ $(\text{Vol}(V) = 1)$ is*

$$BM(P_k, G) = \sum_{a=1}^{k} \frac{1}{|V_a|} \sum_{i,j \in V_a} (w_{ij} - d_i d_j) = \sum_{a=1}^{k} \left[\frac{w(V_a, V_a)}{|V_a|} - \frac{\text{Vol}^2(V_a)}{|V_a|} \right]$$

and the balanced k-module Newman-Girvan modularity of G is

$$BM_k(G) = \max_{P_k \in \mathcal{P}_k} BM(P_k, G).$$

Definition 2.4.3 *The normalized Newman–Girvan modularity corresponding to the k-partition $P_k = (V_1, \dots, V_k)$ of the vertex-set of $G = (V, \boldsymbol{W})$ $(\text{Vol}(V) = 1)$ is*

$$NM(P_k, G) = \sum_{a=1}^{k} \frac{1}{\text{Vol}(V_a)} \sum_{i,j \in V_a} (w_{ij} - d_i d_j) = \sum_{a=1}^{k} \frac{w(V_a, V_a)}{\text{Vol}(V_a)} - 1$$

and the normalized k-module Newman-Girvan modularity of G is

$$NM_k(G) = \max_{P_k \in \mathcal{P}_k} NM(P_k, G).$$

Here we used the fact that $\sum_{a=1}^{k} \text{Vol}(V_a) = 1$. In view of (2.22), minimizing the normalized cut of G over k-partitions of its vertices is equivalent to maximizing $\sum_{a=1}^{k} \frac{w(V_a, V_a)}{\text{Vol}(V_a)}$. Hence, maximizing the normalized Newman–Girvan modularity can be solved with the same spectral method (using the normalized Laplacian) as the normalized cut problem. However, we introduce another method based on the normalized modularity matrix.

We also want to show another insight into the problem of the choice of k from the point of view of computational demand and by using the linear algebraic structure of our objective function. In this way, we will prove that for the selected k, maximizing the above adjusted modularities is equivalent to minimizing the k-variance of the vertex representatives by choosing an appropriate representation; hence, the k-means algorithm is applicable.

2.4.1 Maximizing the balanced Newman–Girvan modularity

The k-partition P_k is uniquely defined by the $n \times k$ balanced partition matrix $\boldsymbol{Z}_k = (\mathbf{z}_1, \dots, \mathbf{z}_k)$ defined in (2.11), that is,

$$BM(P_k, G) = \sum_{a=1}^{k} \mathbf{z}_a^T \boldsymbol{M} \mathbf{z}_a = \text{tr}\big(\boldsymbol{Z}_k^T \boldsymbol{M} \boldsymbol{Z}_k\big)$$

where $\boldsymbol{M}$ is the modularity matrix introduced in (1.19). We want to maximize $\text{tr}(\boldsymbol{Z}_k^T \boldsymbol{M} \boldsymbol{Z}_k)$ over balanced k-partition matrices $\boldsymbol{Z}_k \in \mathcal{Z}_k^B$. Since $\boldsymbol{Z}_k$ is a suborthogonal matrix, $\boldsymbol{Z}_k^T \boldsymbol{Z}_k = \boldsymbol{I}_k$.

Let $\beta_1 \geq \cdots \geq \beta_n$ denote the eigenvalues of the modularity matrix $\boldsymbol{M}$ with corresponding unit-norm, pairwise orthogonal eigenvectors $\mathbf{u}_1, \dots, \mathbf{u}_n$. Let p denote the number of its positive eigenvalues; thus, $\beta_{p+1} = 0$ and $\mathbf{u}_{p+1} = \mathbf{1}/\sqrt{n}$. Now let $\boldsymbol{Y} = (\mathbf{y}_1, \dots, \mathbf{y}_k)$ be an arbitrary $n \times k$ suborthogonal matrix ($k \leq p$). Then using Proposition A.3.12 of the Appendix,

$$\max_{\boldsymbol{Z}_k \in \mathcal{Z}_k^B} BM(\boldsymbol{Z}_k, G) \leq \max_{\boldsymbol{Y}^T\boldsymbol{Y}=\boldsymbol{I}_k} \mathrm{tr}(\boldsymbol{Y}^T\boldsymbol{M}\boldsymbol{Y}) = \sum_{a=1}^{k} \beta_a \leq \sum_{a=1}^{p+1} \beta_a. \tag{2.49}$$

Both inequalities can be attained with equality only in the $k = 1$, $p = 0$ case, when our underlying graph is the complete graph. In this case there is only one cluster with partition vector of equal coordinates (balanced eigenvector corresponding to the single 0 eigenvalue). For $k > 1$, no graph has partition vectors which can coincide with eigenvectors corresponding to positive eigenvalues, since their coordinates do not sum to zero, which would be necessary to be orthogonal to the vector corresponding to the 0 eigenvalue.

It is also obvious that the maximum with respect to k of the maximum in (2.49) is attained with the choice of $k = p + 1$. In Newman (2006), for the non-penalized case, the author shows how $p + 1$ clusters can be constructed by applying a vector partitioning algorithm for $\mathbf{u}_1, \dots, \mathbf{u}_p$. However, in case of large networks, p can also be large, and computation of the positive eigenvalues together with eigenvectors is time-consuming. As a compromise, it will be shown that choosing a $k < p$ such that there is a noticeable gap between β_{k-1} and β_k will also suffice. Further, even for a fixed 'small' $k < p$, finding the true maxim over k-partitions cannot be solved in polynomial time in n, but due to our estimations, spectral partitioning algorithms can be constructed like spectral clustering based on Laplacian eigenvectors, see Section 2.1. Now, we are going to discuss this issue in detail.

We expand $BM(P_k, G)$ with respect to the eigenvalues and eigenvectors of the modularity matrix:

$$BM(P_k, G) = \mathrm{tr}\left(\boldsymbol{Z}_k^T\boldsymbol{M}\boldsymbol{Z}_k\right) = \sum_{a=1}^{k} \mathbf{z}_a^T \left(\sum_{i=1}^{n} \beta_i \mathbf{u}_i \mathbf{u}_i^T\right) \mathbf{z}_a = \sum_{i=1}^{n} \beta_i \sum_{a=1}^{k} \left(\mathbf{u}_i^T \mathbf{z}_a\right)^2.$$

We can increase the last sum if we neglect the terms corresponding to the negative eigenvalues, hence, the outer summation stops at p, or equivalently, at $p + 1$. In this case the inner sum is the largest if $k = p + 1$, when the partition vectors $\mathbf{z}_1, \dots, \mathbf{z}_{p+1}$ are 'close' to the eigenvectors $\mathbf{u}_1, \dots, \mathbf{u}_{p+1}$, respectively. As both systems consist of orthonormal sets of vectors, the two subspaces spanned by them should be close to each other. The subspace $F_{p+1} = \mathrm{Span}\{\mathbf{z}_1, \dots, \mathbf{z}_{p+1}\}$ consists of piecewise constant vectors on $p + 1$ steps, therefore $\mathbf{u}_{p+1} \in F_{p+1}$, and it suffices to process only the first p eigenvectors. The notation $Q_{p+1,p}$ will be used for the increased objective function based on the first p eigenvalue–eigenvector pairs and looking for $p+1$ clusters:

$$BM(P_{p+1}, G) \leq Q_{p+1,p}(\mathbf{Z}_{p+1}, \boldsymbol{M}) := \sum_{i=1}^{p} \beta_i \sum_{a=1}^{p+1} (\mathbf{u}_i^T \mathbf{z}_a)^2.$$

In the sequel, for given $\boldsymbol{M}$, we want to maximize $Q_{p+1,p}(\mathbf{Z}_{p+1}, \boldsymbol{M})$ over $\mathcal{Z}_{p+1}^B$.

For this purpose, let us project the vectors $\sqrt{\beta_i}\mathbf{u}_i$ onto the subspace F_{p+1}:

$$\sqrt{\beta_i}\mathbf{u}_i = \sum_{a=1}^{p+1} [(\sqrt{\beta_i}\mathbf{u}_i)^T \mathbf{z}_a]\mathbf{z}_a + \text{ort}_{\mathcal{F}_{p+1}}(\sqrt{\beta_i}\mathbf{u}_i), \quad i = 1, \ldots, p. \tag{2.50}$$

The first term is the component in the subspace, and the second is orthogonal to it. In fact, the projected copies will be in a p-dimensional subspace of F_{p+1} orthogonal to the $\mathbf{1}$ vector (scalar multiple of $\mathbf{u}_{p+1}$). They will be piecewise constant vectors on $p+1$ steps, and their coordinates sum to 0. This is why one less eigenvectors are used than the number of clusters looked for.

By the Pythagorean theorem, for the squared lengths of the vectors in the decomposition (2.50) we get that

$$\beta_i = \|\sqrt{\beta_i}\mathbf{u}_i\|^2 = \sum_{a=1}^{p+1} [(\sqrt{\beta_i}\mathbf{u}_i)^T \mathbf{z}_a]^2 + \text{dist}^2(\sqrt{\beta_i}\mathbf{u}_i, F_{p+1}), \quad i = 1, \ldots, p.$$

By summing for $i = 1, \ldots, p$, the cumulated second term will turn out to be the sum of inner variances of the vertex representatives in the representation, defined as follows. For a given positive integer $d \leq p$, let the d-dimensional representatives $\mathbf{x}_1, \ldots, \mathbf{x}_n$ of the vertices be row vectors of the $n \times d$ matrix $\boldsymbol{X}_d = (\sqrt{\beta_1}\mathbf{u}_1, \ldots, \sqrt{\beta_d}\mathbf{u}_d)$. Their k-variance is $S_k^2(\boldsymbol{X}_d)$. Since F_k consists of piecewise constant vectors on the partition $(V_1, \ldots, V_k)$, by the ANOVA argument of Section 2.1 (see the Proof of Proposition 2.1.8) it follows that

$$S_k^2(\boldsymbol{X}_d) = \sum_{i=1}^{d} \text{dist}^2(\sqrt{\beta_i}\mathbf{u}_i, F_k).$$

Hence,

$$\begin{aligned}\sum_{i=1}^{p} \beta_i &= \sum_{i=1}^{p}\sum_{a=1}^{p+1} [(\sqrt{\beta_i}\mathbf{u}_i)^T \mathbf{z}_a]^2 + \sum_{i=1}^{p} \text{dist}^2(\sqrt{\beta_i}\mathbf{u}_i, F_{p+1}) \\ &= Q_{p+1,p}(\mathbf{Z}_{p+1}, \boldsymbol{M}) + S_{p+1}^2(\boldsymbol{X}_p),\end{aligned}$$

where the rows of $\boldsymbol{X}_p = (\sqrt{\beta_1}\mathbf{u}_1, \ldots, \sqrt{\beta_p}\mathbf{u}_p)$ are p-dimensional representatives of the vertices. We could as well take $(p+1)$-dimensional representatives as the last coordinates are zeros, and hence, $S_{p+1}^2(\boldsymbol{X}_p) = S_{p+1}^2(\boldsymbol{X}_{p+1})$. Thus, maximizing $Q_{p+1,p}$ is equivalent to minimizing $S_{p+1}^2(\boldsymbol{X}_p)$ that can be obtained by applying the k-means algorithm for the p-dimensional representatives with $p+1$ clusters.

More generally, if there is a gap between β_d and $\beta_{d+1} > 0$, then we may look for k clusters based on d-dimensional representatives of the vertices. Analogously to the above calculations, for $d < k \le p+1$ we have that

$$\sum_{i=1}^{d} \beta_i = \sum_{i=1}^{d} \sum_{a=1}^{k} [(\sqrt{\beta_i}\mathbf{u}_i)^T \mathbf{z}_a]^2 + \sum_{i=1}^{d} \mathrm{dist}^2(\sqrt{\beta_i}\mathbf{u}_i, F_k)$$
$$=: Q_{k,d}(\mathbf{Z}_k, \boldsymbol{M}) + S_k^2(\boldsymbol{X}_d). \tag{2.51}$$

If β_d is much greater than β_{d+1}, the k-variance $S_k^2(\boldsymbol{X}_{d+1})$ is not significantly greater than $S_k^2(\boldsymbol{X}_d)$, since $\boldsymbol{X}_{d+1}$'s last column, $\sqrt{\beta_{d+1}}\mathbf{u}_{d+1}$, will not increase too much the k-variance of the d-dimensional representatives, its norm being much less than that of the first d columns. As the left hand side of (2.51) is not increased significantly by adding β_{d+1}, the quantity $Q_{k,d+1}(\mathbf{Z}_k, \boldsymbol{M})$ is not much greater than $Q_{k,d}(\mathbf{Z}_k, \boldsymbol{M})$ is. Neither the classification nor the value of the modularity is changed much compared to the cost of taking one more eigenvector into consideration. After d has been selected, we can process the k-means algorithm with $k = d+1, \dots, p+1$ clusters. By an easy argument, $S_{k+1}^2(\boldsymbol{X}_d) \le S_k^2(\boldsymbol{X}_d)$, but we can stop, if it is much less. These considerations would minimize computational demand and prove good for randomly generated graphs from different block structures: see Chapter 3.

Calculating eigenvectors is costly; the Lánczos method (see e.g., Golub and van Loan (1989)) performs well if we calculate only eigenvectors corresponding to some leading eigenvalues followed by a spectral gap. In Alpert and Yao (1995); Kannan *et al.* (2004), the authors suggest using as many eigenvectors as possible. In fact, using more eigenvectors (up to p) is better from the point of view of accuracy, but using less eigenvectors (up to a gap in the positive part of the spectrum) is better from the computational point of view, see Newman (2004b); Ng *et al.* (2001). We have to compromise. By these arguments, a local maximum of the modularity can be expected at $k = d+1$.

The advantage of the modularity matrix versus the Laplacian is that here 0 is a watershed, and for small graphs, the $d = p$, $k = p+1$ choice is feasible; for large graphs we look for gaps (as in case of the Laplacian) in the positive part of the modularity spectrum, and the number of clusters is one more than the number of the largest positive eigenvalues with corresponding eigenvectors entered into the classification.

2.4.2 Maximizing the normalized Newman–Girvan modularity

The k-partition P_k is also uniquely defined by the $n \times k$ normalized partition matrix $\mathbf{Z}_k = (\mathbf{z}_1, \dots, \mathbf{z}_k)$ introduced in (2.25). With it,

$$NM(P_k, G) = \sum_{a=1}^{k} \mathbf{z}_a^T \boldsymbol{M} \mathbf{z}_a = \mathrm{tr}[(\boldsymbol{D}^{1/2}\mathbf{Z}_k)^T \boldsymbol{M}_D (\boldsymbol{D}^{1/2}\mathbf{Z}_k)]$$

where $\boldsymbol{M}_D$ is the normalized modularity matrix defined in (1.20). Since the matrix $\boldsymbol{D}^{1/2}\boldsymbol{Z}_k$ is suborthogonal, the maximization here happens with respect to $\boldsymbol{Z}_k^T\boldsymbol{D}\boldsymbol{Z}_k = \boldsymbol{I}_k$, that is over normalized k-partition matrices $\boldsymbol{Z}_k \in \mathcal{Z}_k^N$.

Let $\beta_1' \geq \cdots \geq \beta_n'$ denote the eigenvalues of the symmetric normalized modularity matrix $\boldsymbol{M}_D$ with corresponding unit-norm, pairwise orthogonal eigenvectors $\mathbf{u}_1', \dots, \mathbf{u}_n'$. In Section 1.3 we saw that the spectrum of $\boldsymbol{M}_D$ is in $[-1, 1]$ and includes the 0.

Let p denote the number of positive eigenvalues of $\boldsymbol{M}_D$. Now let $\boldsymbol{Y} = (\mathbf{y}_1, \dots, \mathbf{y}_k)$ be an arbitrary $n \times k$ matrix ($k \leq p$) such that $\boldsymbol{Y}^T\boldsymbol{D}\boldsymbol{Y} = \boldsymbol{I}_k$. With the same linear algebra as used in Section 2.4.1,

$$\max_{\boldsymbol{Z}_k \in \mathcal{Z}_k^n} NM(P_k, G) \leq \max_{\boldsymbol{Y}^T\boldsymbol{D}\boldsymbol{Y}=\boldsymbol{I}_k} \operatorname{tr}(\boldsymbol{Y}^T\boldsymbol{M}\boldsymbol{Y}) \leq \sum_{a=1}^{k} \beta_a' \leq \sum_{a=1}^{p+1} \beta_a'.$$

For further investigation, we expand our objective function with respect to the eigenvectors:

$$NM(P_k, G) = \sum_{i=1}^{n} \beta_i' \sum_{a=1}^{k} \left[(\mathbf{u}_i')^T (\boldsymbol{D}^{1/2}\mathbf{z}_a) \right]^2.$$

We can increase this sum if we neglect the terms corresponding to the negative eigenvalues, hence, the outer summation stops at p, or equivalently, at $p + 1$. The inner sum is the largest in the $k = p + 1$ case, when the unit-norm, pairwise orthogonal vectors $\boldsymbol{D}^{1/2}\mathbf{z}_1, \dots, \boldsymbol{D}^{1/2}\mathbf{z}_{p+1}$ are close to the eigenvectors $\mathbf{u}_1', \dots, \mathbf{u}_{p+1}'$, respectively. In fact, the two subspaces spanned by them should be close to each other. Now the subspace $F_{p+1} = \operatorname{Span}\{\boldsymbol{D}^{1/2}\mathbf{z}_1, \dots, \boldsymbol{D}^{1/2}\mathbf{z}_{p+1}\}$ does not consist of piecewise constant vectors, but the following argument is valid. By the notation $Q_{p+1,p}'(\boldsymbol{Z}_{p+1}, \boldsymbol{M})$ for the increased objective function based on the first p eigenvalue–eigenvector pairs and looking for $p + 1$ clusters we get that

$$NM(P_{p+1}, G) \leq Q_{p+1,p}'(\boldsymbol{Z}_{p+1}, \boldsymbol{M}) := \sum_{i=1}^{p} \beta_i' \sum_{a=1}^{p+1} \left[(\mathbf{u}_i')^T (\boldsymbol{D}^{1/2}\mathbf{z}_a) \right]^2.$$

In the sequel, for given $\boldsymbol{M}$, we want to maximize $Q_{p+1,p}'(\boldsymbol{Z}_{p+1}, \boldsymbol{M})$ over $\mathcal{Z}_{p+1}^N$.

With the argument of Section 2.4.1, now the vectors $\sqrt{\beta_i'}\,\mathbf{u}_i'$ are projected onto the subspace F_{p+1}:

$$\sqrt{\beta_i'}\,\mathbf{u}_i' = \sum_{a=1}^{p+1} \left[\left(\sqrt{\beta_i'}\,\mathbf{u}_i' \right)^T \boldsymbol{D}^{1/2}\mathbf{z}_a \right] \boldsymbol{D}^{1/2}\mathbf{z}_a + \operatorname{ort}_{\mathcal{F}_{p+1}} \left(\sqrt{\beta_i'}\,\mathbf{u}_i' \right), \quad i = 1, \dots, p.$$

As $\sqrt{\beta_{p+1}'}\,\mathbf{u}_{p+1}' = \mathbf{0}$, there is no use in projecting it.

By the Pythagorean theorem, for the squared lengths of the vectors in the above orthogonal decomposition we get that

$$\beta_i' = \left\| \sqrt{\beta_i'}\, \mathbf{u}_i' \right\|^2 = \sum_{a=1}^{p+1} \left[\left(\sqrt{\beta_i'}\, \mathbf{u}_i' \right)^T \boldsymbol{D}^{1/2} \mathbf{z}_a \right]^2 + \text{dist}^2 \left(\sqrt{\beta_i'}\, \mathbf{u}_i', F_{p+1} \right), \quad i = 1, \dots, p.$$

Let the vertex representatives $\mathbf{x}_1', \dots, \mathbf{x}_n' \in \mathbb{R}^p$ be the row vectors of the $n \times p$ matrix $\boldsymbol{X}_p' = (\sqrt{\beta_1'}\, \boldsymbol{D}^{-1/2}\mathbf{u}_1', \dots, \sqrt{\beta_p'}\, \boldsymbol{D}^{-1/2}\mathbf{u}_p')$. Then

$$\text{dist}^2 \left(\sqrt{\beta_i'}\, \mathbf{u}_i', F_{p+1} \right) = \sum_{j=1}^{n} d_j \left(x_{ji}' - c_{ji} \right)^2, \quad i = 1, \dots, p$$

where x_{ji}' is the ith coordinate of the vector $\mathbf{x}_j'$ and c_{ji} is the same for the vector $\mathbf{c}_j \in \mathbb{R}^p$, where there are at most $p + 1$ different ones among the centers $\mathbf{c}_1, \dots, \mathbf{c}_n$ assigned to the vertex representatives. Namely,

$$c_{ji} = \frac{1}{\sum_{\ell \in V_a} d_\ell} \sum_{\ell \in V_a} d_\ell x_{\ell i}', \quad j \in V_a, \quad i = 1, \dots, p.$$

In other words, the column vectors of the $n \times p$ matrix of rows $\mathbf{c}_1, \dots, \mathbf{c}_n$ are stepwise constant vectors on the same $p + 1$ steps corresponding to the $(p + 1)$-partition of the vertices encoded into the partition matrix $\boldsymbol{Z}_{p+1}$.

By summing for $i = 1, \dots, p$, in view of the ANOVA argument of Section 2.4.1, the cumulated second term will turn out to be the weighted $(p + 1)$-variance of the vertex representatives in the $(p + 1)$-partition designated by the partition matrix $\boldsymbol{Z}_{p+1}$:

$$\tilde{S}_{p+1}^2(\boldsymbol{X}_p') = \sum_{i=1}^{p} \text{dist}^2 \left(\sqrt{\beta_i'}\, \mathbf{u}_i', F_{p+1} \right) = \sum_{j=1}^{n} d_j \|\mathbf{x}_j - \mathbf{c}_j\|^2.$$

Therefore,

$$\sum_{i=1}^{p} \beta_i' = Q_{p+1,p}'(\boldsymbol{Z}_{p+1}, \boldsymbol{M}) + \tilde{S}_{p+1}^2(\boldsymbol{X}_p').$$

This applies to a given $(p + 1)$-partition of the vertices. We are looking for the $(p + 1)$-partition maximizing the first term. In view of the above formula, increasing $Q_{p+1,p}'$ can be achieved by decreasing $\tilde{S}_{p+1}^2(\boldsymbol{X}_p')$; latter one is obtained by applying the weighted k-means algorithm with $p + 1$ clusters for the p-dimensional representatives $\mathbf{x}_1', \dots, \mathbf{x}_n'$ with respective weights $d_1, \dots, d_n$.

Analogously, for $d < k \le p+1$:

$$\sum_{i=1}^{d} \beta_i' = \sum_{i=1}^{d}\sum_{a=1}^{k}\left[\left(\sqrt{\beta_i'}\,\mathbf{u}_i'\right)^T \boldsymbol{D}^{1/2}\mathbf{z}_a\right]^2 + \sum_{i=1}^{d} \mathrm{dist}^2\left(\sqrt{\beta_i'}\,\mathbf{u}_i', F_k\right)$$
$$= Q_{k,d}'(\boldsymbol{Z}_k, \boldsymbol{M}) + \tilde{S}_k^2(\boldsymbol{X}_d')$$

where the row vectors of the $n \times d$ matrix $\boldsymbol{X}_d' = (\sqrt{\beta_1'}\,\boldsymbol{D}^{-1/2}\mathbf{u}_1', \dots, \sqrt{\beta_d'}\,\boldsymbol{D}^{-1/2}\mathbf{u}_d')$ are d-dimensional representatives of the vertices. Hence, in the presence of a spectral gap between β_d' and $\beta_{d+1}' > 0$ – in the miniature world of the [0,1] interval – neither $\sum_{i=1}^{d} \beta_i'$ nor $\tilde{S}_k^2(\boldsymbol{X}_d')$ can be increased significantly by introducing one more eigenvalue-eigenvector pair (by using $(d+1)$-dimensional representatives instead of d-dimensional ones). Consequently, $Q_{k,d}'(\boldsymbol{Z}_k, \boldsymbol{M})$ would not change much, and by the argument of Section 2.4.1, $k = d+1$ clusters based on d-dimensional representatives will suffice.

In their paper Karrer and Newman (2011) introduce a model that takes into consideration the heterogeneity in the degrees of vertices. While the usual blockmodel is biased towards placing vertices of similar degrees in the same cluster, the new model is capable of finding clusters of vertices of heterogeneous degrees. Similar stochastic blockmodels will be discussed in Chapter 3.

2.4.3 Anti-community structure and some examples

Given the weighted graph $G = (V, \boldsymbol{W})$, instead of taking the maximum, we take the minimum of $Q_B(P_k, \boldsymbol{W}) = Q_B(\boldsymbol{Z}_k, \boldsymbol{M})$ over balanced k-partition matrices $\boldsymbol{Z}_k$. For fixed $k \le m$, analogously to the inference of Section 2.4.1,

$$\min_{P_k \in \mathcal{P}_k} BM(P_k, G) = \min_{\boldsymbol{Z}_k^T\boldsymbol{Z}_k = \boldsymbol{I}_k} \mathrm{tr}\left(\boldsymbol{Z}_k^T \boldsymbol{M} \boldsymbol{Z}_k\right) \ge \min_{\boldsymbol{Y}^T\boldsymbol{Y}=\boldsymbol{I}_k} \mathrm{tr}(\boldsymbol{Y}^T\boldsymbol{M}\boldsymbol{Y})$$
$$\ge \sum_{a=1}^{k} \beta_{n+1-a} \ge \sum_{a=1}^{m+1} \beta_{n+1-a}$$

where m is the number of negative eigenvalues of $\boldsymbol{M}$ $(m+p<n)$. For the classification, here we use the scaled (by the square root of the absolute value of the corresponding eigenvalue) eigenvectors corresponding to the negative eigenvalues for the representation to find $m+1$ clusters. For large n, it suffices to choose $d < m$ structural negative eigenvalues such that there is a noticeable spectral gap between β_{n+1-d} and β_{n-d}. Then with $\boldsymbol{X}_d = (\sqrt{|\beta_n|}\mathbf{u}_n, \dots, \sqrt{|\beta_{n+1-d}|}\mathbf{u}_{n+1-d})$, we find the minimum of $S_{d+1}^2(\boldsymbol{X}_d)$ by the k-means algorithm with $d+1$ clusters.

The same can be done by minimizing the normalized Newman–Girvan modularity based on the largest absolute value negative eigenvalues and the corresponding eigenvectors of the normalized modularity matrix. The following examples illustrate that large positive eigenvalues of the modularity matrix are indications of a community, while large absolute value negative ones, of an anti-community structure. In Rohe *et al.* (2011), these structures are called 'homophilic' or 'heterophilic'. For the heterophilic structure an example is the network of dating relationships in a high school, where two persons of opposite sex are more likely to date than persons of the same sex.

(i) Let G be the disjoint union of k complete graphs $K_{n_1}, \ldots, K_{n_k}$, respectively, with total number of vertices $n = \sum_{i=1}^{k} n_i$ and corresponding partition $V_1, \ldots, V_k$ of the vertices. This is sometimes called *pure community structure*, since there are no inter-community edges, but all possible intra-community edges are present. The modularity matrix of G has $k - 1$ positive eigenvalues, $\beta_k = 0$ with corresponding eigenvector $\mathbf{1}/\sqrt{n}$, and there is only one negative eigenvalue with multiplicity $n - k$. (In the $n_1 = \cdots = n_k$ special case $\beta_1 = \cdots = \beta_{k-1}$ is a multiple positive eigenvalue.) Here k communities are detected by the k-means algorithm applied for the $(k - 1)$-dimensional representatives based on the eigenvectors corresponding to the positive eigenvalues. As these eigenvectors themselves have piecewise constant structures over the clusters $V_1, \ldots, V_k$, the k-variance of the representatives is 0, and the maximum $BM(P_k, G)$ is a slightly smaller positive number than the maximum $Q_{k,k-1}(\boldsymbol{Z}_k, \boldsymbol{M})$, latter one being the sum of the positive eigenvalues. In the $k = 1$ case the modularity matrix is negative semidefinite, and both the maximum $BM(P_k, G)$ and $Q_{k,k-1}(\boldsymbol{Z}_k, \boldsymbol{M})$ are zeros. The normalized modularity matrix $\boldsymbol{M}_D$ of G has the eigenvalue 1 with multiplicity $k - 1$, one 0 eigenvalue and all the other eigenvalues are in the $(-1, 0)$ interval taking on at most $k - 1$ different values. (In the $n_1 = \cdots = n_k$ case there is only one negative eigenvalue with multiplicity $n - k$.)

(ii) Let G be the complete k-partite graph $K_{n_1,\ldots,n_k}$, with total number of vertices $n = \sum_{i=1}^{k} n_i$ and corresponding partition $V_1, \ldots, V_k$ of the vertices. This is sometimes called *pure anti-community structure* as the empty modules may model hub-authorities and correspond to perfectly disassortative mixing. The modularity matrix of G has $k - 1$ negative eigenvalues, all the other eigenvalues are zeros. (In the $n_1 = \cdots = n_k$ special case there is one negative eigenvalue with multiplicity $k - 1$.) Here k communities are detected by the k-means algorithm applied for the $(k - 1)$-dimensional representatives based on the eigenvectors corresponding to the negative eigenvalues. As these eigenvectors themselves have piecewise constant structures over the clusters $V_1, \ldots, V_k$, the k-variance of the representatives is 0, the minimum $BM(P_k, \boldsymbol{G})$ is negative, but slightly larger than the minimum $Q_{k,k-1}(\boldsymbol{Z}_k, \boldsymbol{M})$, latter one being the sum of the negative eigenvalues. The normalized modularity matrix $\boldsymbol{M}_D$ of G has $k - 1$ negative eigenvalues in the $[-1, 0)$ interval, all the other eigenvalues are zeros. (In the $n_1 = \cdots = n_k$ case the only negative eigenvalue has multiplicity $k - 1$.)

Note that the complete graph ((i) with $k = 1$) and complete k-partite graphs ((ii) with $k \geq 2$) have the zero as the largest modularity eigenvalue. Dragan Stevanovic conjectures that these are the only graphs with a negative semidefinite modularity matrix (of which zero is always an eigenvalue). An extensive numerical search also supports this fact, though it is an open question yet.

We remark that strategic interaction games also provide examples for both community and anti-community structures. Here the agents are vertices of a graph and agents connected with an edge (of non-zero weight) influence each other's actions, see for example Ballester *et al.* (2006); Jackson and Zenou (forthcoming) for economic networks and simple graphs. The theory can be extended to edge-weighted graphs, where the impact of an agent's action on another agent's action depends on the edge-weight between them. The actions themselves are nonnegative real numbers that – taking into consideration the graph structure – maximize the joint quadratic payoff which is of the following two types. In games of so-called *strategic complements*, an increase of the actions of other players leads a given player's higher actions to have relatively higher payoffs compared to that player's lower actions. In games of so-called *strategic substitutes* the opposite is true: an increase in other players' actions leads to relatively lower payoffs under higher actions of a given player. If we classify the agents such that those exhibiting similar actions belong to the same cluster, then a community and an anti-community structure is likely to develop in the two above cases, respectively. We may extend the theory to multiple actions, where the action vectors play a similar role to the representatives of the vertices in the quadratic placement problems of Chapter 1. For genetic applications consult Wilkinson and Huberman (2004).

Now we present two applications of the balanced Newman–Girvan modularity for real-life networks.

- *Zachary's karate club data*: When we maximized the balanced modularity by means of applying the k-means algorithm (with $k = 2$ clusters) for the coordinates of the eigenvector corresponding to the leading positive eigenvalue of the modularity matrix of the karate club's network, our algorithm gave exactly the same partition of the club members that found in the original paper Zachary (1977), see Figure 2.1.

- *The bottlenose dolphin community of Doubtful Sound*: We investigated the graph of social connections of 40 bottlenose dolphins retained for association analysis by Lusseau *et al.* (2003). The authors found three groups with individuals most frequently seen together (see Lusseau's Figure 5), though the groups were not separated clearly by their hierarchical clustering algorithm. Based on one and two leading positive eigenvalues and the corresponding eigenvectors of the modularity matrix, by the k-means algorithm, we found two and three clusters, respectively. Though we processed the algorithm for the two- and three-cluster cases separately, one cluster of the three turned out to be the same as one of the clusters of the two-cluster case. In our Figure 2.2, squares and circles represent individuals of the two main clusters obtained by our spectral clustering algorithm maximizing the balanced modularity with $k = 2$, and the shaded and open circles denote the separation of the second cluster when we processed our algorithm with $k = 3$. These three communities are practically the same as discussed on page 401 of Lusseau *et al.* (2003). Squares correspond to a male

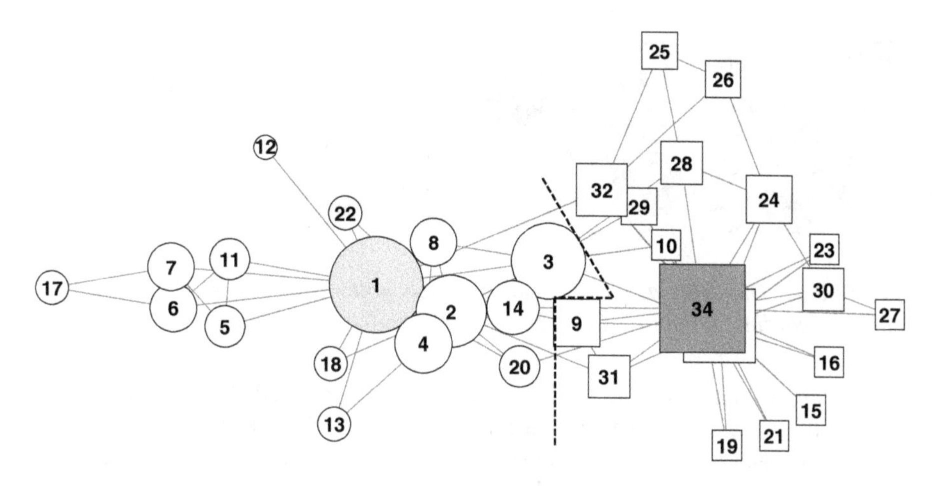

Figure 2.1 The network of social connections in the karate club network of Zachary. Circles and squares represent nodes of the two clusters with sizes proportional to their degrees. The shaded nodes are the administrator (1) and the instructor (34, covering 33). The separation of the two clusters found by our spectral algorithm maximizing the balanced modularity coincides with the real-life separation of the club members found in the original paper and denoted by the dashed line. Reproduced based on Bolla (2011b) with permission of the APS.

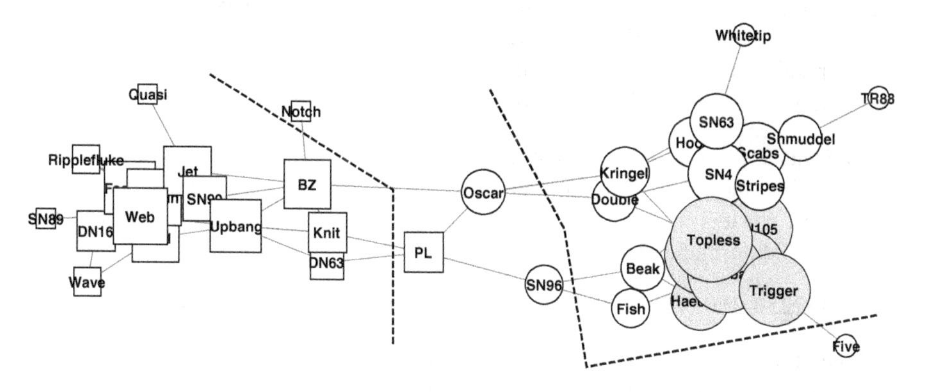

Figure 2.2 The network of social connections between 40 bottlenose dolphins retained for association analysis by Lusseau et al. (2003). Squares and circles represent individuals of the two main clusters obtained by our spectral clustering algorithm maximizing the balanced modularity with $k = 2$. In the $k = 3$ case, the squares remained in the same cluster, while circles separated into the shaded and open ones. The dense parts of these clusters coincide with the three communities described in Lusseau et al. (2003). The two main communities observed in the original paper are separated by dashed lines, while the intermediate low degree nodes are not classified uniquely by the original paper. Reproduced based on Bolla (2011b) with permission of the APS.

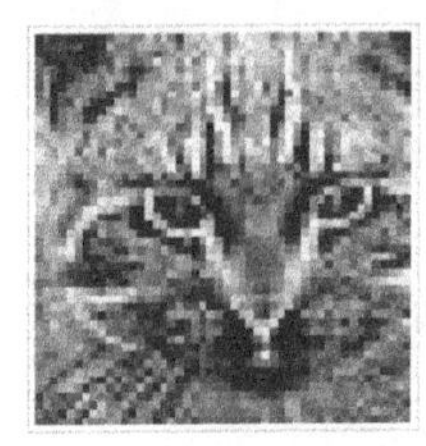 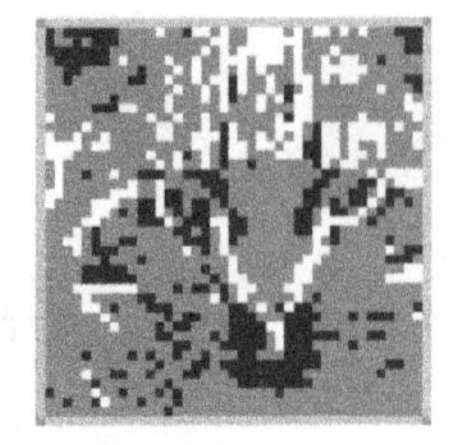 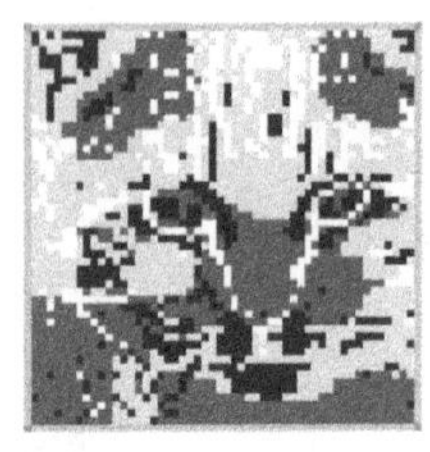 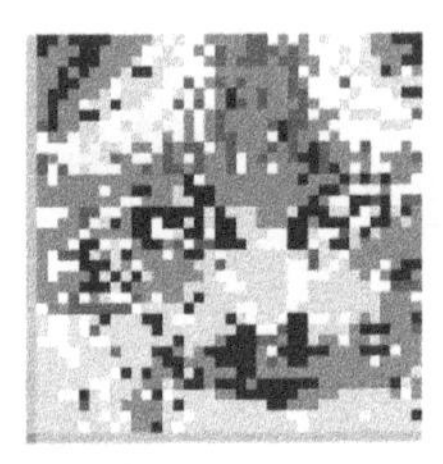

Figure 2.3 The original picture and the pixels colored with 3, 4, and 5 different colors according to their cluster memberships.

group with an unknown sex individual at the bottom, shaded circles correspond to the group of six males and one female (Trigger) at the left upper corner, finally, open squares represent the female band at the top right of Lusseau's Figure 5; there were loose connections between these two kinds of circles in accord with the fact that they corresponded to one cluster in the two-cluster situation. The two main communities observed in the original paper are separated by dashed lines, while the intermediate low-degree nodes – corresponding to the middle part of Lusseau's Figure 5 – are not classified uniquely in the original paper.

As for the normalized modularity matrix, we used the eigenvectors corresponding to its $k-1$ largest eigenvalues to find k clusters of $48^2 = 2304$ pixels based on a Gaussian kernel, see Section 1.5 (the distances of the pixels depended not only on their spatial distances). More precisely, we assigned the points $\mathbf{x}_1, \dots, \mathbf{x}_{2304} \in \mathbb{R}^3$ to the pixels, the coordinates of which characterize the spatial location, color, and brightness of the pixels. With the positive parameter σ, the similarity between pixels i and j was $w_{ij} = e^{-\frac{\|\mathbf{x}_i - \mathbf{x}_j\|^2}{2\sigma^2}}$ for $i \neq j$. Figure 2.3 shows the original picture, and the picture when the pixels were colored according to their cluster memberships with number of clusters 3, 4, and 5. Since the largest absolute value eigenvalues of the 2304×2304 normalized modularity matrix are

$$0.137259,\ 0.0142548,\ 0.000925228,\ -0.000670733,\ -0.000670674,\ \dots$$

and the number of the positive eigenvalues is three, with a gap after the second one, the 3- or 4-cluster solution seems the most reasonable. It is an intriguing question – unsolved so far – whether the dimension k of the original data points can be detected in the spectrum of $\boldsymbol{M}_D$ when n is large.

2.5 Normalized bicuts of contingency tables

We are given an $m \times n$ contingency table $\boldsymbol{C}$ on row set Row and column set Col as introduced in Section 1.2. For a fixed integer k, $0 < k \leq r = \text{rank}(\boldsymbol{C})$, we want to simultaneously partition the rows and columns of $\boldsymbol{C}$ into disjoint, nonempty subsets

$$Row = R_1 \cup \cdots \cup R_k, \quad Col = C_1 \cup \cdots \cup C_k$$

such that the cuts $c(R_a, C_b) = \sum_{i \in R_a} \sum_{j \in C_b} c_{ij}, \quad a, b = 1, \dots, k$ between the row-column cluster pairs be as homogeneous as possible.

Definition 2.5.1 *The normalized bicut of the contingency table* $\boldsymbol{C}$ *with respect to the k-partitions* $P_{row} = (R_1, \dots, R_k)$ *and* $P_{col} = (C_1, \dots, C_k)$ *of its rows and columns and the collection of signs* σ *is defined as follows:*

$$\nu_k(P_{row}, P_{col}, \sigma) = \sum_{a=1}^{k} \sum_{b=1}^{k} \left(\frac{1}{\text{Vol}(R_a)} + \frac{1}{\text{Vol}(C_b)} + \frac{2\sigma_{ab}\delta_{ab}}{\sqrt{\text{Vol}(R_a)\text{Vol}(C_b)}} \right) c(R_a, C_b), \tag{2.52}$$

where

$$\text{Vol}(R_a) = \sum_{i \in R_a} d_{row,i} = \sum_{i \in R_a} \sum_{j=1}^{n} c_{ij}, \quad \text{Vol}(C_b) = \sum_{j \in C_b} d_{col,j} = \sum_{j \in C_b} \sum_{i=1}^{m} c_{ij}$$

are volumes of the clusters (also see formulas (1.13) and (1.14)), δ_{ab} *is the Kronecker delta, and the sign* σ_{ab} *is equal to 1 or −1 (it only has relevance in the* $a = b$ *case), and* $\sigma = (\sigma_{11}, \dots, \sigma_{kk})$ *is the collection of the relevant signs.*

The normalized k-way bicut of the contingency table $\boldsymbol{C}$ *is the minimum of (2.52) over all possible k-partitions* $\mathcal{P}_{row,k}$ *and* $\mathcal{P}_{col,k}$ *of its rows and columns, and over all possible collections of signs* σ*:*

$$\nu_k(\boldsymbol{C}) = \min_{P_{row}, P_{col}, \sigma} \nu_k(P_{row}, P_{col}, \sigma).$$

Note that $\nu_k(\boldsymbol{C})$ penalizes row- and column clusters of extremely different volumes in the $a \neq b$ case, whereas in the $a = b$ case σ_{aa} moderates the balance between $\text{Vol}(R_a)$ and $\text{Vol}(C_a)$.

Theorem 2.5.2 *Let* $1 = s_0 \geq s_1 \geq \dots \geq s_{r-1} > 0$ *be the positive singular values of the normalized contingency table* $\boldsymbol{C}_{corr} = \boldsymbol{D}_{row}^{-1/2} \boldsymbol{C} \boldsymbol{D}_{col}^{-1/2}$ *belonging to* $\boldsymbol{C}$*. Then for any positive integer* $k \leq r$*, such that* $s_{k-1} > s_k$*,*

$$\nu_k(\boldsymbol{C}) \geq 2k - \sum_{i=0}^{k-1} s_i.$$

Proof. We will show that $\nu_k(P_{row}, P_{col}, \sigma)$ is the objective function Q_k of (1.15) in the special representation, where the column vectors of $\boldsymbol{X}$ and $\boldsymbol{Y}$ are normalized partition vectors corresponding to P_{row} and P_{col}, respectively. From here, the statement follows, since in view of Theorem 1.2.1, the overall minimum is $2k - \sum_{i=0}^{k-1} s_i$. Indeed, denoting by x_{ia} the ith coordinate of the ath column of $\boldsymbol{X}$, in view of (2.25),

$$x_{ia} = \begin{cases} \frac{1}{\sqrt{\text{Vol}(R_a)}} & \text{if} \quad i \in R_a \\ 0 & \text{otherwise} \end{cases} \qquad (a = 1, \dots, k).$$

Likewise, denoting by y_{jb} the jth coordinate of the bth column of $\boldsymbol{Y}$,

$$y_{jb} = \begin{cases} \frac{1}{\sqrt{\mathrm{Vol}(C_b)}} & \text{if} \quad j \in C_b \\ 0 & \text{otherwise} \end{cases}, \quad (b = 1, \dots, k).$$

With this, the matrices $\boldsymbol{X}$ and $\boldsymbol{Y}$ satisfy the conditions imposed on the representatives. Actually, the row and column represenatives, $\mathbf{r}_1, \dots \mathbf{r}_m \in \mathbb{R}^k$ and $\mathbf{q}_1, \dots, \mathbf{q}_n \in \mathbb{R}^k$, are the row vectors of $\boldsymbol{X}$ and $\boldsymbol{Y}$, and it is easy to verify that

$$\|\mathbf{r}_i - \mathbf{q}_j\|^2 = \frac{1}{\mathrm{Vol}(R_a)} + \frac{1}{\mathrm{Vol}(C_b)} + \frac{2\sigma_{bb}\delta_{ab}}{\sqrt{\mathrm{Vol}(R_a)\mathrm{Vol}(C_b)}}, \quad \text{if} \quad i \in R_a,\ j \in C_b.$$

Note that in Section 1.2 we introduced $\boldsymbol{C}_{corr}$ as correspondence matrix belonging to $\boldsymbol{C}$. This matrix is used intensively in correspondence analysis. For further reading about correspondence analysis we recommend Section C.3 of the Appendix and Faust (2005); Faust *et al.* (2002) and Bolla *et al.* (2010). If $\boldsymbol{C}$, or equaivalently, $\boldsymbol{C}_{corr}$ is non-decomposable, then 1 is a single singular value.

Observe, that in the case of a symmetric contingency table, we get the same result with the representation, based on the eigenvectors corresponding to the largest absolute value eigenvalues of the normalized modularity matrix. However, $\nu_k(P_{row}, P_{col}, \sigma)$ cannot always be directly related to the normalized cut, except the following two special cases.

- When the $k-1$ largest absolute value eigenvalues of the normalized modularity matrix are all positive, or equivalently, if the k smallest eigenvalues (including the zero) of the normalized Laplacian matrix are farther from 1 than any other eigenvalue which is greater than 1. In this case, the $k-1$ largest singular values (apart from the 1) of the correspondence matrix are identical to the $k-1$ largest eigenvalues of the normalized modularity matrix, and the left and right singular vectors are identical to the corresponding eigenvector with the same orientation. Consequently, for the k-dimensional (in fact, $(k-1)$-dimensional) row- and column-representatives $\mathbf{r}_i = \mathbf{q}_i$ $(i = 1, \dots, n = m)$ holds. With the choice $\sigma_{bb} = 1$ $(b = 1, \dots, k)$, the corresponding $\nu_k(\boldsymbol{C})$ is twice the normalized cut of our weighted graph in which weights of edges within the clusters do not count. In this special situation, the normalized bicut also favors k-partitions with low inter-cluster edge-densities (therefore, intra-cluster densities tend to be large, as they do not count in the objective function).

- When the $k-1$ largest absolute value eigenvalues of the normalized modularity matrix are all negative, then $\mathbf{r}_i = -\mathbf{q}_i$ for all $(k-1)$-dimensional row and column representatives, and any (but only one) of them can be the corresponding vertex representative. Now $\nu_k(\boldsymbol{C})$, which is attained with the choice $\sigma_{bb} = -1$ $(b = 1, \dots, k)$, differs from the normalized cut in that it also counts the edge-weights within the clusters. Indeed, in the $a = b$, $R_a = C_a = V_a$ case

$$\|\mathbf{r}_i - \mathbf{q}_j\|^2 = \frac{1}{\mathrm{Vol}(V_a)} + \frac{1}{\mathrm{Vol}(V_b)} + \frac{2}{\sqrt{\mathrm{Vol}(V_a)\mathrm{Vol}(V_b)}} = \frac{4}{\mathrm{Vol}(V_a)}$$

if $i, j \in V_a$. Here, by minimizing the normalized k-way cut, rather a so-called anti-community structure is detected in that $c(R_a, C_a) = c(V_a, V_a)$ is suppressed to compensate for the term $\frac{4}{\mathrm{Vol}(V_a)}$.

We remark that Ding *et al.* (2003) treats this problem for two row- and column-clusters and minimizes another objective function such that it favors 2-partitions where $c(R_1, C_2)$ and $c(R_2, C_1)$ are small compared to $c(R_1, C_1)$ and $c(R_2, C_2)$. The solution is also given by the transformed left- and right-hand side singular vector pair corresponding to s_1. However, it is the objective function ν_k which best complies with the SVD of the correspondence matrix, and hence, gives the continuous relaxation of the normalized bicut minimization problem. The idea of Ding *et al.* could be naturally extended to the case of several, but the same number of row and column clusters, and it may work well in the keyword-document classification problem. Though, in some real-life problems, for example clustering genes and conditions of microarrays, we rather want to find clusters of similarly functioning genes that equally (not especially weakly or strongly) influence conditions of the same cluster; this issue is discussed in detail in Section 3.3. For further reading about microarrays, we recommend Higham *et al.* (2005); Kluger *et al.* (2003); Omberg *et al.* (2007). Dhillon (2001) also suggests a multipartition algorithm that runs the k-means algorithm simultaneously for the row- and column representatives.

References

Ajtai M, Komlos J and Szemerédi E 1983 Sorting in $c \log n$ parallel steps. *Combinatorica* **3**, 1–9.

Aldous DJ 1989 Lower bounds for covering times for reversible Markov chains and random walks on graphs. *J. Theor. Probab.* **2**, 91–100.

Alon N 1986 Eigenvalues and expanders. *Combinatorica* **6** (2), 83–96.

Alon N and Milman VD 1985 λ_1, isoperimetric inequalities for graphs and superconcentrators. *J. Comb. Theory Ser. B* **38**, 73–88.

Alon N and Milman VD 1987 Better expanders and superconcentrators. *J. Algorithms* **8**, 337–347.

Alon N and Sudakov B 2000 Bipartite subgraphs and the smallest eigenvalue. *Comb. Probab. Comput.* **9**, 1–12.

Alpert CJ and Yao S.-Z 1995 Spectral partitioning: the more eigenvectors, the better, in *Proc. 32nd ACM/IEEE International Conference on Design Automation* (eds Preas BT, Karger PG, Nobandegani BS and Pedram M), Association of Computer Machinery, New York, pp. 195–200.

Arora S, Barak B and Steurer D 2010 Subexponential algorithms for unique games and related problems, in *Proc. 51st IEEE Symposium on Foundations of Computer Science (FOCS 2010)*, Washington DC, USA, IEEE Computer Society, pp. 563–572.

Azran A and Ghahramani Z 2006 Spectral methods for automatic multiscale data clustering, in *Proc. IEEE Computer Society Conference on Computer Vision and Pattern Recognition (CVPR 2006), New York NY* (eds Fitzgibbon A, Taylor CJ and Lecun Y), IEEE Computer Society, Los Alamitos, California, pp. 190–197.

Babai L and Szegedy M 1992 Local expansion of symmetric graphs. *Comb. Probab. Comput.* **1**, 1–11.

Ballester C, Calvó-Armengol A and Zenou Y 2006 Who's who in networks. Wanted: The key player. *Econometrica* **74** (5), 1403–1417.

Bolla M 1989 Spectra, Euclidean representation, and vertex-colourings of hypergraphs, Preprint 78/1989, Mathematical Institute of the Hungarian Academy of Sciences.

Bolla M 1993 Spectra and Euclidean representation of hypergraphs. *Discret. Math.* **117**, 19–39.

Bolla M and M.-Sáska G 2002 Isoperimetric properties of weighted graphs related to the Laplacian spectrum and canonical correlations. *Stud. Sci. Math. Hung.* **39**, 425–441.

Bolla M and M.-Sáska G 2004 Optimization problems for weighted graphs and related correlation estimates. *Discret. Math.* **282**, 23–33.

Bolla M and Tusnády G 1990 Spectra and colourings of weighted graphs, Preprint 52/1990, Mathematical Institute of the Hungarian Academy of Sciences.

Bolla M and Tusnády G 1994 Spectra and Optimal Partitions of Weighted Graphs. *Discret. Math.* **128**, 1–20.

Bolla M, Friedl K and Krámli A 2010 Singular value decomposition of large random matrices (for two-way classification of microarrays). *J. Multivariate Anal.* **101**, 434–446.

Bolla M 2011 Beyond the expanders. *Int. J. Comb.*, 787596.

Bolla M 2011 Penalized versions of the Newman–Girvan modularity and their relation to multi-way cuts and k-means clustering. *Phys. Rev. E* **84**, 016108.

Buser P 1984 On the bipartition of graphs. *Discret. Appl. Math.* **9**, 105–109.

Cheeger J 1970 A lower bound for the smallest eigenvalue of the Laplacian, in *Problems in Analysis* (ed. R. C. Gunning), Princeton Univ. Press, Princeton NJ, pp. 195–199.

Chung F 1997 *Spectral Graph Theory*, CBMS Regional Conference Series in Mathematics **92**. American Mathematical Society, Providence RI.

Clauset A, Newman MEJ and Moore C 2004 Finding community structure in very large networks. *Phys. Rev. E* **70**, 066111.

Dhillon IS 2001 Co-clustering documents and words using bipartite spectral graph partitioning, in *Proc. ACM Int. Conf. Knowledge Disc. Data Mining (KDD 2001)* (eds Provost FJ and Srikant R), Association for Computer Machinery, New York, pp. 269–274.

Diaconis P and Stroock D 1991 Geometric bounds for eigenvalues of Markov chains. *Ann. Appl. Probab.* **1**, 36–62.

Ding C, He X, Zha H, Gu M and Simon HD 2003 A minmax cut method for data clustering and graph partitioning, Technical Report 54111, Lawrence Berkeley National Laboratory.

Donath WE and Hoffman AJ 1973 Lower bounds for the partitioning of graphs. *IBM J. Res. Develop.* **17**, 420–425.

Duch J and Arenas A 2005 Community detection in complex networks using extremal optimization. *Phys. Rev. E* **72**, 027104.

Faust K 2005 Using correspondence analysis for joint displays of affiliation networks, in *Models and Methods in Social Network Analysis* Vol. 7, (eds Carrington PJ and Scott J), Cambridge Univ. Press, Cambridge, pp. 117–147.

Faust K, Willert KE, Rowlee DD and Skvoretz J 2002 Scaling and statistical models for affiliation networks: Patterns of participation among Soviet politicians during the Brezhnev era. *Social Networks* **24**, 231–259.

Fiedler M 1972 Bounds for eigenvalues of doubly stochastic matrices. *Linear Algebra Appl.* **5** (98), 299–310.

Fiedler M 1973 Algebraic connectivity of graphs. *Czech. Math. J.* **23** (98), 298–305.

Fortunato S 2010 Community detection in graphs. *Phys. Rep.* **486**, 75–174.

Golub GH and van Loan CF 1996 *Matrix Computations*, 3rd edn Johns Hopkins University Press, Baltimore, MD.

Hagen L and Kahng AB 1992 New spectral methods for ratio cut partitioning and clustering. *IEEE Trans. Comput. Aided Des.* **11**, 1074–1085.

Higham DJ, Kalna G and Vass JK 2005 Analysis of the singular value decomposition as a tool for processing microarray expression data, in *Proceedings of ALGORITMY 2005* (eds Handlovicova A, Kriva Z and Sevcovic D), Slovak University of Technology, Bratislava.

Hoffman AJ 1972 Eigenvalues and partitionings of the edges of a graph. *Linear Algebra Appl.* **5**, 137–146.

Hoory S, Linial N and Widgerson A 2006 Expander graphs and their applications. *Bull. Amer. Math. Soc. (N. S.)* **43** (4), 439–561.

Jackson MO and Zenou Y 2014 Games on Networks, to appear in the *Handbook of Game Theory* **4** (eds Young P and Zamir S), Elsevier Science.

Juhász F and Mályusz K 1980 Problems of cluster analysis from the viewpoint of numerical analysis. In *Numerical Methods, Coll. Math. Soc. J. Bolyai* (ed. Rózsa P), North-Holland, Amsterdam, Vol 22, pp. 405–415.

Kannan R, Vempala S and Vetta A 2004 On clusterings: Good, bad and spectral. *J. ACM* **51**, 497–515.

Karrer B and Newman MEJ 2011 Stochastic blockmodels and community structure in networks. *Phys. Rev. E* **83**, 016107.

Kluger Y, Basri R, Chang JT and Gerstein M 2003 Spectral biclustering of microarray data: coclustering genes and conditions. *Genome Res.* **13**, 703–716.

Lee JR, Gharan SO and Trevisan L 2012 Multi-way spectral partitioning and higher-order Cheeger inequalities, in *Proc. 44th Annual ACM Symposium on the Theory of Computing (STOC 2012)*, ACM, New York NY, pp. 1117–1130.

Louis A, Raghavendra P, Tetali P and Vempala S 2011 Algorithmic extension of Cheeger's inequality to higher eigenvalues and partitions, in *Approximation, Randomization and Combinatorial Optimization. Algorithms and Techniques. Lecture Notes in Computer Science* (eds Goldborg LA, Jansen K, Ravi R and Rolim JDP), Springer, vol. 6845, pp. 315–326.

Louis A, Raghavendra P, Tetali P and Vempala S 2012 Many sparse cuts via higher eigenvalues, in *Proc. 44th Annual ACM Symposium on the Theory of Computing (STOC 2012)*, ACM, New York NY, pp. 1131–1140.

Lovász L 1993 *Combinatorial Problems and Exercises*. ACM, Akadémiai Kiadó–North Holland, Budapest–Amsterdam.

Lovász L 1993 Random walks on graphs: a survey, in *Combinatorics, Paul Erdős is Eighty. János Bolyai Society, Mathematical Studies* Keszthely, Hungary, vol. 2, pp. 1–46.

Lovász L and Simonovits M 1993 Random walks in a convex body and an improved volume algorithm. *Random Struct. Algorithms* **4**, 359–412.

Lovász L and Winkler P 1995 Exact mixing in an unknown Markov chain. *Electron. J. Comb.* **2**, paper R15, 1–14.

Lubotzky A 1994 Discrete graphs, expanging graphs, and invariant measures. (With an appendix by J. D. Rogawski), in *Progress in Mathematics* **125**, Birkhäuser Verlag, Basel.

Lusseau D, Schneider K, Boisseau OJ *et al.* 2003 The bottlenose dolphin community of Doubtful Sound features a large proportion of long-lasting associations. *Behav. Ecol. Sociobiol.* **54**, 396–405.

Meila M and Shi J 2001 Learning segmentation by random walks, in *Proc. 13th Neural Information Processing Systems Conference (NIPS 2001)* (eds Leen TK, Dietterich TG and Tresp V), MIT Press, Cambridge, USA, pp. 873–879.

Mohar B 1988 Isoperimetric inequalities, growth and the spectrum of graphs. *Linear Algebra Appl.* **103**, 119–131.

Mohar B 1989 Isoperimetric numbers of graphs. *J. Comb. Theor, Ser. B* **47**, 274–291.

Mohar B and Poljak S 1990 Eigenvalues and the max-cut problem. *Czech. Math. J.* **40**, 343–352.

Newman MEJ 2004a Detecting community structure in networks. *Eur. Phys. J. B* **38**, 321–333.

Newman MEJ 2004b Fast algorithm for detecting community structure in networks. *Phys. Rev. E* **69**, 066133.

Newman MEJ 2004c Analysis of weighted networks. *Phys. Rev. E* **70**, 056131.

Newman MEJ 2006 Finding community structure in networks using the eigenvectors of matrices. *Phys. Rev. E* **74**, 036104.

Newman MEJ 2010 *Networks, An Introduction*, Oxford University Press.

Newman MEJ and Girvan M 2004 Finding and evaluating community structure in networks. *Phys. Rev. E* **69**, 026113.

Ng AY Jordan MI and Weiss Y 2001 On spectral clustering: analysis and an algorithm, in *Proc. 14th Neural Information Processing Systems Conference (NIPS 2001)* (eds Dietterich TG, Becker S and Ghahramani Z), MIT Press, Cambridge, USA, pp. 849–856.

Omberg L, Golub GH and Alter O 2007 A tensor higher-order singular value decomposition for integrative analysis of DNA microarray data from different studies. *Proc. Natl. Acad. Sci. USA* **104** (47), 18371–18376.

Pólya G and Szegő S 1951 Isoperimetric inequalities in mathematical physics, *Ann. Math. Studies* **27**, Princeton University Press.

Reichardt J. and Bornholdt S 2007 Partitioning and modularity of graphs with arbitrary degree distribution. *Phys. Rev. E* **76**, 015102(R).

Rohe K, Chatterjee S and Yu B 2011 Spectral clustering and the high-dimensional stochastic blockmodel. *Ann. Stat.* **39** (4), 1878–1915.

Sinclair A and Jerrum M 1989 Approximate counting, uniform generation and rapidly mixing Markov chains. *Inf. Comput.* **82**, 93–133.

Telcs A 2006 *The Art of Random Walks*, Lecture Notes in Mathematics 1885, Springer.

Tetali P 1991 Random walks and effective resistance of networks. *J. Theor. Probab.* **1**, 101–109.

Thomassen C 1990 Resistances and currents in infinite electrical networks. *J. Comb. Theory* **49**, 87–102.

Trevisan L 2009 Max cut and the smallest eigenvalue, in *Proc. 41th Annual ACM Symposium on the Theory of Computing (STOC 2009)*, ACM, Bethesda, Maryland USA, pp. 1117–1130.

von Luxburg U 2006 A tutorial on spectral clustering. *Stat. Comput.* **17**, 395–416.

White S and Smyth P 2005 A spectral clustering approach to find communities in graphs, in *Proc. 5th SIAM International Conference on Data Mining* (eds Kargupta H, Srivastava J, Kamath C and Goodman A), SIAM (Society for Industrial and Applied Mathematics), Newport Beach, California, pp. 274–285.

Wilkinson DM and Huberman BA 2004 A method for finding communities of related genes. *Proc. Natl. Acad. Sci. USA* **101** (50), 5241–5248.

Zachary WW 1977 An information flow model for conflict and fission in small groups. *J. Antropol. Res.* **33**, 452–473.

3

Large networks, perturbation of block structures

In this chapter we apply the results of the previous chapters for the spectral clustering of large networks. Networks are modeled either by edge-weighted graphs or contingency tables, and usually subject to random errors due to their evolving and flexible nature. Asymptotic properties of SD and SVD of the involved matrices are discussed when not only the number of graph vertices or the number of rows and columns of the contingency table tends to infinity, but the cluster sizes also grow proportionally with them. Mostly, perturbation results for the SD and SVD of blown-up matrices burdened with a Wigner-type error matrix are investigated. Conversely, given a weight-matrix or rectangular array of nonnegative entries, we are looking for the underlying block-structure. We will show that under very general circumstances, clusters of the vertices of a large graph, and simultaneously those of the rows and columns of a large contingency table, can be identified with high probability. In this framework, so-called volume regular cluster pairs are also considered with homogeneous information flow within the clusters and between the pairs of them.

3.1 Symmetric block structures burdened with random noise

First the notion of a very general kind of random noise will be introduced.

Definition 3.1.1 *Let w_{ij} $(1 \le i \le j \le n)$ be independent, real-valued random variables defined on the same probability space, $w_{ji} = w_{ij}$ (for $j < i$), $\mathbb{E}(w_{ij}) = 0$ $(\forall i, j)$,*

Spectral Clustering and Biclustering: Learning Large Graphs and Contingency Tables, First Edition. Marianna Bolla.

and the w_{ij}'s are uniformly bounded, that is, there is a constant $K > 0$ – that does not depend of n – such that $|w_{ij}| \leq K$, $\forall i, j$. Then the $n \times n$ real symmetric random matrix $\boldsymbol{W}_n = (w_{ij})_{1 \leq i,j \leq n}$ is called symmetric Wigner-noise.

This random matrix is the generalization of that introduced by E. Wigner when he formulated his famous semicircle law: see Theorem B.2.2. Note that the condition of uniform boundedness of the entries could be relaxed: the entries may have Gaussian distribution or sub-Gaussian moments. However, most of the subsequent results and the frequently used Theorem B.2.3 rely on Definition 3.1.1. This theorem implies that for the spectral norm of an $n \times n$ symmetric Wigner noise

$$\|\boldsymbol{W}_n\| = \max_{1 \leq i \leq n} |\lambda_i(\boldsymbol{W}_n)| \leq 2\sigma\sqrt{n} + \mathcal{O}(n^{1/3}\log n) \tag{3.1}$$

holds with probability tending to 1 as $n \to \infty$, where σ^2 is the uniform bound for the variances of the w_{ij}'s.

Definition 3.1.2 *The $n \times n$ matrix $\boldsymbol{B}$ is a symmetric blown-up matrix if there is a positive integer $k < n$, a $k \times k$ symmetric pattern matrix $\boldsymbol{P}$ with entries p_{ij} $(0 < p_{ij} < 1)$, and there are positive integers $n_1, \dots, n_k$, $\sum_{i=1}^{k} n_i = n$ such that – after rearranging its rows and columns in the same way – the matrix $\boldsymbol{B}$ can be divided into k^2 blocks, where the block (i, j) is an $n_i \times n_j$ matrix with entries all equal to p_{ij} $(1 \leq i, j \leq k)$.*

Let us fix $\boldsymbol{P}$, blow it up to an $n \times n$ matrix $\boldsymbol{B}_n$, and consider the noisy matrix $\boldsymbol{A}_n = \boldsymbol{B}_n + \boldsymbol{W}_n$ as $n_1, \dots, n_k \to \infty$, roughly speaking, at the same rate. For this purpose, an exact growth rate condition is formulated as follows.

Definition 3.1.3 *Under Growth Rate Condition 1, briefly,* **GC1**, *the following is understood: $n = \sum_{i=1}^{k} n_i \to \infty$ in such a way that $\frac{n_i}{n} \geq c$ with some constant $0 < c \leq \frac{1}{k}$.*

While perturbing $\boldsymbol{B}_n$ by $\boldsymbol{W}_n$, assume that for the uniform bound of the entries of $\boldsymbol{W}_n$ the condition

$$K \leq \min\left\{ \min_{i,j \in \{1,\dots,k\}} p_{ij},\ 1 - \max_{i,j \in \{1,\dots,k\}} p_{ij} \right\} \tag{3.2}$$

is satisfied. In this way, the entries of $\boldsymbol{A}_n$ are in the [0,1] interval, and $\boldsymbol{A}_n$ defines an edge-weighted graph $G_n = (V, \boldsymbol{A}_n)$ on n vertices. We are interested in the spectral properties of the expanding random graph sequence G_n. In fact, G_n is the noisy version of the deterministic edge-weighted graph $(V, \boldsymbol{B}_n)$ the vertices of which are partitioned into clusters $V_1, \dots, V_k$ with $|V_a| = n_a$ $(a = 1, \dots k)$ such that vertices of any pair V_a, V_b are connected with an edge of the same weight p_{ab}. Hence, loops are also present.

With an appropriate Wigner-noise we can achieve that the noisy matrix $\boldsymbol{A}_n = \boldsymbol{B}_n + \boldsymbol{W}_n$ contains 1's in the (a, b)-th block with probability p_{ab}, and 0's otherwise. Indeed, for indices $1 \le a < b \le k$ and $i \in V_a$, $j \in V_b$ let

$$w_{ij} := \begin{cases} 1 - p_{ab} & \text{with probability} \quad p_{ab} \\ -p_{ab} & \text{with probability} \quad 1 - p_{ab} \end{cases} \tag{3.3}$$

be independent random variables; further, for $a = 1, \dots, k$ and $i, j \in V_a$ $(i \le j)$ let

$$w_{ij} := \begin{cases} 1 - p_{aa} & \text{with probability} \quad p_{aa} \\ -p_{aa} & \text{with probability} \quad 1 - p_{aa} \end{cases} \tag{3.4}$$

be also independent, otherwise $\boldsymbol{W}$ is symmetric. This $\boldsymbol{W}$ satisfies the conditions of Definition 3.1.1 with uniformly bounded entries of zero expectation and variance bounded by

$$\sigma^2 = \max_{1 \le i \le j \le k} p_{ij}(1 - p_{ij}) \le \frac{1}{4}.$$

So, the noisy weighted graph $G_n = (V, \boldsymbol{A}_n)$ becomes a *generalized random graph* on the *planted partition* $V_1, \dots, V_k$ of the vertices such that vertices of V_a and V_b are connected independently, with probability p_{ab}, $1 \le a \le b \le k$; see for example Bickel and Chen (2009); Coja-Oghlan and Lanka (2009b); Karrer and Newman (2011); McSherry (2001); Rohe *et al.* (2011); Söderberg (2003). This so-called *stochastic block-model* was first mentioned in Holland *et al.* (1983), and discussed much later in Bollobás *et al.* (2007) as a special case of an inhomogeneous random graph. Leskovec *et al.* (2009) show that the best cluster sizes are not very large in a diverse set of real-life networks. This fact motivates the idea to increase the number of clusters with the number of vertices. In the stochastic block-model of Choi *et al.* (2012); Rohe *et al.* (2011) this idea is also used. Note that this model is the generalization of the classical Erdős–Rényi random graph (the first random graph of the history, introduced in Erdős and Rényi (1960) and also discussed in Bollobás (2001)) which corresponds to the $k = 1$ case.

Definition 3.1.4 *Let $\mathbf{A}_n$ be an $n \times n$ symmetric matrix and $\mathcal{P}_n$ a property which mostly depends on the SD of $\mathbf{A}_n$. Then $\mathbf{A}_n$ can have the property $\mathcal{P}_n$ in the two following senses. If $\mathbf{A}_n$ is the weight matrix of an edge-weighted graph G_n, then we equivalently say that G_n has the property $\mathcal{P}_n$, and denote this fact by $\mathbf{A}_n \in \mathcal{P}_n$ or $G_n \in \mathcal{P}_n$.*

WP1 *The property $\mathcal{P}_n$ holds for $\mathbf{A}_n$ with probability tending to 1 if*

$$\lim_{n \to \infty} \mathbb{P}(\mathbf{A}_n \in \mathcal{P}_n) = 1.$$

AS *The property $\mathcal{P}_n$ holds for $\mathbf{A}_n$ almost surely if*

$$\mathbb{P}(\exists\, n_0 \in \mathbb{N}\, \text{s.t. for}\, n \ge n_0 \,:\, \mathbf{A}_n \in \mathcal{P}_n) = 1.$$

Here we may assume **GC1** *of Definition 3.1.3 for the growth of n together with the cluster sizes.*

Note that **AS** always implies **WP1**. Conversely, if in addition to **WP1**

$$\sum_{n=1}^{\infty} \mathbb{P}(\boldsymbol{A}_n \notin \mathcal{P}_n) < \infty$$

also holds, then, by the Borel–Cantelli lemma, $\boldsymbol{A}_n$ has $\mathcal{P}_n$ **AS**.

To established almost sure properties, Theorem B.2.4 of the Appendix (sharp concentration of the eigenvalues) plays a crucial role. In view of this, for the spectral norm, that is, the largest absolute value eigenvalue, of an $n \times n$ Wigner-noise $\boldsymbol{W}_n$, the following relation holds with every positive real number t:

$$\mathbb{P}(|\|\boldsymbol{W}_n\| - \mathbb{E}(\|\boldsymbol{W}_n\|)| > t) \le \exp\left(-\frac{(1-o(1))t^2}{32K^2}\right),$$

where K is the uniform bound for the entries of $\boldsymbol{W}_n$.

This inequality and the fact that, in view of (3.1), $\|\boldsymbol{W}_n\| = \mathcal{O}(\sqrt{n})$ together imply that $\mathbb{E}\|\boldsymbol{W}_n\| = \mathcal{O}(\sqrt{n})$. Therefore, there exist positive constants c_1 and c_2 such that they do not depend on n (they only depend on K) and

$$\mathbb{P}\left(\|\boldsymbol{W}_n\| > c_1\sqrt{n}\right) \le e^{-c_2 n}. \tag{3.5}$$

As the right-hand side of (3.5) is the general term of a convergent series, by the Borel–Cantelli lemma it follows that the spectral norm of $\boldsymbol{W}_n$ is of order $\sqrt{n}$, **AS**. This observation will provide the base of the almost sure results of this chapter. Note that in combinatorics literature, sometimes **WP1** is called **AS**. However, from a probabilistic point of view, **AS** is much stronger than **WP1**, and it makes a difference in practice too: an almost sure property means that no matter how $\boldsymbol{A}_n$ is selected, it must have property $\mathcal{P}_n$ if n is large enough.

3.1.1 General blown-up structures

The spectrum of a symmetric blown-up matrix is characterized as follows.

Proposition 3.1.5 *Under the growth rate condition* **GC1**, *all the non-zero eigenvalues of the $n \times n$ blown-up matrix $\boldsymbol{B}_n$ of the $k \times k$ symmetric pattern matrix $\boldsymbol{P}$ are of order n in absolute value.*

Proof. As there are at most k linearly independent rows of $\boldsymbol{B}_n$, $r = \text{rank}(\boldsymbol{B}_n) \le k$. Let $\beta_1, \dots, \beta_r > 0$ be the non-zero eigenvalues of $\boldsymbol{B}_n$ with corresponding orthonormal eigenvectors $\mathbf{u}_1, \dots, \mathbf{u}_r \in \mathbb{R}^n$. For notational convenience, we discard the subscripts: let $\beta \neq 0$ be an eigenvalue with corresponding eigenvector $\mathbf{u}$, $\|\mathbf{u}\| = 1$. It is easy to see that $\mathbf{u}$ has a piecewise constant structure: it has n_i coordinates equal to

$u(i)$ $(i = 1, \ldots, k)$, where $n_1, \ldots, n_k$ are the blow-up sizes. Then, with these coordinates, the eigenvalue–eigenvector equation

$$\boldsymbol{B}\mathbf{u} = \beta\mathbf{u}$$

has the form

$$\sum_{j=1}^{k} n_j p_{ij} u(j) = \beta u(i), \quad i = 1, \ldots, k. \tag{3.6}$$

With the notation

$$\tilde{\mathbf{u}} = (u(1), \ldots, u(k))^T, \qquad \boldsymbol{N} = \text{diag}(n_1, \ldots, n_k), \tag{3.7}$$

(3.6) can be rewritten in the form

$$\boldsymbol{P}\boldsymbol{N}\tilde{\mathbf{u}} = \beta\tilde{\mathbf{u}}. \tag{3.8}$$

Further, introducing the transformation

$$\mathbf{v} = \boldsymbol{N}^{1/2}\tilde{\mathbf{u}}, \tag{3.9}$$

Equation (3.8) is equivalent to

$$\boldsymbol{N}^{1/2}\boldsymbol{P}\boldsymbol{N}^{1/2}\mathbf{v} = \beta\mathbf{v}. \tag{3.10}$$

It is easy to see that the transformation (3.9) results in a unit-norm vector. Furthermore, applying the transformation (3.9) to the $\tilde{\mathbf{u}}_i$ vectors obtained from the $\mathbf{u}_i$ $(i = 1, \ldots, r)$, the orthogonality is also preserved. Consequently, $\mathbf{v}_i = \boldsymbol{N}^{1/2}\tilde{\mathbf{u}}_i$ is an eigenvector corresponding to the eigenvalue β_i of the $k \times k$ matrix $\boldsymbol{N}^{1/2}\boldsymbol{P}\boldsymbol{N}^{1/2}$, $i = 1, \ldots, r$. With the shrinking

$$\widetilde{\boldsymbol{N}} = \frac{1}{n}\boldsymbol{N}, \tag{3.11}$$

(3.10) is also equivalent to

$$\widetilde{\boldsymbol{N}}^{1/2}\boldsymbol{P}\widetilde{\boldsymbol{N}}^{1/2}\mathbf{v} = \frac{\beta}{n}\mathbf{v},$$

that is the $k \times k$ matrix $\widetilde{\boldsymbol{N}}^{1/2}\boldsymbol{P}\widetilde{\boldsymbol{N}}^{1/2}$ has nonzero eigenvalues $\frac{\beta_i}{n}$ with orthonormal eigenvectors $\mathbf{v}_i$ $(i = 1, \ldots, r)$.

Now we want to establish relations between the eigenvalues of $\boldsymbol{P}$ and $\widetilde{\boldsymbol{N}}^{1/2}\boldsymbol{P}\widetilde{\boldsymbol{N}}^{1/2}$. Since we are interested in the absolute values of the nonzero eigenvalues, we will use singular values (recall that the singular values of a symmetric matrix are the absolute values of its real eigenvalues). Also, we are interested only in the r nonzero eigenvalues, where $r = \text{rank}(\boldsymbol{B}) = \text{rank}(\widetilde{\boldsymbol{N}}^{1/2}\boldsymbol{P}\widetilde{\boldsymbol{N}}^{1/2})$, therefore, it suffices to consider vectors $\mathbf{x}$, for which $\widetilde{\boldsymbol{N}}^{1/2}\boldsymbol{P}\widetilde{\boldsymbol{N}}^{1/2}\mathbf{x} \neq \mathbf{0}$ and apply the minimax principle of

Theorem A.3.16 of the Appendix. In view of this theorem, for $i \in \{1, \dots, r\}$ and an arbitrary i-dimensional subspace $H \subset \mathbb{R}^n$:

$$\min_{\mathbf{x}\in H} \frac{\|\widetilde{\boldsymbol{N}}^{1/2}\boldsymbol{P}\widetilde{\boldsymbol{N}}^{1/2}\mathbf{x}\|}{\|\mathbf{x}\|} = \min_{\mathbf{x}\in H} \frac{\|\widetilde{\boldsymbol{N}}^{1/2}\boldsymbol{P}\widetilde{\boldsymbol{N}}^{1/2}\mathbf{x}\|}{\|\boldsymbol{P}\widetilde{\boldsymbol{N}}^{1/2}\mathbf{x}\|} \cdot \frac{\|\boldsymbol{P}\widetilde{\boldsymbol{N}}^{1/2}\mathbf{x}\|}{\|\widetilde{\boldsymbol{N}}^{1/2}\mathbf{x}\|} \cdot \frac{\|\widetilde{\boldsymbol{N}}^{1/2}\mathbf{x}\|}{\|\mathbf{x}\|}$$

$$\geq s_k(\widetilde{\boldsymbol{N}}^{1/2}) \cdot \min_{\mathbf{x}\in H} \frac{\|\boldsymbol{N}\widetilde{\boldsymbol{N}}^{1/2}\mathbf{x}\|}{\|\widetilde{\boldsymbol{N}}^{1/2}\mathbf{x}\|} \cdot s_k(\widetilde{\boldsymbol{N}}^{1/2}) \geq c \cdot \min_{\mathbf{x}\in H} \frac{\|\boldsymbol{P}\widetilde{\boldsymbol{N}}^{1/2}\mathbf{x}\|}{\|\widetilde{\boldsymbol{N}}^{1/2}\mathbf{x}\|},$$

with the constant c of the growth rate condition **GC1** (see Definition 3.1.3). Now taking the maximum for all possible i-dimensional subspace H we obtain that $|\lambda_i(\widetilde{\boldsymbol{N}}^{1/2}\boldsymbol{P}\widetilde{\boldsymbol{N}}^{1/2})| \geq c|\lambda_i(\boldsymbol{P})| > 0$. On the other hand,

$$|\lambda_i(\widetilde{\boldsymbol{N}}^{1/2}\boldsymbol{P}\widetilde{\boldsymbol{N}}^{1/2})| \leq \|\widetilde{\boldsymbol{N}}^{1/2}\boldsymbol{P}\widetilde{\boldsymbol{N}}^{1/2}\| \leq \|\widetilde{\boldsymbol{N}}^{1/2}\| \cdot \|\boldsymbol{P}\| \cdot \|\widetilde{\boldsymbol{N}}^{1/2}\| \leq \|\boldsymbol{P}\| \leq k.$$

These together imply that $\lambda_i(\widetilde{\boldsymbol{N}}^{1/2}\boldsymbol{P}\widetilde{\boldsymbol{N}}^{1/2})$ can be bounded from below and from above with a positive constant that does not depend on n and n_i's, it only depends on c and $\|\boldsymbol{P}\|$. Hence, because of $\lambda_i(\widetilde{\boldsymbol{N}}^{1/2}\boldsymbol{P}\widetilde{\boldsymbol{N}}^{1/2}) = \frac{\beta_i}{n}$, we obtain that $\beta_1, \dots, \beta_r = \Theta(n)$.

For simplicity, in the sequel, we will assume that $\operatorname{rank}(\boldsymbol{P}) = k$, consequently, $\operatorname{rank}(\boldsymbol{B}) = k$ too.

Theorem 3.1.6 *Let $\boldsymbol{B}_n$ be an $n \times n$ blown-up matrix of the $k \times k$ symmetric pattern matrix $\boldsymbol{P}$ with non-zero eigenvalues $\beta_1, \dots, \beta_k$, and $\boldsymbol{W}_n$ be an $n \times n$ Wigner-noise. Then there are k eigenvalues $\lambda_1, \dots, \lambda_k$ of the noisy random matrix $\boldsymbol{A}_n = \boldsymbol{B}_n + \boldsymbol{W}_n$ such that*

$$|\lambda_i - \beta_i| \leq 2\sigma\sqrt{n} + \mathcal{O}(n^{1/3}\log n), \quad i = 1, \dots, k \tag{3.12}$$

and for the other $n - k$ eigenvalues

$$|\lambda_j| \leq 2\sigma\sqrt{n} + \mathcal{O}(n^{1/3}\log n), \quad j = k+1, \dots, n \tag{3.13}$$

holds **AS** *as $n \to \infty$ under* **GC1**.

Proof. The statement immediately follows by applying the Weyl's perturbation theorem (see Theorem A.3.19) for the spectrum of the symmetric matrix $\boldsymbol{B}_n$ characterized in Proposition 3.1.5, where the spectral norm of the perturbation $\boldsymbol{W}_n$ is estimated by (3.1). This proves the order of eigenvalues **WP1**. In view of (3.5) and the Borel–Cantelli lemma, it implies that this is an **AS** property as well.

Consequently, taking into consideration the order $\Theta(n)$ of the non-zero eigenvalues of $\boldsymbol{B}_n$, there is a spectral gap between the k largest absolute value and the other eigenvalues of $\boldsymbol{A}_n$, and this is of order $\Delta - 2\varepsilon$, where

$$\varepsilon = 2\sigma\sqrt{n} + \mathcal{O}(n^{1/3}\log n) \quad \text{and} \quad \Delta = \min_{1\leq i\leq k} |\beta_i|. \tag{3.14}$$

Then Theorem 3.1.6 guarantees the existence of k protruding, so-called *structural* eigenvalues of $\boldsymbol{A}_n = \boldsymbol{B}_n + \boldsymbol{W}_n$. With the help of this theorem we are also able to estimate the distances between the corresponding eigen-subspaces of the matrices $\boldsymbol{B}_n$ and $\boldsymbol{A}_n$.

Let us denote the unit-norm eigenvectors corresponding to the largest eigenvalues $\beta_1, \dots, \beta_k$ of $\boldsymbol{B}_n$ by $\mathbf{u}_1, \dots, \mathbf{u}_k$ and those corresponding to the largest eigenvalues $\lambda_1, \dots, \lambda_k$ of $\boldsymbol{A}_n$ by $\mathbf{x}_1, \dots, \mathbf{x}_k$. Let $F := \mathrm{Span}\{\mathbf{u}_1, \dots, \mathbf{u}_k\} \subset \mathbb{R}^n$ be the k-dimensional eigen-subspace, and let $\mathrm{dist}(\mathbf{x}, F)$ denote the Euclidean distance between the vector $\mathbf{x} \in \mathbb{R}^n$ and the subspace F.

Proposition 3.1.7 *With the above notation, the following estimate holds* **AS** *for the sum of the squared distances between* $\mathbf{x}_1, \dots, \mathbf{x}_k$ *and* F*:*

$$\sum_{i=1}^{k} \mathrm{dist}^2(\mathbf{x}_i, F) \le k \frac{\varepsilon^2}{(\Delta - \varepsilon)^2} = \mathcal{O}\left(\frac{1}{n}\right). \tag{3.15}$$

Proof. Let us choose one of the eigenvectors $\mathbf{x}_1, \dots, \mathbf{x}_k$ of $\boldsymbol{A}_n$ and denote it simply by $\mathbf{x}$ with corresponding eigenvalue λ. To estimate the distance between $\mathbf{x}$ and F, we expand $\mathbf{x}$ in the basis $\mathbf{u}_1, \dots, \mathbf{u}_n$ with coefficients $t_1, \dots, t_n \in \mathbb{R}$:

$$\mathbf{x} = \sum_{i=1}^{n} t_i \mathbf{u}_i.$$

The eigenvalues $\beta_1, \dots, \beta_k$ of the matrix $\boldsymbol{B}_n$ corresponding to $\mathbf{u}_1, \dots, \mathbf{u}_k$ are of order n (by Proposition 3.1.5), whereas the other eigenvalues are zeros.

Then, on the one hand

$$\boldsymbol{A}_n \mathbf{x} = (\boldsymbol{B}_n + \boldsymbol{W}_n)\mathbf{x} = \sum_{i=1}^{k} t_i \beta_i \mathbf{u}_i + \boldsymbol{W}_n \mathbf{x}, \tag{3.16}$$

and on the other hand

$$\boldsymbol{A}_n \mathbf{x} = \lambda \mathbf{x} = \sum_{i=1}^{n} t_i \lambda \mathbf{u}_i. \tag{3.17}$$

Equating the right-hand sides of (3.16) and (3.17) we get that

$$\sum_{i=1}^{k} t_i (\lambda - \beta_i) \mathbf{u}_i + \sum_{i=k+1}^{n} t_i \lambda \mathbf{u}_i = \boldsymbol{W}_n \mathbf{x}.$$

The Pythagorean theorem yields

$$\sum_{i=1}^{k} t_i^2 (\lambda - \beta_i)^2 + \sum_{i=k+1}^{n} t_i^2 \lambda^2 = \|\boldsymbol{W}_n \mathbf{x}\|^2 = \mathbf{x}^T \boldsymbol{W}_n^T \boldsymbol{W}_n \mathbf{x} \le \varepsilon^2, \tag{3.18}$$

since $\|\mathbf{x}\| = 1$ and the largest eigenvalue of $\boldsymbol{W}_n^T \boldsymbol{W}_n$ is ε^2.

The squared distance between $\mathbf{x}$ and F is $\text{dist}^2(\mathbf{x}, F) = \sum_{i=k+1}^{n} t_i^2$. In view of $|\lambda| \geq \Delta - \varepsilon$,

$$(\Delta - \varepsilon)^2 \text{dist}^2(\mathbf{x}, F) = (\Delta - \varepsilon)^2 \sum_{i=k+1}^{n} t_i^2 \leq \sum_{i=k+1}^{n} t_i^2 \lambda^2$$
$$\leq \sum_{i=1}^{k} t_i^2 (\lambda - \beta_i)^2 + \sum_{i=k+1}^{n} t_i^2 \lambda^2 \leq \varepsilon^2$$

where in the last inequality we used (3.18). From here,

$$\text{dist}^2(\mathbf{x}, F) \leq \frac{\varepsilon^2}{(\Delta - \varepsilon)^2} = \mathcal{O}\left(\frac{1}{n}\right) \tag{3.19}$$

where the order of the estimate follows from the order of ε and Δ of (3.14).

Applying (3.19) for the eigenvectors $\mathbf{x}_1, \ldots, \mathbf{x}_k$ of $\boldsymbol{A}_n$, and adding the k inequalities together, we obtain the same order of magnitude for the sum of the squared distances, which finishes the proof.

Let $G_n = (V, \boldsymbol{A}_n)$ be the random edge-weighted graph on the n-element vertex set and edge-weights in $\boldsymbol{A}_n$. Denote by $V_1, \ldots, V_k$ the partition of V with respect to the blow-up of $\boldsymbol{B}_n$ (it defines a clustering of the vertices). Proposition 3.1.7 implies the well-clustering property of the representatives of the vertices of G_n in the following representation. Let $\boldsymbol{X}$ be the $n \times k$ matrix containing the eigenvectors $\mathbf{x}_1, \ldots, \mathbf{x}_k$ in its columns. Let the k-dimensional representatives of the vertices be the row vectors of $\boldsymbol{X}$ and $S_k^2(P_k; \boldsymbol{X})$ denote the k-variance – see (C.5) of the Appendix – of these representatives in the clustering $P_k = (V_1, \ldots, V_k)$.

Theorem 3.1.8 *For the k-variance of the above representation of the noisy weighted graph $G_n = (V, \boldsymbol{A}_n)$, the relation*

$$S_k^2(\boldsymbol{X}) = \mathcal{O}\left(\frac{1}{n}\right)$$

holds **AS** *as $n \to \infty$ under* **GC1**.

Proof. By the considerations of the Proof of Proposition 2.1.8, $S_k^2(P_k; \boldsymbol{X})$ is equal to the left-hand side of (3.15), therefore, it is of order $\mathcal{O}(1/n)$. This is also inherited to $S_k^2(\boldsymbol{X}) = \min_{P_k' \in \mathcal{P}_k} S_k^2(P_k'; \boldsymbol{X})$.

Consequently, the addition of any kind of a Wigner-noise to a weight matrix that has a blown-up structure will not change the order of its protruding eigenvalues, and the block structure of it can be concluded from the vertex representatives of the noisy matrix, where the representation is performed by means of the corresponding eigenvectors.

Let us characterize the Laplacian spectra of the above edge-weighted graphs with blown-up weight matrix $\boldsymbol{B}_n$. We will follow the notation of Bolla (2008). For our convenience, the Laplacian of the edge-weighted graph with weight matrix $\boldsymbol{A}$ will be denoted by

$$\boldsymbol{L}(\boldsymbol{A}) = \boldsymbol{D}(\boldsymbol{A}) - \boldsymbol{A}, \tag{3.20}$$

where $\boldsymbol{D}(\boldsymbol{A})$ is the diagonal degree-matrix.

Let $\boldsymbol{B}_n$ be the blown-up matrix of the $k \times k$ pattern matrix $\boldsymbol{P}$ of rank k, and blow-up sizes $n_1, \dots, n_k$. Let $V_1, \dots, V_k$ denote the corresponding vertex-clusters of the graph $(V, \boldsymbol{B}_n)$, $|V_i| = n_i$, $i = 1, \dots, k$. With a general pattern-matrix, our graph is connected, therefore its Laplacian $L(\boldsymbol{B}_n)$ has one zero eigenvalue ($\lambda_0 = 0$) with corresponding eigenvector $\mathbf{u}_0 = \mathbf{1}_n$, and the positive eigenvalues are as follows. There are $k-1$ eigenvalues, say, $\lambda_1, \dots, \lambda_{k-1}$ with piecewise constant eigenvectors. Let λ be one of these eigenvalues with corresponding eigenvector $\mathbf{u}$ which has equal coordinates, say, $u(1), \dots, u(k)$ within the blocks $V_1, \dots, V_k$. In terms of them, the eigenvalue–eigenvector equation for the coordinates looks like

$$\sum_{j \neq i} n_j p_{ij} u(i) - \sum_{j \neq i} n_j p_{ij} u(j) = \lambda u(i), \quad i = 1, \dots, k.$$

With the notation of (3.7), the above system of equations is equivalent to

$$\boldsymbol{D}(\boldsymbol{N}^{1/2}\boldsymbol{P}\boldsymbol{N}^{1/2}) \cdot \tilde{\mathbf{u}} - \boldsymbol{N}^{1/2}\boldsymbol{P}\boldsymbol{N}^{1/2}\tilde{\mathbf{u}} = \lambda \tilde{\mathbf{u}},$$

which is the eigenvalue–eigenvector equation of the $k \times k$ Laplacian $\boldsymbol{L}(\boldsymbol{N}^{1/2}\boldsymbol{P}\boldsymbol{N}^{1/2})$. After dividing both sides by n, with the notation of (3.11), it also becomes the eigenvalue–eigenvector equation of the $k \times k$ Laplacian $\boldsymbol{L}(\widetilde{\boldsymbol{N}}^{1/2}\boldsymbol{P}\widetilde{\boldsymbol{N}}^{1/2})$:

$$\boldsymbol{L}(\widetilde{\boldsymbol{N}}^{1/2}\boldsymbol{P}\widetilde{\boldsymbol{N}}^{1/2}) \cdot \tilde{\mathbf{u}} = \frac{\lambda}{n} \tilde{\mathbf{u}}.$$

Due to this coincidence, with the same considerations as in the proof of Proposition 3.1.5, we can see that the $k-1$ positive eigenvalues of $\boldsymbol{L}(\widetilde{\boldsymbol{N}}^{1/2}\boldsymbol{P}\widetilde{\boldsymbol{N}}^{1/2})$ are bounded from below and from above with a constant that does not depend on n and n_i's, it only depends on the spectral norm of $\boldsymbol{P}$ and the constant c of **GC1**. Hence, $\lambda_i = \Theta(n)$, $i = 1, \dots, k-1$, and the corresponding eigenvectors are piecewise constant vectors of coordinates $u(1), \dots, u(k)$ with multiplicities $n_1, \dots, n_k$, also satisfying $\sum_{i=1}^{k} n_i u(i) = 0$ (due to the orthogonality to the vector $\mathbf{1}_n$). Consequently, they form a $(k-1)$-dimensional subspace in $\mathbb{R}^n$. (Note that the piecewise constant vectors on k steps form a k-dimensional subspace, but $\text{Span}\{\mathbf{u}_1, \dots, \mathbf{u}_{k-1}\}$ is merely a subspace of this, namely, the one orthogonal to the vector $\mathbf{1}_n$.

It is easy to check that the other positive eigenvalues of $L(\boldsymbol{B}_n)$ are the following numbers γ_i with multiplicity $n_i - 1$:

$$\gamma_i = \sum_{j \neq i} n_j p_{ij} = n \sum_{j \neq i} \frac{n_j}{n} p_{ij} = \Theta(n), \quad i = 1, \dots, k,$$

since the summation is only over $k-1$ indices and we also use **GC1**. It can easily be seen that an eigenvector $\mathbf{v}_i$ corresponding to γ_i with coordinates $v_i(j)$ $(j = 1, \dots, n)$ is such that $v_i(j) = 0$ if $j \notin V_i$ and $\sum_{j \in V_i} v_i(j) = 0$, due to the orthogonality to $\mathbf{u}_1, \dots, \mathbf{u}_{k-1}$ and $\mathbf{1}_n$. Consequently, they form $(n_i - 1)$-dimensional subspaces. Further, one can check that $\sum_{i=1}^{k-1} \lambda_i = \sum_{i=1}^{k} \gamma_i$.

Summarizing, the non-zero eigenvalues of $\boldsymbol{L}(\boldsymbol{B}_n)$ are all of order $\Theta(n)$, therefore the eigenvectors corresponding to the non-zero eigenvalues above are capable of revealing the underlying block-structure of the blown-up graph, as they have either piecewise constant structure or zero coordinates except one cluster. However, Wigner-type perturbations cannot be treated in this case for the following reasons. Obviously, Laplacians of edge-disjoint simple graphs are added together; moreover, Laplacians of edge-weighted graphs are also added together. However, no edge-weighted graph corresponds to a Wigner-noise, which usually has negative entries, and cannot be the weight-matrix of an edge-weighted graph. Even if we define the formal Laplacian $\boldsymbol{L}(\boldsymbol{W})$ for the Wigner-noise $\boldsymbol{W}$ by (3.20), $\boldsymbol{L}(\boldsymbol{W})$ is not a Wigner-noise any more. Nonetheless, perturbation results, analogous to those of Theorem 3.1.6 for the adjacency spectrum, can be proved for the normalized Laplacian spectrum of the noisy graph in the miniature world of the $[0, 2]$ interval.

Proposition 3.1.9 *Let $\boldsymbol{B}_n$ be the blown-up matrix of the $k \times k$ symmetric pattern matrix $\boldsymbol{P}$ of rank k. Under the growth rate condition* **GC1**, *there exists a constant $\delta \in (0, 1)$, independent of n, such that there are k eigenvalues of the normalized Laplacian of the edge-weighted graph $(V, \boldsymbol{B}_n)$ that are not equal to 1, and they are localized in the union of intervals $[0, 1-\delta]$ and $[1+\delta, 2]$.*

This statement, as well as the following results are not proved here, as they are proved more generally in the next section for rectangular correspondence matrices. In Proposition 3.2.9 we will prove that the correspondence matrix $\boldsymbol{D}(\boldsymbol{B}_n)^{-1/2} \cdot \boldsymbol{B}_n \cdot \boldsymbol{D}(\boldsymbol{B}_n)^{-1/2}$ is also a blown-up matrix and has k non-zero singular values within the interval $[\delta, 1]$. It implies that it has k non-zero eigenvalues within $[-1, -\delta] \cup [\delta, 1]$. Consequently, $\boldsymbol{I}_n - \boldsymbol{D}(\boldsymbol{B}_n)^{-1/2} \cdot \boldsymbol{B}_n \cdot \boldsymbol{D}(\boldsymbol{B}_n)^{-1/2}$ has k non-1 eigenvalues within $[0, 1-\delta]$ and $[1+\delta, 2]$.

Theorem 3.1.9 states that the normalized Laplacian eigenvalues of $\boldsymbol{B}_n$, which are not equal to 1, are bounded away from 1. Equivalently, the non-zero eigenvalues of the normalized modularity matrix belonging to the edge-weighted graph $(V, \boldsymbol{B}_n)$ are bounded away from zero (there are only $k-1$ such eigenvalues as 1 is not an eigenvalue of this matrix if the underlying graph is connected, see Section 1.3). We claim that this property is inherited to the the normalized Laplacian of the noisy graph $G_n = (V, \boldsymbol{A}_n)$, where $\boldsymbol{A}_n = \boldsymbol{B}_n + \boldsymbol{W}_n$ with a Wigner-noise $\boldsymbol{W}_n$. More precisely, the following statement is formulated.

Theorem 3.1.10 *Let $G_n = (V, \boldsymbol{A}_n)$ be random edge-weighted graph with $\boldsymbol{A}_n = \boldsymbol{B}_n + \boldsymbol{W}_n$, where $\boldsymbol{B}_n$ is the blown-up matrix of the $k \times k$ pattern matrix $\boldsymbol{P}$ of rank k, and $\boldsymbol{W}_n$ is Wigner-noise. Then there exists a positive constant $\delta \in (0, 1)$, independent of n, such that for every $0 < \tau < 1/2$ the following statement holds* **AS** *as $n \to \infty$ un-*

der the growth rate condition **GC1***: there are exactly k eigenvalues of the normalized Laplacian of* G_n *that are located in the union of intervals* $[-n^{-\tau}, 1-\delta+n^{-\tau}]$ *and* $[1+\delta-n^{-\tau}, 2+n^{-\tau}]$*, while all the others are in the interval* $(1-n^{-\tau}, 1+n^{-\tau})$.

This statement also follows from the the analogous Theorem 3.2.10 stated for rectangular matrices. Here $m = n$, and hence, the so-called **GC2** introduced and required there, is automatically satisfied here. Note that if the uniform bound of the entries of $\boldsymbol{W}_n$ satisfies (3.2), then the random matrix $\boldsymbol{A}_n$ has nonnegative entries and its normalized Laplacian spectrum is in the [0, 2] interval. In this case the eigenvalues bounded away from 1 are either in the $[0, 1-\delta+n^{-\tau}]$ or in the $[1+\delta-n^{-\tau}, 2]$ interval.

Let $\mathbf{u}'_0, \dots, \mathbf{u}'_{k-1}$ be unit-norm, pairwise orthogonal eigenvectors corresponding to the non-one eigenvalues (including the 0) of the weighted Laplacian of $\boldsymbol{B}_n$. The n-dimensional vectors obtained by the transformations

$$\mathbf{x}_i = \boldsymbol{D}(\boldsymbol{B}_n)^{-1/2} \cdot \mathbf{u}'_i \quad (i = 0, \dots, k-1)$$

($\mathbf{x}_0 = \mathbf{1}$) are vector components of the optimal k-dimensional representation of the weighted graph $(V, \boldsymbol{B}_n)$, see Theorem 1.3.3. The $n \times k$ matrix $\boldsymbol{X}^* = (\mathbf{x}_0, \dots, \mathbf{x}_{k-1})$ contains the optimum vertex representatives in its rows.

Let $0 < \tau < 1/2$ be arbitrary and $\epsilon := n^{-\tau}$. Let us also denote the unit-norm, pairwise orthogonal eigenvectors corresponding to the k eigenvalues of the normalized Laplacian of $G_n = (V, \boldsymbol{A}_n)$, separated from 1, by $\mathbf{v}_0, \dots, \mathbf{v}_{k-1} \in \mathbb{R}^n$ (their existence is guaranteed by Theorem 3.1.10). Further, set

$$F' := \mathrm{Span}\{\mathbf{u}'_0, \dots, \mathbf{u}'_{k-1}\}.$$

Proposition 3.1.11 *With the above notation, for the distance between* $\mathbf{v}_i$ *and* F'*, the following estimate holds* **AS** *as* $n \to \infty$ *under* **GC1***:*

$$\mathrm{dist}(\mathbf{v}_i, F) \le \frac{\epsilon}{(\delta - \epsilon)} = \frac{1}{(\frac{\delta}{\epsilon} - 1)}, \quad i = 0, \dots, k-1. \tag{3.21}$$

Observe that the statement is similar to that of Proposition 3.1.7 with δ instead of Δ and ϵ instead of ε. The right-hand side of (3.21) is of order $n^{-\tau}$ that tends to zero, as $n \to \infty$. For the proof see the upcoming Section 3.2.

Proposition 3.1.11 implies the well-clustering property of the vertex representatives by means of the transformed eigenvectors

$$\mathbf{y}_i = \boldsymbol{D}(\boldsymbol{A}_n)^{-1/2} \cdot \mathbf{v}_i, \quad i = 0, \dots, k-1.$$

The optimal k-dimensional representatives of the random edge-weighted graph $G_n = (V, \boldsymbol{A}_n)$ are row vectors of the $n \times k$ matrix $\boldsymbol{Y}^* = (\mathbf{y}_0, \dots, \mathbf{y}_{k-1})$. The weighted k-variance of the representatives, defined by (C.6) of the Appendix, is $\tilde{S}_k^2(P_k; \boldsymbol{Y}^*)$ with respect to the k-partition $P_k = (V_1, \dots, V_k)$ of the vertices corresponding to the blow-up.

Proposition 3.1.12 *With the above notation,*

$$\tilde{S}_k^2(\boldsymbol{Y}^*) \le \frac{k}{(\frac{\delta}{\epsilon} - 1)^2}$$

holds **AS** *as* $n \to \infty$ *under* **GC1**.

Proof. An easy calculation shows that

$$\tilde{S}_k^2(P_k; \boldsymbol{Y}^*) = \sum_{i=0}^{k-1} \text{dist}^2(\mathbf{v}_i, F')$$

and hence, $\tilde{S}_k^2(\boldsymbol{Y}^*) \le \tilde{S}_k^2(P_k; \boldsymbol{Y}^*)$.

Hereby, we enlist the eigenvalues and illustrate the clusters of some generalized random graphs based on the following pattern matrices.

- *Type 1*: Edges come into existence with large probability in the diagonal, and small probability in the off-diagonal blocks:

$$k = 3, \ \boldsymbol{P} = \begin{pmatrix} 0.8 & 0.1 & 0.15 \\ 0.1 & 0.75 & 0.2 \\ 0.15 & 0.2 & 0.7 \end{pmatrix}, \ \frac{n_1}{n} = 0.30, \ \frac{n_2}{n} = 0.33, \ \frac{n_3}{n} = 0.37.$$

Figures 3.1 and 3.2 show the noisy blown-up matrices with the special noise of (3.3) and (3.4), that is the adjacency matrices of generalized random graphs,

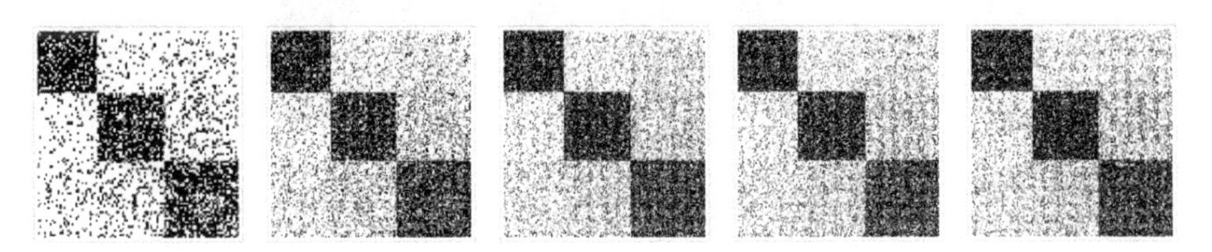

Figure 3.1 Adjacency matrices of sizes $n \times n$ *(*$n = 100, 200, 300, 400, 500$*) of Type 1 generalized random graphs, where edges (black) come into existence between blocks* i *and* j *with probability* p_{ij} *(entries of the pattern matrix* $\boldsymbol{P}$*).*

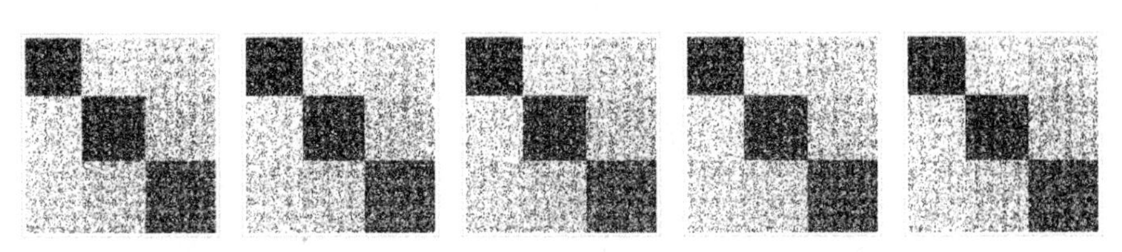

Figure 3.2 Adjacency matrices of sizes $n \times n$ *(*$n = 600, 700, 800, 900, 1000$*) of Type 1 generalized random graphs with the same pattern matrix.*

where black and white squares correspond to the 1 and 0 entries of the adjacency matrix. The largest absolute value eigenvalues of the 1000×1000 normalized modularity matrix are

$$0.64791,\ 0.50964,\ -0.21923,\ 0.21511,\ -0.20884,\ 0.20768,\ 0.19999$$

with a gap (in absolute value) after the second one, indicating the existence of 3 underlying clusters. The vertices are sorted according to their cluster memberships, where the clusters are obtained by the k-means algorithm.

- *Type 2*: Edges come into existence with small probability in the diagonal, and large probability in the off-diagonal blocks:

$$k = 3,\ \boldsymbol{P} = \begin{pmatrix} 0.1 & 0.8 & 0.75 \\ 0.8 & 0.15 & 0.7 \\ 0.75 & 0.7 & 0.2 \end{pmatrix},\ \frac{n_1}{n} = 0.30,\ \frac{n_2}{n} = 0.33,\ \frac{n_3}{n} = 0.37.$$

Figures 3.3 and 3.4 show the noisy blown-up matrices. The largest absolute value eigenvalues of the 1000×1000 normalized modularity matrix are

$$-0.38458,\ -0.33551,\ 0.06739,\ 0.06672,\ -0.06653,\ -0.06591,\ 0.06502$$

with a gap (in absolute value) after the second one, indicating the existence of 3 underlying clusters.

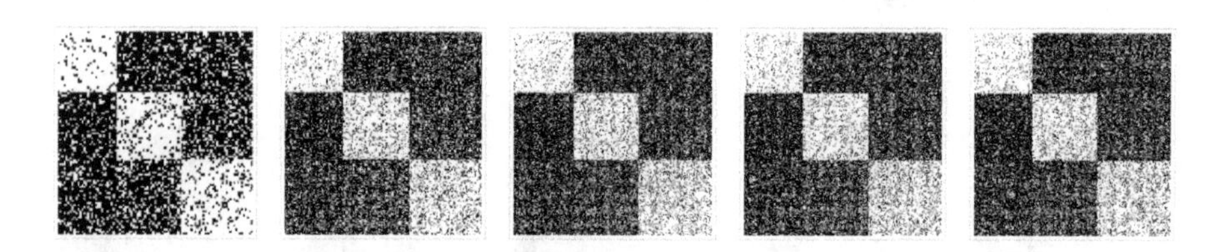

Figure 3.3 Adjacency matrices of sizes $n \times n$ ($n = 100, 200, 300, 400, 500$) of Type 2 generalized random graphs, where edges (black) come into existence between blocks i and j with probability p_{ij} (entries of the pattern matrix $\boldsymbol{P}$).

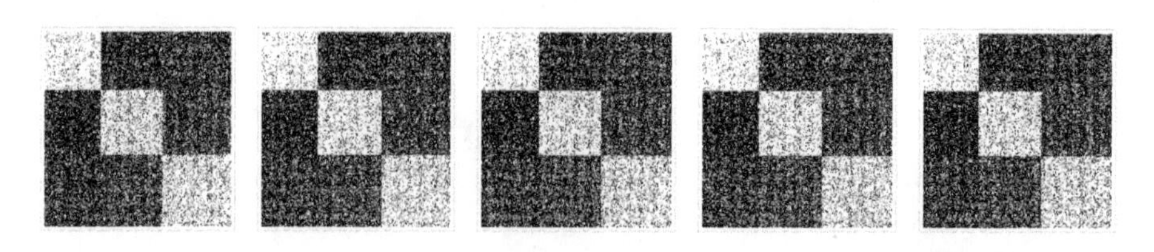

Figure 3.4 Adjacency matrices of sizes $n \times n$ ($n = 600, 700, 800, 900, 1000$) of Type 2 generalized random graphs with the same pattern matrix.

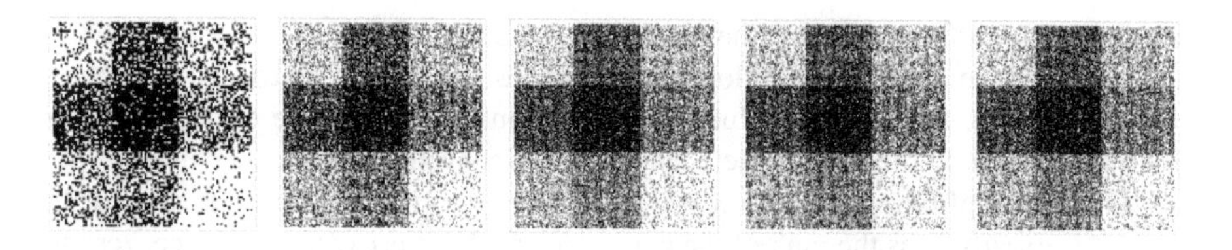

Figure 3.5 Adjacency matrices of sizes $n \times n$ ($n = 100, 200, 300, 400, 500$) of Type 3 generalized random graphs, where edges (black) come into existence between blocks i and j with probability p_{ij} (entries of the pattern matrix $\boldsymbol{P}$).

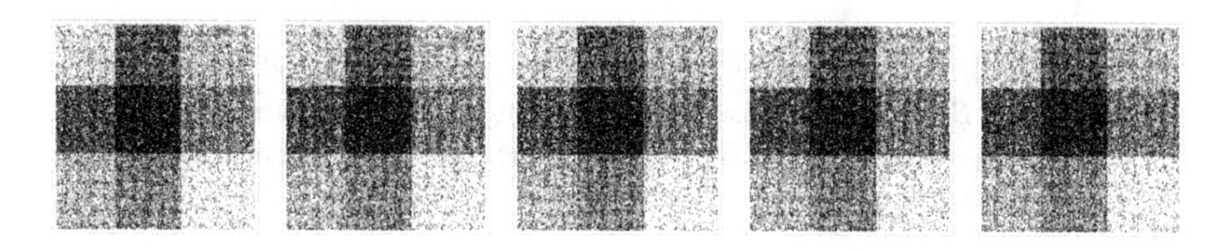

Figure 3.6 Adjacency matrices of sizes $n \times n$ ($n = 600, 700, 800, 900, 1000$) of Type 3 generalized random graphs with the same pattern matrix.

- *Type 3*: General pattern matrix of probabilities of inter-cluster connections:

$$k = 3,\ \boldsymbol{P} = \begin{pmatrix} 0.2 & 0.7 & 0.3 \\ 0.7 & 0.9 & 0.5 \\ 0.3 & 0.5 & 0.1 \end{pmatrix},\ \frac{n_1}{n} = 0.30,\ \frac{n_2}{n} = 0.33,\ \frac{n_3}{n} = 0.37.$$

Figures 3.5 and 3.6 show the noisy blown-up matrices. The largest absolute value eigenvalues of the 1000×1000 normalized modularity matrix are

$$-0.16481,\ 0.14511,\ -0.06924,\ 0.06853,\ -0.06810,\ 0.06736,\ -0.06736$$

with a gap (in absolute value) after the second one, indicating the existence of 3 underlying clusters.

3.1.2 Blown-up multipartite structures

As for the complete k-partite graph as a blown-up graph, we can prove a little more, concerning the allocation of its eigenvalues, see also Bolla (2005).

Proposition 3.1.13 *Let the entries of the $k \times k$ pattern matrix $\boldsymbol{P}$ be the following: $p_{ii} = 0$ $(i = 1, \dots, k)$ and $p_{ij} = p_{ji} = p \in [0, 1]$ $(1 \le i < j \le k)$. Let $\boldsymbol{B}_n$ be the blown-up matrix of $\boldsymbol{P}$ with block sizes $n_1 \le n_2 \le \cdots \le n_k$, $n := \sum_{i=1}^k n_i$. Then $\boldsymbol{B}_n$ has exactly $n - k$ zero eigenvalues, whereas the negative eigenvalues of $\boldsymbol{B}_n$ are in the interval $[-pn_k, -pn_1]$ and the positive ones in $[p(n - n_k), p(n - n_1)]$.*

Proof. It suffices to prove the statement for $p = 1$. In the case $0 \le p < 1$ the statement of the proposition follows from this, as the pattern matrix is multiplied by p, therefore, all the eigenvalues of $\boldsymbol{P}$ and consequently, those of $\boldsymbol{B}_n$, are also multiplied by p. In the trivial case $p = 0$ all the eigenvalues are zeros.

For a general blown-up matrix we have already seen that its rank is at most k. Now it is exactly k, as the rank of the matrix $\boldsymbol{N}^{1/2}\boldsymbol{P}\boldsymbol{N}^{1/2}$ is exactly k. Hence, zero is an eigenvalue of $\boldsymbol{B}_n$ with multiplicity $n - k$ and corresponding eigenspace

$$\left\{ \mathbf{x} = (x_1, \dots, x_n)^T \neq \mathbf{0} : \sum_{j \in V_i} x_j = 0, \quad i = 1, \dots, k \right\} \subset \mathbb{R}^n. \tag{3.22}$$

Due to the orthogonality to (3.22), any eigenvector $\mathbf{u}$ corresponding to an eigenvalue $\beta \neq 0$ of $\boldsymbol{B}_n$ has a piecewise constant structure: n_1 coordinates equal to $u(1)$, $\dots, n_k$ coordinates equal to $u(k)$. The corresponding eigenvalue–eigenvector equation $\boldsymbol{B}_n\mathbf{u} = \beta\mathbf{u}$ gives that

$$\sum_{j \neq i} n_j u(j) = \beta u(i), \quad i = 1, \dots, k,$$

consequently,

$$\sum_{j=1}^{k} n_j u(j) = (n_i + \beta) u(i), \quad i = 1, \dots, k, \tag{3.23}$$

that is – with regard to the left-hand side – independent of i.

If $\beta = -n_i$ for some index i, then β is in the desired range, and there is nothing to prove. If $\beta_i \neq -n_i$ $(i = 1, \dots, k)$, then none of the $u(i)$'s can be zero, otherwise – due to (3.23) – all the $u(i)$'s are zeros, but the zero vector cannot be an eigenvector. Let i be an arbitrary integer in $\{1, \dots, k\}$. As $u(i) \neq 0$, $\mathbf{u}$ can be scaled such that $u(i) = 1$. Therefore, (3.23) becomes

$$\sum_{j=1}^{k} n_j u(j) = n_i + \beta. \tag{3.24}$$

Equating (3.24) to (3.23) applied for the other indices implies that

$$u(j) = \frac{n_i + \beta}{n_j + \beta}, \quad j \neq i.$$

Summing for $j = 1, \dots, k$ yields

$$\sum_{j=1}^{k} n_j u(j) = (n_i + \beta) \sum_{j=1}^{k} \frac{n_j}{n_j + \beta}$$

and in view of (3.24) it is also equal to $n_i + \beta$, therefore,

$$\sum_{j=1}^{k} \frac{n_j}{n_j + \beta} = 1. \tag{3.25}$$

Since $\mathrm{tr}(\boldsymbol{B}_n) = 0$, $\boldsymbol{B}_n$ must have both negative and positive eigenvalues. Let us assume that there is an eigenvalue $\beta < -n_k$. Then on the left-hand side of (3.25) all the terms are negative, and their sum could not be 1. Consequently, all the eigenvalues must be at least $-n_k$. Now let us assume that there is a negative eigenvalue with $-n_1 < \beta < 0$. Then for all the terms on the left-hand side of (3.25)

$$\frac{n_j}{n_j + \beta} > 1, \quad j = 1, \dots, k$$

holds, therefore their sum cannot be 1. So, for the negative eigenvalues

$$-n_k \le \beta \le -n_1$$

is proved.

For the positive eigenvalues, we will use

$$0 < n_1 + \beta \le n_j + \beta \le n_k + \beta, \quad j = 1, \dots, k.$$

After taking the reciprocals, multiplying by n_j, and summing for $j = 1, \dots, k$, we obtain that

$$\sum_{j=1}^{k} \frac{n_j}{n_1 + \beta} \ge \sum_{j=1}^{k} \frac{n_j}{n_j + \beta} \ge \sum_{j=1}^{k} \frac{n_j}{n_k + \beta},$$

that is, in view of (3.25),

$$\frac{n}{n_1 + \beta} \ge 1 \ge \frac{n}{n_k + \beta},$$

which implies

$$n - n_k \le \beta \le n - n_1,$$

that was to be proved for the positive eigenvalues in the case of $p = 1$.

We remark the following:

- In the special case of $n_1 = \cdots = n_k = \frac{n}{k}$, all the negative eigenvalues of $\boldsymbol{B}_n$ are equal to $-p\frac{n}{k}$, and all the positive ones to $p(n - \frac{n}{k})$. As the sum of the eigenvalues of $\boldsymbol{B}_n$ is zero, $-p\frac{n}{k}$ is an eigenvalue with multiplicity $k - 1$, whereas $p(n - \frac{n}{k})$ is a single eigenvalue.

- If n_i is a block-size with multiplicity k_i ($\sum_{i=1}^{k} k_i = k$), then $-pn_i$ is an eigenvalue of $\boldsymbol{B}_n$ with multiplicity $k_i - 1$. Accordingly, if n_i is a single block-size, then $-pn_i$ cannot be an eigenvalue of $\boldsymbol{B}_n$. If especially, $k_1 = k$, then $-pn/k$ is an eigenvalue with multiplicity $k-1$, in accordance with the previous remark.

- In the case of $p = 1$, our matrix $\boldsymbol{B}_n$ is the adjacency matrix of $K_{n_1,\dots,n_k}$, the complete k-partite graph on disjoint, edge-free vertex-subsets $V_1, \dots, V_k$ with $|V_i| = n_i$ $(i = 1, \dots, k)$.

The Laplacian spectrum of the complete k-partite graph $K_{n_1,\dots,n_k}$ is completely characterized in Example (d) of Section 1.1.3. In that example we also discussed the structure-revealing representation which is obtained by means of the eigenvectors corresponding to the $k-1$ largest Laplacian eigenvalues. Here representatives of vertices in the same cluster V_i coincide, $i = 1, \dots, k$.

The normalized Laplacian eigenvalues of the edge-weighted graph with weight-matrix $\boldsymbol{B}_n$ of Proposition 3.1.13 are the same as the normalized Laplacian eigenvalues of $K_{n_1,\dots,n_k}$, and they are characterized by Proposition 3.1.9 as follows: 1 is an eigenvalue with multiplicity $n-k$, and – under **GC1** – there is a δ (independent on n) that the other eigenvalues are localized in the union of intervals $[0, 1-\delta]$ and $[1+\delta, 2]$. We remark that actually here, 0 is a single eigenvalue, and there are $k-1$ eigenvalues within the interval $[1+\delta, 2]$. In the special case of $n_1 = \dots, n_k = \frac{n}{k}$, our complete k-partite graph is $(n - \frac{n}{k})$-regular, hence its normalized Laplacian eigenvalues are the adjacency eigenvalues divided by $n - \frac{n}{k}$, that is, a single 0, the number 1 with multiplicity $n-k$ and the number $n/(n-\frac{n}{k}) = \frac{k}{k-1}$ with multiplicity $k-1$. Consequently, 2 is an eigenvalue only if $k = 2$, in accord with the fact that 2 is an eigenvalue of the normalized Laplacian of a connected graph if and only if the graph is bipartite.

3.1.3 Weak links between disjoint components

Let $k < n$ be a fixed integer. Now the underlying structure is the following: our edge-weighted graph consists of k disjoint components on $n_1, \dots, n_k$ vertices, respectively. With $n = \sum_{i=1}^{k} n_i$, let $\boldsymbol{B}$ denote the $n \times n$ symmetric weight matrix. In this model, $\boldsymbol{B}$ is the Kronecker-sum of the weight matrices of the components, that is, $\boldsymbol{B} = \boldsymbol{B}^{(1)} \oplus \dots \oplus \boldsymbol{B}^{(k)}$, where $\boldsymbol{B}^{(i)}$ is $n_i \times n_i$ symmetric matrix with non-diagonal entries μ_i's and diagonal ones ν_i's ($\mu_i > 0$ and ν_i are real numbers), $i = 1, \dots, k$. This means that within the connected components of the edge-weighted graph $(V, \boldsymbol{B})$ each pair of vertices is connected with an edge of the same weight, and loops are also allowed (when $\nu_i \neq 0$). The spectrum of $\boldsymbol{B}$ is the union of the spectra of $\boldsymbol{B}^{(i)}$'s. It is easy to verify that the eigenvalues of $\boldsymbol{B}^{(i)}$ are $(n_i - 1)\mu_i + \nu_i$ with eigen-direction $\mathbf{1}_{n_i}$ and $\nu_i - \mu_i$ with multiplicity $n_i - 1$ and corresponding eigen-subspace $\mathbf{1}_{n_i}^{\perp}$ $(i = 1, \dots, k)$.

Now $\boldsymbol{B}$ is not a blown-up matrix, unless $\nu_i = \mu_i$ $(i = 1, \dots, k)$. However, keeping μ_i's and ν_i's fixed, we can increase the size of $\boldsymbol{B}$ in such a way that $n_1, \dots, n_k \to \infty$ under the growth rate condition **GC1**. In the sequel, we use the notation $\boldsymbol{B}_n$ for the expanding $\boldsymbol{B}$. We put a Wigner-noise $\boldsymbol{W}_n$ on $\boldsymbol{B}_n$. Concerning the spectral properties of the weight matrix $\boldsymbol{A}_n = \boldsymbol{B}_n + \boldsymbol{W}_n$ of the random edge-weighted graph $G_n = (V, \boldsymbol{A}_n)$, the following result can be stated.

Theorem 3.1.14 *Let $\mathbf{W}_n$ be an $n \times n$ Wigner-noise (with uniform bound and variance of the entries K and σ^2) and the matrix $\boldsymbol{B}_n$ be defined as above. The numbers k, K, σ, μ_i, and ν_i ($i = 1, \dots, k$) are kept fixed as $n_1, \dots, n_k$ tend to infinity under* **GC1**. *Then, for the eigenvalues $\lambda_1, \dots, \lambda_n$ of $\boldsymbol{A}_n = \boldsymbol{B}_n + \mathbf{W}_n$ the following inequalities hold* **AS**. *There is an ordering of the k largest eigenvalues $\lambda_1, \dots, \lambda_k$ such that*

$$|\lambda_i - [(n_i - 1)\mu_i + \nu_i]| \le 2\sigma\sqrt{n} + O(n^{1/3}\log n), \quad i = 1, \dots, k;$$

among the other eigenvalues, for $i = 1, \dots, k$ there are $n_i - 1$ λ_j's with

$$|\lambda_j - [\nu_i - \mu_i]| \le 2\sigma\sqrt{n} + O(n^{1/3}\log n).$$

The proof is similar to that of Theorem 3.1.6 if we take into consideration the spectrum of $\boldsymbol{B}_n$, the Weyl's perturbation theory (Theorem A.3.19 of the Appendix) and the bound (3.1) for the spectral norm of a Wigner-noise. The complete proof is found in Bolla (2004), where *asymptotic k-variate normality* for the random vector $(\lambda_1, \dots, \lambda_k)$ was also proved with covariance matrix $2\sigma^2 \boldsymbol{I}_k$, and it was shown that the k-variance of the vertices of $G_n = (V, \boldsymbol{A}_n)$ – in the Euclidean representation defined by the corresponding eigenvectors – is $\mathcal{O}(1/n)$.

Theorem 3.1.14 implies that the k largest eigenvalues of the random matrix $\boldsymbol{A}_n$ are of order $\Theta(n)$, and there must be a spectral gap between the k largest and the remaining eigenvalues **AS** as $n \to \infty$ under **GC1**. In view of the asymptotic normality, the k largest eigenvalues are highly concentrated on their expectation of order n, independently, with finite variance. For instance, such data structures occur, when the n objects come from k loosely connected strata ($k < n$). Note that in Granovetter (1973) the importance of so-called weak links between social strata is emphasized. In our model, the weak links correspond to the entries of the Wigner-noise. For a detailed description about social network analysis see Wasserman and Faust (1994).

The Laplacian spectrum of the graph $(V, \boldsymbol{B})$ can easily be characterized. Because of the relation $\boldsymbol{B} = \boldsymbol{B}^{(1)} \oplus \dots \oplus \boldsymbol{B}^{(k)}$, the spectrum of $\boldsymbol{L}(\boldsymbol{B})$ is the union of the spectra of $\boldsymbol{L}(\boldsymbol{B}^{(i)})$'s. But

$$L(\boldsymbol{B}^{(i)}) = [(n_i - 1)\mu_i + \nu_i]\boldsymbol{I}_{n_i} - \boldsymbol{B}^{(i)}, \quad i = 1, \dots, k,$$

therefore the spectrum of $\boldsymbol{L}(\boldsymbol{B}^{(i)})$ is obtained from that of $\boldsymbol{B}^{(i)}$ by a simple linear transformation. Consequently, the spectrum of $\boldsymbol{L}(\boldsymbol{B}_n)$ is as follows: the zero with multiplicity k and corresponding eigenspace of partition vectors defined in (2.11); the numbers $n_i\mu_i$ with multiplicity $n_i - 1$ and eigen-subspace described in (3.22), $i = 1, \dots, k$. Note that the diagonal entries, ν_i's, cancel here in accord with our former observation that loops do not contribute to the Laplacian.

The normalized Laplacian spectrum of the edge-weighted graph $(V, \boldsymbol{B})$ is again the union of those of the blocks. The normalized Laplacian matrix belonging to block i is

$$\begin{aligned} &\{[(n_i - 1)\mu_i + \nu_i]\boldsymbol{I}_{n_i}\}^{-1/2}\boldsymbol{L}(\boldsymbol{B}^{(i)})\{[(n_i - 1)\mu_i + \nu_i]\boldsymbol{I}_{n_i}\}^{-1/2} \\ &= \frac{1}{(n_i - 1)\mu_i + \nu_i}\boldsymbol{L}(\boldsymbol{B}^{(i)}), \quad i = 1, \dots, k. \end{aligned}$$

Hence, the normalized Laplacian spectrum $(V, \boldsymbol{B})$ is as follows: the zero with multiplicity k and the numbers $\frac{n_i \mu_i}{(n_i - 1)\mu_i + \nu_i}$ with multiplicity $n_i - 1, i = 1, \ldots, k$. Note that the latter ones tend to 1 as $n_i \to \infty$ $(i = 1, \ldots, k)$. Here the loops do contribute to the spectrum.

Observe that adding a general Wigner-noise to the structures of Subsections 3.1.2 and 3.1.3 will not always result in a weight-matrix of nonnegative weights. Only special Wigner-noises – having zero entries in the positions of the zero blocks of the block-matrix, otherwise satisfying(3.2) – can be treated here.

In Table 3.1 we summarize the adjacency, Laplacian, normalized Laplacian and modularity spectra and spectral subspaces of the three main types of block- and blown-up matrices based on Subsections 3.1.1, 3.1.2, and 3.1.3. Through this table we want to send the following message to the reader. Whenever the rank of the $k \times k$ pattern matrix $\boldsymbol{P}$ is k (even if it contains some entries equal to each other), the blown-up matrix (under the usual conditions for the blow-up sizes) will asymptotically have k (or $k - 1$ in the modularity case) outstanding eigenvalues (separated from the others) with corresponding eigen-subspace such that the derived vertex-representatives will reveal the k underlying clusters. The latter fact is based on the piecewise constant structure of the (not necessarily) unique eigenvectors. What is only important is that the eigen-subspace corresponding to structural eigenvalues has dimension k ($k - 1$ in the modularity case) and is separated from the eigen-subspace corresponding to the eigenvalues in the remainder of the spectrum. If the pattern matrix has rank k, however small the difference between its entries is (some of them may have the same value), then the differences are magnified when the blow-up sizes tend to infinity (under the growth rate conditions). Of course, the speed of the clusters' separation depends on the relative values of the pattern matrix's entries and that of the blown-up sizes.

3.1.4 Recognizing the structure

In the previous section we investigated how the addition of a completely random Wigner-noise influences the behavior of the outstanding, in other words, *structural eigenvalues* of the underlying matrix of a deterministic structure. Wigner-type matrices became important in quantum mechanics, whereas in the case of macroscopic matrices they are merely landmarks of a random noise added to the underlying linear structure of the edge-weight matrix of some communication, social, or biological networks. However hard it is to recognize the structure concealed by the noise, in a number of models it is possible by means of spectral techniques and large deviations principles. When we perform spectral clustering, it is a crucial issue how many structural eigenvalues – with corresponding eigenvectors – to retain for the vertex-representation.

Note that numerical algorithms for the SD of a matrix with a size of millions are not immediately applicable, and some newly developed randomized algorithms are to be used instead, for example Achlioptas and McSherry (2007). These algorithms exploit the randomness of the underlying matrix, and rely on the fact that a random noise will not change the order of magnitude of the structural eigenvalues. Sometimes, instead of depriving the matrix of the noise, rather a noise is added (by digitalizing the

Table 3.1 Spectra and spectral subspaces of some special block- and blown-up matrices

Graph $G = (V, \boldsymbol{B})$	Adjacency matrix $\boldsymbol{B}$	Laplacian matrix $\boldsymbol{D} - \boldsymbol{B}$	Normalized Laplacian $\boldsymbol{I} - \boldsymbol{D}^{-1/2}\boldsymbol{B}\boldsymbol{D}^{-1/2}$	Normalized modularity $\boldsymbol{D}^{-1/2}\boldsymbol{B}\boldsymbol{D}^{-1/2} - \sqrt{\mathbf{d}}\sqrt{\mathbf{d}}^T$
$\boldsymbol{B} = \oplus_{i=1}^{k} \boldsymbol{B}_i$, where the $n_i \times n_i$ $\boldsymbol{B}_i$ has diagonal ν_i and off-diagonal μ_i $V = (V_1, \dots, V_k)$ $\lvert V_i \rvert = n_i$ $(i = 1, \dots, k)$	$\lambda_i = (n_i - 1)\mu_i + \nu_i$ $(i = 1, \dots, k)$ with piecewise constant eigenvectors over V_i's; $\nu_i - \mu_i$ with multiplicity $n_i - 1$, and eigenvectors with 0-sum coordinates over V_i $(i = 1, \dots, k)$	0 with multiplicity k and piecewise constant eigenvectors over V_i's; $n_i\mu_i$ with multiplicity $n_i - 1$, and eigenvectors with 0-sum coordinates over V_i $(i = 1, \dots, k)$	0 with multiplicity k and stepwise constant eigenvectors over V_i's; $\frac{n_i\mu_i}{\nu_i + (n_i - 1)\mu_i}$ with multiplicity $n_i - 1$, and eigenvectors with 0-sum coordinates over V_i $(i = 1, \dots, k)$	0 single; 1 with multiplicity $k - 1$ and stepwise constant eigenvectors over V_i's; $1 - \frac{n_i\mu_i}{\nu_i + (n_i - 1)\mu_i}$ with multiplicity $n_i - 1$, and eigenvectors with 0-sum coordinates over V_i $(i = 1, \dots, k)$
$G = K_{n_1,\dots,n_k}$ with independent sets V_i's, $\lvert V_i \rvert = n_i$ $(i = 1, \dots, k)$. w.l.g. assume that $n_1 \le \dots \le n_k$ $(n = \sum_{i=1}^{k} n_i)$	0 with multiplicity $n - k$ with eigenvectors of 0-sum coordinates over V_i's; the other k eigenvalues are in $[-n_k, -n_1] \cup [n - n_k, n - n_1]$ with piecewise constant eigenvectors	0 single; n with multiplicity $k - 1$ and piecewise constant eigenvectors over V_i's; $n - n_i$ with multiplicity $n_i - 1$ and eigenvectors with 0-sum coordinates over V_i's	0 single; 1 with multiplicity $n - k$; $k - 1$ eigenvalues in $[1 + \delta, 2]$, where δ does not depend on n under $\frac{n_i}{n} \ge c$ $(i = 1, \dots, k)$	0 with multiplicity $n - k + 1$; $k - 1$ eigenvalues in $[-1, -\delta]$, where δ does not depend on n under $\frac{n_i}{n} \ge c$ $(i = 1, \dots, k)$
$\boldsymbol{B}$ is the blown-up matrix of $\boldsymbol{P} = (p_{ij})$ $(i, j = 1, \dots, k)$, with blow-up sizes $n_1, \dots, n_k$ and clusters $V_1, \dots, V_k$; $\lvert V_i \rvert = n_i$ $(n = \sum_{i=1}^{k} n_i)$ $\mathrm{rank}(\boldsymbol{P}) = k$, $\frac{n_i}{n} \ge c$	0 with multiplicity $n - k$ with eigenvectors of 0-sum coordinates over V_i's; k non-zero eigenvalues $\lambda_1, \dots, \lambda_k = \Theta(n)$ with piecewise constant eigenvectors over $V_1, \dots, V_k$	0 single; $\lambda_1, \dots, \lambda_{k-1} = \Theta(n)$ with piecewise constant eigenvectors over V_i's; $\gamma_i = \sum_{j \ne i} n_j p_{ij}$ with multiplicity $n_i - 1$ and zero-sum coordinates over V_i $(i = 1, \dots, k)$; $\sum_{i=1}^{k-1} \lambda_i = \sum_{i=1}^{k} \gamma_i$	$\exists 0 < \delta < 1$ s.t. there are k eigenvalues (including the 0) in $[0, 1 - \delta] \cup [1 + \delta, 2]$ with piecewise constant eigenvectors over V_i's, and the 1 with multiplicity $n - k$	$\exists 0 < \delta < 1$ s.t. there are $k - 1$ eigenvalues (excluding the 1) in $[-1, -\delta] \cup [\delta, 1)$ with piecewise constant eigenvectors over V_i's, and the 0 with multiplicity $n - k + 1$

Based on Subsections 3.1.3, 3.1.2, and 3.1.1

entries of or sparsifying the underlying matrix with an appropriate randomization) to make the matrix more easily decomposable by means of the classical methods (e.g., the Lánczos method, see Golub and van Loan (1996)). These algorithms are only capable of finding a low-rank approximation of our matrix, where this rank cannot exceed the number of the structural eigenvalues of the original matrix. If there are no such eigenvalues, this property is also inherited to the randomized matrix, so the worst that can happen is that we discover, unfortunately, that there is no linear structure in our matrix at all, but it is a noise itself. In all the other cases we obtain a good approximation for the part of the spectrum needed, exploiting the randomness in our original data.

Both the number of eigenvalues to be retained and algorithmic questions can be analyzed by means of the results of this section. Also note that Wigner-type noises over a remarkable structure are not only numerically tractable, but have significance in real-life networks too. For example, sociologist Granovetter (1973) shows that sometimes weak ties better help people to find job than strong (historical or family) relations in which they are stuck.

We investigate the opposite question too: what kind of random matrices have a blown-up matrix as a skeleton, except of a 'small' perturbation? The following proposition states that under very general conditions an $n \times n$ random symmetric matrix with nonnegative, uniformly bounded entries (so that it can be the weight matrix of an edge-weighted graph) has at least one eigenvalue greater than of order $\sqrt{n}$.

Proposition 3.1.15 *Let $\boldsymbol{A}$ be an $n \times n$ random symmetric matrix such that $0 \le a_{ij} \le 1$ and the entries are independent for $i \le j$. Further, let us assume that there are positive constants c_1 and c_2 and $0 < \delta \le \Delta \le 1/2$ such that, with the notation $X_i = \sum_{j=1}^{n} a_{ij}$,*

$$\mathbb{E}(X_i) \ge c_1 n^{\frac{1}{2}+\delta} \quad \text{and} \quad \operatorname{Var}(X_i) \le c_2 n^{\frac{1}{2}+\Delta}, \quad i = 1, \dots, n.$$

Then for every $0 < \varepsilon < \delta$:

$$\lim_{n\to\infty} \mathbb{P}\left(\lambda_{\max}(\boldsymbol{A}) \ge c_1 n^{\frac{1}{2}+\varepsilon}\right) = 1.$$

Remark that the above conditions automatically hold true if there is a constant $0 < \mu_0 < 1$ such that $\mathbb{E}(a_{ij}) \ge \mu_0$ for all i, j pairs. This is the case in the theorems of Juhász and Mályusz (1980) and Füredi and Komlós (1981). In our case there can be a lot of zero entries; we require only that in each row there are at least $c_1 n^{1/2+\delta}$ entries with expectation greater than or equal to any small fixed positive constant μ_0. As the matrix is symmetric, this also holds for its columns. Therefore, among the n^2 entries there must be at least $\Theta(n^{1+2\delta})$ ones (but not anyhow) with expectation of at least a fixed $0 < \mu_0 < 1$, all the others can be zeros. Also note that by the Perron–Frobenius theory (see Theorems A.3.30 and A.3.32), the largest eigenvalue of $\boldsymbol{A}$ is always positive.

To prove Proposition 3.1.15, the following lemma is needed (see e.g., Hoeffding (1963)).

Lemma 3.1.16 (Chernoff inequality for large deviations) *Let $X_1, \dots, X_n$ be independent random variables, $|X_i| \le K$, $X := \sum_{i=1}^n X_i$. Then for every $a > 0$:*

$$\mathbb{P}\left(|X - \mathbb{E}(X)| > a\right) \le e^{-\frac{a^2}{2(\mathrm{Var}(X)+Ka/3)}}.$$

Proof. [**Proposition** 3.1.15] As a consequence of Proposition A.3.33 of the Appendix, $\lambda_{\max}(\boldsymbol{A}) \ge \min_i X_i$, hence

$$\mathbb{P}\left(\lambda_{\max}(\boldsymbol{A}) \ge c_1 n^{\frac{1}{2}+\varepsilon}\right) \ge \mathbb{P}\left(\min_i X_i \ge c_1 n^{\frac{1}{2}+\varepsilon}\right),$$

and it is enough to prove that the latter probability tends to 1 as $n \to \infty$. We will prove that the probability of the complementary event tends to 0:

$$\mathbb{P}\left(\text{for at least one } i:\ X_i < c_1 n^{\frac{1}{2}+\varepsilon}\right) \le n\mathbb{P}\left(\text{for a general } i:\ X_i < c_1 n^{\frac{1}{2}+\varepsilon}\right). \tag{3.26}$$

From now on, we will drop the suffix i and X denotes the sum of the entries in an arbitrary row of $\boldsymbol{A}$. As X is the sum of n independent random variables satisfying the conditions of Lemma 3.1.16 with $K = 1$,

$$\begin{aligned}
\mathbb{P}\left(X < c_1 n^{\frac{1}{2}+\varepsilon}\right) &= \mathbb{P}\left(\mathbb{E}(X) - X > \mathbb{E}(X) - c_1 n^{\frac{1}{2}+\varepsilon}\right)\\
&\le \mathbb{P}\left(|X - \mathbb{E}(X)| > \mathbb{E}(X) - c_1 n^{\frac{1}{2}+\varepsilon}\right)\\
&\le \mathbb{P}\left(|X - \mathbb{E}(X| > c_1 n^{\frac{1}{2}}(n^{\delta} - n^{\varepsilon})\right)\\
&\le e^{-\frac{c_1^2 n (n^{\delta}-n^{\varepsilon})^2}{2(c_2 n^{\frac{1}{2}+\Delta} + n^{\frac{1}{2}}(n^{\delta}-n^{\varepsilon})/3)}}\\
&\le e^{-c_3 n^{\frac{1}{2}} \frac{(n^{\delta}-n^{\varepsilon})^2}{n^{\Delta}}}\\
&= e^{-c_3 n^{\frac{1}{2}-\Delta}(n^{\delta}-n^{\varepsilon})^2}
\end{aligned}$$

with some positive constant c_3, in view of the inequalities $0 < \varepsilon < \delta \le \Delta \le 1/2$. Thus, the right-hand side of (3.26) can be estimated from above by

$$\frac{n}{e^{c_3 n^{\frac{1}{2}-\Delta}(n^{\delta}-n^{\varepsilon})^2}} \le \frac{n}{e^{c_4 n^{\gamma}}}$$

with some constants $c_4 > 0$ and $\gamma > 0$, because of $0 < \varepsilon < \delta \le \Delta \le 1/2$. The last term above tends to 0 as $n \to \infty$, which finishes the proof.

Note that the constants δ and Δ were only responsible for the speed of the convergence. Now we will use Proposition 3.1.15 to deprive a random symmetric matrix of the noise.

Theorem 3.1.17 *Let $\boldsymbol{A}_n$ be a sequence of $n \times n$ symmetric matrices, where n tends to infinity. Assume that $\boldsymbol{A}_n$ has exactly k eigenvalues of order greater than $\sqrt{n}$ (k is fixed),* *and there is a k-partition of the vertices of* $G_n = (V, \boldsymbol{A}_n)$ *such that the k-variance of the representatives – in the representation with the corresponding eigenvectors – is* $\mathcal{O}(1/n)$. *Then there is an explicit construction for a symmetric blown-up matrix* $\boldsymbol{B}_n$ *such that* $\boldsymbol{A}_n = \boldsymbol{B}_n + \boldsymbol{E}_{m\times n}$, *with* $\|\boldsymbol{E}_n\| = \mathcal{O}(\sqrt{n})$.

Instead of the complete proof we only describe the construction, since the estimations are special cases of those performed more generally, for rectangular matrices, when we prove Theorem 3.2.13. However, we now prove the following lemma which will frequently be used in the sequel.

Lemma 3.1.18 *Let* $\mathbf{x}_1, \dots, \mathbf{x}_k \in \mathbb{R}^n$ *and* $\mathbf{y}_1, \dots, \mathbf{y}_k \in \mathbb{R}^n$ *be orthonormal sets* ($k \le n$). *Then another orthonormal set of vectors* $\mathbf{v}_1, \dots, \mathbf{v}_k$ *within* $F = \mathrm{Span}\{\mathbf{y}_1, \dots, \mathbf{y}_k\}$ *can be found such that*

$$\sum_{i=1}^{k} \|\mathbf{x}_i - \mathbf{v}_i\|^2 \le 2 \sum_{i=1}^{k} \mathrm{dist}^2(\mathbf{x}_i, F).$$

Proof. In the proof we follow the ideas of Bolla (2004). We will give a construction for such $\mathbf{v}_i$'s which are 'close' to the individual $\mathbf{x}_i$'s, respectively. Note that $\boldsymbol{X} = (\mathbf{x}_1, \dots, \mathbf{x}_k)$ and $\boldsymbol{Y} = (\mathbf{y}_1, \dots, \mathbf{y}_k)$ are $n \times k$ suborthogonal matrices. Since the vectors $\mathbf{v}_i$'s also form an orthonormal set within F, they can be obtained by applying a rotation (within F) for the $\mathbf{y}_i$'s. That is, we are looking for a $k \times k$ orthogonal matrix $\boldsymbol{R}$ such that $(\mathbf{v}_1, \dots, \mathbf{v}_k) = \boldsymbol{YR}$ and, with it,

$$\sum_{i=1}^{k} \|\mathbf{x}_i - \mathbf{v}_i\|^2 = \mathrm{tr}[(\boldsymbol{X} - \boldsymbol{YR})^T(\boldsymbol{X} - \boldsymbol{YR})] \le 2 \sum_{i=1}^{k} d^2(\mathbf{x}_i, F) \tag{3.27}$$

holds. By the properties of the trace operator (discussed in Section A of the Appendix),

$$\begin{aligned} \mathrm{tr}[(\boldsymbol{X} - \boldsymbol{YR})^T(\boldsymbol{X} - \boldsymbol{YR})] &= \mathrm{tr}(\boldsymbol{X}^T\boldsymbol{X}) + \mathrm{tr}(\boldsymbol{R}^T\boldsymbol{Y}^T\boldsymbol{YR}) - 2\mathrm{tr}(\boldsymbol{X}^T\boldsymbol{YR}) \\ &= \mathrm{tr}(\boldsymbol{X}^T\boldsymbol{X}) + \mathrm{tr}[(\boldsymbol{Y}^T\boldsymbol{Y})(\boldsymbol{RR}^T)] - 2\mathrm{tr}(\boldsymbol{X}^T\boldsymbol{YR}) \\ &= 2[k - \mathrm{tr}(\boldsymbol{X}^T\boldsymbol{YR})] \end{aligned} \tag{3.28}$$

is obtained, where we used that $\boldsymbol{X}^T\boldsymbol{X} = \boldsymbol{Y}^T\boldsymbol{Y} = \boldsymbol{RR}^T = \boldsymbol{I}_k$. The expression in (3.28) is minimal if and only if $\mathrm{tr}(\boldsymbol{X}^T\boldsymbol{YR})$ is maximal as a function of $\boldsymbol{R}$. At this point, we use Proposition A.3.24 of the Appendix, according to which $\mathrm{tr}(\boldsymbol{X}^T\boldsymbol{Y})\boldsymbol{R}$ is maximal if $(\boldsymbol{X}^T\boldsymbol{Y})\boldsymbol{R}$ is symmetric, and the maximum is $\sum_{i=1}^{k} s_i$, with s_i's being the singular values of $\boldsymbol{X}^T\boldsymbol{Y}$. Indeed, let $\boldsymbol{VSU}^T$ the SVD of $\boldsymbol{X}^T\boldsymbol{Y}$, where $\boldsymbol{V}$ and $\boldsymbol{U}$ are $k \times k$ orthogonal matrices and $\boldsymbol{S}$ is a $k \times k$ diagonal matrix with the singular values in its main diagonal (see Theorem A.3.5 and Equation (A.12) of the Appendix). Note that the singular values s_i's are the cosines of the *principal (canonical) angles* between the subspaces $\mathrm{Span}\{\mathbf{x}_1, \dots, \mathbf{x}_k\}$ and F.

By the above SVD, $(\boldsymbol{X}^T\boldsymbol{Y})\boldsymbol{R} = \boldsymbol{V}\boldsymbol{S}\boldsymbol{U}^T\boldsymbol{R}$ is symmetric if $\boldsymbol{U}^T\boldsymbol{R} = \boldsymbol{V}^T$, that is, $\boldsymbol{R} = \boldsymbol{U}\boldsymbol{V}^T$. Consequently, the minimum that can be attained in (3.28) is equal to

$$2\sum_{i=1}^{k}(1 - s_i). \tag{3.29}$$

Eventually, the sum of the distances in (3.27) can also be written in terms of the singular values $s_1, \dots, s_k$. Since $\boldsymbol{Y}\boldsymbol{Y}^T$ is the matrix of the orthogonal projection onto F,

$$\sum_{i=1}^{k} d^2(\mathbf{x}_i, F) = \operatorname{tr}[(\boldsymbol{X} - \boldsymbol{Y}\boldsymbol{Y}^T\boldsymbol{X})^T(\boldsymbol{X} - \boldsymbol{Y}\boldsymbol{Y}^T\boldsymbol{X})] = \operatorname{tr}(\boldsymbol{X}^T\boldsymbol{X}) - \operatorname{tr}(\boldsymbol{X}^T\boldsymbol{Y}\boldsymbol{Y}^T\boldsymbol{X})$$

$$= k - \sum_{i=1}^{k} s_i^2 = \sum_{i=1}^{k}(1 - s_i^2), \tag{3.30}$$

where we also used that the $k \times k$ symmetric matrix $\boldsymbol{X}^T\boldsymbol{Y}\boldsymbol{Y}^T\boldsymbol{X}$ has eigenvalues $s_1^2, \dots, s_k^2$.

Comparing (3.29) and (3.30), it remains to show that $\sum_{i=1}^{k}(1 - s_i) \le \sum_{i=1}^{k}(1 - s_i^2)$. But s_i's are the singular values of the matrix $\boldsymbol{X}^T\boldsymbol{Y}$, therefore denoting by $s_{\max}(.)$ the maximum singular value of the matrix in the argument, in view of Proposition A.3.26 of the Appendix, we have

$$s_i \le s_{\max}(\boldsymbol{X}^T\boldsymbol{Y}) \le s_{\max}(\boldsymbol{X}) \cdot s_{\max}(\boldsymbol{Y}) = 1,$$

as all positive singular values of the matrices $\boldsymbol{X}$ and $\boldsymbol{Y}$ – being suborthogonal matrices – are equal to 1. Hence, $s_i \ge s_i^2$ implies the desired relation (3.27).

Proof. [**Theorem** 3.1.17, the construction part.] Let $\mathbf{x}_1, \dots, \mathbf{x}_k$ denote the eigenvectors corresponding to $\lambda_1, \dots, \lambda_k$, the k largest absolute value (of order larger than $\sqrt{n}$) eigenvalues of $\boldsymbol{A}_n$. The representatives – that are row vectors of the $n \times k$ matrix $\boldsymbol{X} = (\mathbf{x}_1, \dots, \mathbf{x}_k)$ – by the assumption of the theorem, form k clusters in $\mathbb{R}^k$ with k-variance less than c/n with some constant c. Let $V_1, \dots, V_k$ denote the clusters (properly reordering the rows of $\boldsymbol{X}$, together they give the index set $\{1, \dots, n\}$). Let $\mathbf{x}^1, \dots, \mathbf{x}^n \in \mathbb{R}^k$ be the Euclidean representatives of the vertices (the rows of $\boldsymbol{X}$), and let $\bar{\mathbf{x}}^1, \dots, \bar{\mathbf{x}}^k$ denote the cluster centers. Now let us choose the following representation of the vertices. The representatives are row vectors of the $n \times k$ matrix $\tilde{\boldsymbol{X}}$ such that the first n_1 rows of $\tilde{\boldsymbol{X}}$ are equal to $\bar{\mathbf{x}}^1$, the next n_2 rows of $\tilde{\boldsymbol{X}}$ are equal to $\bar{\mathbf{x}}^2$, ..., and so on; the last n_k rows of $\tilde{\boldsymbol{X}}$ are equal to $\bar{\mathbf{x}}^k$. Finally, let $\mathbf{y}_1, \dots, \mathbf{y}_k \in \mathbb{R}^n$ be the column vectors of $\tilde{\boldsymbol{X}}$. By the considerations of the Proof of Proposition 2.1.8 and Theorem 3.1.8,

$$S_k^2(\boldsymbol{X}) = \sum_{i=1}^{k} \operatorname{dist}^2(\mathbf{x}_i, F),$$

where the k-dimensional subspace F is spanned by the vectors $\mathbf{y}_1, \dots, \mathbf{y}_k$; further, by the assumption of the theorem, $S_k^2(\boldsymbol{X}) < \frac{c}{n}$.

Then, in view of Lemma 3.1.18, a set $\mathbf{v}_1, \dots, \mathbf{v}_k$ of orthonormal vectors within F can be found such that

$$\sum_{i=1}^{k} \|\mathbf{x}_i - \mathbf{v}_i\|^2 \leq 2\frac{c}{n}$$

holds **AS**. It is important that $\mathbf{v}_i$'s are also piecewise constant vectors over the vertex-clusters $V_1, \dots, V_k$.

Finally, for the matrix $\boldsymbol{A}_n = \sum_{i=1}^{n} \lambda_i \mathbf{x}_i \mathbf{x}_i^T$, the blown-up matrix $\boldsymbol{B}_n = \sum_{i=1}^{k} \lambda_i \mathbf{v}_i \mathbf{v}_i^T$ is constructed. Then the spectral norm of the error matrix $\boldsymbol{E}_n = \boldsymbol{A}_n - \boldsymbol{B}_n$ is $\mathcal{O}(\sqrt{n})$, as it will be proved more generally, in the rectangular case.

In the models discussed in Subsection 3.1.1, a very general kind of random noise was added to the weight matrix of some special edge-weighted graphs. We have shown that if the weight matrix has some structural eigenvalues, then a Wigner-noise cannot essentially disturb this structure: the adjacency matrix of the noisy graph will have the same number of protruding eigenvalues with corresponding eigenvectors revealing the structure of the graph. Vice versa, if – in addition – the representation with the eigenvectors corresponding to the structural eigenvalues shows good classification properties, we have shown how to find the clusters themselves. Now we will briefly discuss some other kind of contemporary random graph models from the point of view of stability. Juhász (1996) investigates the possibly complex eigenvalues of not necessarily symmetric random block matrices under more special conditions on the entries.

Theoretically, for any graph on n vertices, the regularity lemma of Szemerédi (see Komlós *et al.* (2002); Simonovits and T.-Sós (1991); Szemerédi (1976); Tao (2006)) guarantees the existence of a favorable partition of the vertices with maximum number of clusters independent of n. Here we cite the regularity lemma together with the notion of favorable cluster pairs between which the edge-density is homogeneous.

Definition 3.1.19 *We say that the disjoint pair $V_i, V_j \subset V$ $(i \neq j)$ is ε-regular, if for any $A \subset V_i$, $B \subset V_j$, with $|A| > \varepsilon|V_i|$, $|B| > \varepsilon|V_j|$,*

$$|\rho(A, B) - \rho(V_i, V_j)| < \varepsilon$$

holds, where $\rho(A, B)$ denotes the edge-density between the disjoint vertex-subsets A and B. More precisely, denoting by $e(A, B)$ the number of cut-edges between A and B,

$$\rho(A, B) = \frac{e(A, B)}{|A| \cdot |B|}.$$

Theorem 3.1.20 (Szemerédi's regularity lemma) *For every $\varepsilon > 0$ and integer $m > 0$ there are integers $P(\varepsilon, m)$, $Q(\varepsilon, m)$ with the following property: for every* *simple graph $G = (V, E)$ with $n > P(\varepsilon, m)$ vertices there is a partition of V into $k + 1$ classes $V_0, V_1, \dots, V_k$ such that*

- $m \le k \le Q(\varepsilon, m)$,
- $|V_0| \le \varepsilon n$,
- $|V_1| = |V_2| = \cdots = |V_k|$,
- *all but at most εk^2 of the pairs (V_i, V_j) are ε-regular.*

If the graph is sparse – the number of edges $e = o(n^2)$ – then $k = 1$, otherwise k can be arbitrarily large (but it does not depend on n, it merely depends on ε).

If our random graph is a generalized random graph, then $e(V_i, V_j)$ is the sum of $|V_i| \cdot |V_j|$ independent, identically distributed Bernoulli variables with parameter p_{ij} ($1 \le i, j \le k$), where p_{ij}'s are entries of the pattern matrix $\boldsymbol{P}$. Hence $e(A, B)$ is a binomially distributed random variable with expectation $|A| \cdot |B| \cdot p_{ij}$ and variance $|A| \cdot |B| \cdot p_{ij}(1 - p_{ij})$. Therefore, by Lemma 3.1.16 (with the choice of $K = 1$) and with $A \subset V_i$, $B \subset V_j$, $|A| > \varepsilon|V_i|$, $|B| > \varepsilon|V_j|$ we have that

$$\begin{aligned}\mathbb{P}\left(|\rho(A, B) - p_{ij}| > \varepsilon\right) &= \mathbb{P}\left(\left|e(A, B) - |A| \cdot |B| \cdot p_{ij}\right| > \varepsilon \cdot |A| \cdot |B|\right)\\ &\le e^{-\frac{\varepsilon^2|A|^2|B|^2}{2[|A||B|p_{ij}(1-p_{ij})+\varepsilon|A||B|/3]}}\\ &= e^{-\frac{\varepsilon^2|A||B|}{2[p_{ij}(1-p_{ij})+\varepsilon/3]}}\\ &\le e^{-\frac{\varepsilon^4|V_i||V_j|}{2[p_{ij}(1-p_{ij})+\varepsilon/3]}},\end{aligned}$$

that tends to 0 as $|V_i| = n_i \to \infty$ and $|V_j| = n_j \to \infty$. Hence, any pair V_i, V_j is **AS** ε-regular. We note, however, that the regularity lemma does not give a construction for the clusters. Provided the conditions of Theorem 3.1.17 hold, by the cluster centers a similar construction may exist. The algorithmic aspects of the regularity lemma are discussed in Alon *et al.* (1994). A matrix approximation theorem to derive a constructive version of the regularity lemma is introduced in Frieze and Kannan (1999). Further aspects will be discussed in Section 3.3 and Chapter 4.

3.1.5 Random power law graphs and the extended planted partition model

In fact, there are other types of real-world graphs that are more or less vulnerable to random noise, for example scale-free graphs introduced in Barabási and Albert (1999) and more precisely characterized as a random graph process in Bollobás *et al.* (2001). Bollobás and Riordan (2003) investigate the vulnerability of this graph under the effect of removing edges if $n \to \infty$.

In the sequel, we will use the definition of Chung *et al.* (2003) for a random graph on n vertices with given positive expected degree sequence $d_1, \dots, d_n$. Let $b_{ij} := d_i d_j / \sum_{m=1}^{n} d_m$ be the weight of the edge between vertices i and j, where loops are also present and we assume that $\max_i d_i^2 \le \sum_{i=1}^n d_i$. So the deterministic edge-weight matrix $\boldsymbol{B} = (b_{ij})_{i,j=1}^n$, providing the skeleton for the random graph, is a matrix of rank 1, having the eigenvalue zero with multiplicity $n-1$, further the only positive eigenvalue is equal to

$$\mathrm{tr}(\boldsymbol{B}) = \frac{\sum_{i=1}^n d_i^2}{\sum_{i=1}^n d_i}, \tag{3.31}$$

the so-called second order average degree introduced in Chung *et al.* (2003). In our approach, the random noise means the addition of a Wigner-noise to $\boldsymbol{B}$, the effect of which depends on the asymptotic order of the quantity (3.31).

The random power law graph is a special case of this model. Let $\beta > 0$ denote the power in the distribution of the actual degrees: the probability that a vertex has degree x is proportional to $x^{-\beta}$ (x is not necessarily an integer, it can be a generalized degree as well); it is reminiscent of the Pareto distribution. The maximum eigenvalue of our graph is proportional to the squareroot of the maximum degree, see Chung *et al.* (2003). Móri (2005) proves that in case of trees the maximum degree is asymptotically of order $n^{\frac{1}{\beta-1}}$ if n is 'large', and this asymptotic order is also valid for other power law graphs with $\beta > 1$. Hence, with $\frac{1}{2(\beta-1)} > \frac{1}{2}$, that is, with $\beta < 2$, the largest eigenvalue has order greater than $\sqrt{n}$ which is not changed significantly after a Wigner-noise is added.

In view of Chung *et al.* (2003), the following degree sequence gives a power law graph with parameters β (the power) and i_0 (specifies the support of the distribution):

$$d_i = c \cdot i^{-\frac{1}{\beta-1}}, \quad i = i_0, \dots, i_0 + n,$$

where – in accord with the notation of the above paper – the $n+1$ vertices are indexed from i_0 to $i_0 + n$, and c is a normalizing constant that depends on β, n, and i_0.

In order to have a real graph, the following two inequalities must hold:

$$\sum_{i=i_0}^{i_0+n} d_i = 2e \le 2\binom{n+1}{2} = (n+1)n \sim n^2 \tag{3.32}$$

where e denotes the number of edges, and for the minimum degree

$$d_{\min} = d_{i_0+n} = c \cdot (i_0 + n)^{-\frac{1}{\beta-1}} \ge 1. \tag{3.33}$$

For large n the sum $\sum_{i=i_0}^{i_0+n} d_i$ is bounded by means of integration, hence the left-hand side of (3.32) is estimated as

$$\begin{aligned} \sum_{i=i_0}^{i_0+n} d_i &= c \sum_{i=i_0}^{i_0+n} i^{-\frac{1}{\beta-1}} \ge c \int_{i=i_0}^{i_0+n-1} x^{-\frac{1}{\beta-1}}\, dx \\ &= c\frac{\beta-1}{2-\beta}\left[i_0^{-\frac{\beta-1}{2-\beta}} - (i_0+n-1)^{-\frac{\beta-1}{2-\beta}} \right], \end{aligned} \tag{3.34}$$

where $1 < \beta < 2$.

Relations (3.32) – (3.34) give upper and lower estimates for c:

$$(i_0 + n)^{\frac{1}{\beta-1}} \le c \le \frac{n^2}{\frac{\beta-1}{2-\beta}\left[i_0^{-\frac{\beta-1}{2-\beta}} - (i_0 + n - 1)^{-\frac{\beta-1}{2-\beta}} \right]} = \mathcal{O}(n^2)$$

for large n's. This surely holds, if $\frac{1}{\beta-1} \le 2$, that is, if $\beta \ge 1.5$. If, in addition, $\beta < 2$ is also satisfied, the largest eigenvalue is greater than $\sqrt{n}$ in magnitude. Consequently, for $\beta \in [1.5, 2)$ c can be chosen such that the number of edges is $e = \Theta(n^2)$, so our graph is dense enough to have more than one cluster by the regularity lemma. In other words, our graph has a blown-up skeleton and, therefore, it is robust enough. For example, β is 1.5 in the flux distribution examined in Almaas *et al.* (2004). Scale-free graphs (see e.g., Farkas *et al.* (2001)) with $\beta \in [1.5, 2)$ frequently occur in the case of cellular networks. Perhaps, because of this, such metabolic networks can better tolerate a Wigner-noise – that more or less affects each of the edges – than those with $\beta \ge 2$, usual in the case of social and communication networks. For the $2 < \beta < 3$ case, Leskovec *et al.* (2008) prove that in the above defined power law graph, with probability tending to 1 (as $n \to \infty$), there exists a cut of size $\Theta(\log n)$ whose conductance is $\Theta(\frac{1}{\log n})$; further there exist constants c', $\varepsilon > 0$ such that there are no sets of size larger than $c' \log n$ having conductance smaller than ε. Here the conductance of a cut $(S, \overline{S})$ in G is $\max\{\phi(S, G), \phi(\overline{S}, G)\}$ with the expansion ϕ defined in (2.46), the minimum of which maxima over the cuts of G is the isoperimetric number of (2.37).

For the several clusters case, where the number k of the clusters is fixed, Chaudhuri *et al.* (2012) recommend an extended planted partition model. The authors note that the normalized Laplacian-based clustering does not perform well when the graph has a number of low degree vertices. The algorithm discussed in Coja-Oghlan and Lanka (2009b) helps this situation by eliminating the low degree vertices and clustering the rest of the graph. The authors in Coja-Oghlan and Lanka (2009a) also give a lower estimate for the spectral gap of the core of a random graph with a given expected degree distribution $d_1, \dots, d_n$, where the estimation also depends on the minimum expected degree. Their result applies to sparse graphs too. Coja-Oghlan (2010) gives a construction via spectral techniques for partitioning the graph's vertices into k clusters provided k is such that there exists a rank k approximation of the adjacency matrix that plays the role of our blown-up matrix.

The extended planted partition model of Chaudhuri *et al.* (2012) postulates that there is a hidden partition $V_1, \dots, V_k$ of the vertices of the simple graph G and a number d_i associated with each vertex. Then i, j are connected with probability $pd_i d_j$ if they are in the same cluster, and with probability $qd_i d_j$, otherwise. Here $0 < p, q < 1$ are parameters of the model, akin to the constant $\tau > 0$ with which the authors define the degree-corrected normalized Laplacian of the graph in the following way. Let $\boldsymbol{A}$ be the adjacency matrix and $\boldsymbol{D}$ be the degree-matrix of G. The degree-corrected normalized Laplacian of G is $\boldsymbol{I} - (\boldsymbol{D} + \tau \boldsymbol{I})^{-1/2} \boldsymbol{A} (\boldsymbol{D} + \tau \boldsymbol{I})^{-1/2}$. They use the SD of this matrix, or equivalently, the SVD of the degree-corrected random walk Laplacian $\boldsymbol{I} - (\boldsymbol{D} + \tau \boldsymbol{I})^{-1} \boldsymbol{A}$. Then by means of spectral projections onto the subspace spanned by the bottom k right-hand side singular vectors of this matrix, akin to the vertex representation technique, they find a k-partition of the vertices. Note that this method of degree-correction is reminiscent of that of the ridge regression.

For algorithmic issues, see also Kannan *et al.* (2005), and for robustness issues, Karrer *et al.* (2008).

3.2 Noisy contingency tables

Contingency tables are rectangular arrays of nonnegative real entries, corresponding to standardized counts of two categorical variables (see Sections 1.2 and B.1), for example keyword–document matrices or microarrays. In this section, the results of the previous section will be extended to the stability of the SVD of large noisy contingency tables, further to two-way clustering of the row and column categories of the underlying categorical variables. As the categories may be measured in different units, sometimes a normalization is necessary. For this reason, normalized contingency tables and correspondence analysis techniques are used (see Section C.3 of the Appendix). To begin with, the notions of the previous section are adopted to rectangular arrays.

Definition 3.2.1 *The $m \times n$ real matrix $\boldsymbol{W}$ is a Wigner-noise if its entries w_{ij} $(1 \le i \le m,\ 1 \le j \le n)$ are independent random variables, $\mathbb{E}(w_{ij}) = 0$, and the w_{ij}'s are uniformly bounded (that is, there is a constant $K > 0$, independently of m and n, such that $|w_{ij}| \le K,\ \forall i, j$).*

Though the main results of this paper can be extended to w_{ij}'s with any light-tail distribution (especially to Gaussian distributed entries), the subsequent results will be based on the assumptions of Definition 3.2.1, sometimes also using the notation σ^2 for the common bound of the entries' variances. In view of Theorem B.2.6 of the Appendix, the spectral norm of an $m \times n$ Wigner-noise $\boldsymbol{W}_{m\times n}$ is at most of order $\sqrt{m+n}$, with probability tending to 1 as $n, m \to \infty$.

Definition 3.2.2 *The $m \times n$ real matrix $\boldsymbol{B}$ is a blown-up matrix if there is an $a \times b$ so-called pattern matrix $\boldsymbol{P}$ with entries $0 < p_{ij} < 1$, and there are positive integers $m_1, \dots, m_a$ with $\sum_{i=1}^{a} m_i = m$ and $n_1, \dots, n_b$ with $\sum_{i=1}^{b} n_i = n$ such that – after rearranging its rows and columns – the matrix $\boldsymbol{B}$ can be divided into $a \times b$ blocks, where the block (i, j) is an $m_i \times n_j$ matrix with entries all equal to p_{ij} $(1 \le i \le a,\ 1 \le j \le b)$.*

Let us fix $\boldsymbol{P}$, blow it up to an $m \times n$ matrix $\boldsymbol{B}_{m\times n}$, and consider the noisy matrix $\boldsymbol{A}_{m\times n} = \boldsymbol{B}_{m\times n} + \boldsymbol{W}_{m\times n}$ as $m, n \to \infty$ under one or both of the subsequent growth rate conditions.

Definition 3.2.3 *The following growth rate conditions for the growth of the sizes and that of the cluster sizes of an $m \times n$ rectangular array are introduced.*

GC1 *There exists a constant $0 < c \le \frac{1}{a}$ such that $\frac{m_i}{m} \ge c$ ($i = 1, \dots, a$) and a constant $0 < d \le \frac{1}{b}$ such that $\frac{n_i}{n} \ge d$ ($i = 1, \dots, b$).*

GC2 *There exist constants $C \ge 1$, $D \ge 1$, and $C_0 > 0$, $D_0 > 0$ such that $m \le C_0 n^C$ and $n \le D_0 m^D$ for sufficiently large m and n.*

We remark the following:

- **GC1** implies that

$$c \le \frac{m_k}{m_i} \le \frac{1}{c} \quad \text{and} \quad d \le \frac{n_\ell}{n_j} \le \frac{1}{d} \tag{3.35}$$

  hold for any pair of indices $k, i \in \{1, \dots, a\}$ and $\ell, j \in \{1, \dots, b\}$.

- **GC2** implies that for sufficiently large m and n,

$$\left(\frac{1}{D_0}\right)^{\frac{1}{D}} n^{\frac{1}{D}} \le m \le C_0 n^C \quad \text{and} \quad \left(\frac{1}{C_0}\right)^{\frac{1}{C}} m^{\frac{1}{C}} \le n \le D_0 m^D.$$

  Therefore, **GC2** mildly regulates the relation between m and n, but they need not tend to infinity at the same rate.

While perturbing $\boldsymbol{B}_{m \times n}$ by $\boldsymbol{W}_{m \times n}$, assume that for the uniform bound of the entries of $\boldsymbol{W}_{m \times n}$ the condition

$$K \le \min \left\{ \min_{\substack{i \in \{1,\dots,a\} \\ j \in \{1,\dots,b\}}} p_{ij},\ 1 - \max_{\substack{i \in \{1,\dots,a\} \\ j \in \{1,\dots,b\}}} p_{ij} \right\} \tag{3.36}$$

is satisfied. In this way, the entries of $\boldsymbol{A}_{m \times n}$ are in the [0,1] interval, and hence, $\boldsymbol{A}_{m \times n}$ defines a standardized contingency table. We are interested in asymptotic properties of the SVD of this expanding contingency table sequence as $n, m \to \infty$ under the growth rate conditions.

With an appropriate Wigner-noise we can guarantee that the noisy table $\boldsymbol{A}_{m \times n}$ contains 1's in the (a, b)-th block with probability p_{ab}, and 0's otherwise. Indeed, for indices $1 \le a, b \le k$ and $i \in V_a$, $j \in V_b$ let

$$w_{ij} := \begin{cases} 1 - p_{ab} & \text{with probability} \quad p_{ab} \\ -p_{ab} & \text{with probability} \quad 1 - p_{ab} \end{cases} \tag{3.37}$$

be independent random variables. This $\boldsymbol{W}_{m \times n}$ satisfies the conditions of Definition 3.2.1 with uniformly bounded entries of zero expectation. Let $R_1, \dots, R_a$ and $C_1, \dots, C_b$ denote the row- and column clusters induced by the blow-up. In the random 0-1 contingency table $\boldsymbol{A}_{m \times n}$, the row and column categories of R_i and C_j are in interaction with probability p_{ab}. Such schemes are sought for in microarray analysis and they are called chess-board patterns, see Kluger *et al.* (2003) for details. In terms of microarrays, the above property means that genes of the same cluster R_i equally influence conditions of the same cluster C_j.

Definition 3.2.4 *Let $\mathbf{A}_{m\times n}$ be an $m \times n$ contingency table and $\mathcal{P}_{m,n}$ a property which mostly depends on the SVD of $\mathbf{A}_{m\times n}$. Then $\mathbf{A}_{m\times n}$ can have the property $\mathcal{P}_{m,n}$ in the two following senses.*

WP1 *The property $\mathcal{P}_{m,n}$ holds for $\mathbf{A}_{m\times n}$ with probability tending to 1 if $\lim_{m,n\to\infty} \mathbb{P}\left(\mathbf{A}_{m\times n} \in \mathcal{P}_{m,n}\right) = 1$.*

AS *The property $\mathcal{P}_{m,n}$ holds for $\mathbf{A}_{m\times n}$ almost surely if*

$$\mathbb{P}\left(\exists\, m_0, n_0 \in \mathbb{N} \text{ s.t. for } m \geq m_0 \text{ and } n \geq n_0 \,:\, \mathbf{A}_{m\times n} \in \mathcal{P}_{m,n}\right) = 1.$$

Here we may assume **GC1** *and/or* **GC2** *for the simultaneous growth of m and n. In fact,* **GC2** *will only be used for noisy correspondence matrices.*

Analogously to the quadratic case, **AS** always implies **WP1**. Furthermore, if in addition to **WP1**

$$\sum_{m,n=1}^{\infty} \mathbb{P}\left(\mathbf{A}_{m\times n} \notin \mathcal{P}_{m,n}\right) < \infty$$

also holds, then, by the Borel–Cantelli lemma, $\mathbf{A}_{m\times n}$ has $\mathcal{P}_{m,n}$ **AS**.

To find **AS** estimate for the spectral norm, that is, the largest singular value of $\mathbf{W}_{m\times n}$, Theorem B.2.4 of the Appendix can be adapted to rectangular matrices in the following manner. Let $\mathbf{W}_{m\times n}$ be a Wigner-noise with entries uniformly bounded by K. The $(m+n)\times(m+n)$ symmetric matrix

$$\widetilde{\mathbf{W}} = \frac{1}{K}\cdot\begin{pmatrix} \mathbf{0} & \mathbf{W}_{m\times n} \\ \mathbf{W}^T_{m\times n} & \mathbf{0} \end{pmatrix}$$

satisfies the conditions of Theorem B.2.4, and in view of Proposition A.3.29 of the Appendix, its largest and smallest eigenvalues are

$$\lambda_i(\widetilde{\mathbf{W}}) = -\lambda_{n+m-i+1}(\widetilde{\mathbf{W}}) = \frac{1}{K}\cdot s_i(\mathbf{W}_{m\times n}), \quad i = 1, \ldots, \min\{m, n\},$$

while the others are zeros, where $\lambda_i(.)$ and $s_i(.)$ denote the ith largest eigenvalue and singular value of the matrix in the argument, respectively. Therefore

$$\boldsymbol{P}\left(|s_1(\mathbf{W}_{m\times n}) - \mathbb{E}(s_1(\mathbf{W}_{m\times n}))| > t\right) \leq \exp\left(-\frac{(1-o(1))t^2}{32K^2}\right). \tag{3.38}$$

The fact that $s_1(\mathbf{W}_{m\times n}) = \|\mathbf{W}_{m\times n}\| = \mathcal{O}(\sqrt{m+n})$ **WP1** and inequality (3.38) together ensure that $\mathbb{E}(\|\mathbf{W}\|) = \mathcal{O}(\sqrt{m+n})$. Hence, no matter how $\mathbb{E}\|\mathbf{W}_{m\times n}\|$ behaves when $m, n \to \infty$, the following rough estimate holds: there exist positive constants

c_1 and c_2, depending merely on the common bound K of the entries of $\boldsymbol{W}_{m\times n}$, such that

$$\mathbb{P}\left(\|\boldsymbol{W}_{m\times n}\| > c_1\sqrt{m+n} \right) \le e^{-c_2(m+n)}. \tag{3.39}$$

Since the right-hand side of (3.39) forms a convergent series, the spectral norm of a Wigner-noise $\boldsymbol{W}_{m\times n}$ is of order $\sqrt{m+n}$ **AS**. This observation will provide the base of the subsequent **AS** results, which are also **WP1** ones.

3.2.1 Singular values of a noisy contingency table

Proposition 3.2.5 *Under* **GC1**, *all the non-zero singular values of the blown-up contingency table* $\boldsymbol{B}_{m\times n}$ *are of order* $\sqrt{mn}$.

Proof. As there are at most a and b linearly independent rows and linearly independent columns in $\boldsymbol{B}_{m\times n}$, respectively, the rank r of the matrix $\boldsymbol{B}_{m\times n}$ cannot exceed $\min\{a, b\}$. Note that r is also the rank of the pattern matrix $\boldsymbol{P}$. Let $s_1 \ge s_2 \ge \cdots \ge s_r > 0$ be the positive singular values of $\boldsymbol{B}_{m\times n}$. Let $\mathbf{v}_k \in \mathbb{R}^m$, $\mathbf{u}_k \in \mathbb{R}^n$ be a singular vector pair corresponding to s_k, $k = 1, \dots, r$. Without loss of generality, $\mathbf{v}_1, \dots, \mathbf{v}_r$ and $\mathbf{u}_1, \dots, \mathbf{u}_r$ can be chosen unit-norm, pairwise orthogonal vectors in $\mathbb{R}^m$ and $\mathbb{R}^n$, respectively.

For the subsequent calculations we drop the subscript k, and $\mathbf{v}, \mathbf{u}$ denotes a singular vector pair corresponding to the singular value $s > 0$ of the blown-up matrix $\boldsymbol{B}_{m\times n}$, $\|\mathbf{v}\| = \|\mathbf{u}\| = 1$. It is easy to see that they have piecewise constant structures: $\mathbf{v}$ has m_i coordinates equal to $v(i)$, $i = 1, \dots, a$ and $\mathbf{u}$ has n_j coordinates equal to $u(j)$, $j = 1, \dots, b$. With these coordinates, the singular value–singular vector equation

$$\boldsymbol{B}_{m\times n}\mathbf{u} = s\mathbf{v} \tag{3.40}$$

has the form

$$\sum_{j=1}^{b} n_j p_{ij} u(j) = s v(i), \quad i = 1, \dots, a. \tag{3.41}$$

With the notation

$$\tilde{\mathbf{u}} = (u(1), \dots, u(a))^T, \qquad \tilde{\mathbf{v}} = (v(1), \dots, v(b))^T,$$

and

$$\boldsymbol{D}_a = \mathrm{diag}(m_1, \dots, m_a), \qquad \boldsymbol{D}_b = \mathrm{diag}(n_1, \dots, n_b), \tag{3.42}$$

the equations in (3.41) can be written as

$$\boldsymbol{P}\boldsymbol{D}_b\tilde{\mathbf{u}} = s\tilde{\mathbf{v}}. \tag{3.43}$$

Introducing the transformations

$$\mathbf{w} = \boldsymbol{D}_b^{1/2}\tilde{\mathbf{u}}, \qquad \mathbf{z} = \boldsymbol{D}_a^{1/2}\tilde{\mathbf{v}}, \tag{3.44}$$

the Equation 3.43 is equivalent to

$$\boldsymbol{D}_a^{1/2}\boldsymbol{P}\boldsymbol{D}_b^{1/2}\mathbf{w} = s\mathbf{z}. \tag{3.45}$$

Applying the transformation (3.44) for the $\tilde{\mathbf{u}}_k, \tilde{\mathbf{v}}_k$ pairs obtained from the $\mathbf{u}_k, \mathbf{v}_k$ pairs ($k = 1, \ldots, r$), orthonormal systems in $\mathbb{R}^a$ and $\mathbb{R}^b$ are obtained:

$$\mathbf{w}_k{}^T\mathbf{w}_\ell = \sum_{j=1}^{b} n_j u_k(j) u_\ell(j) = \delta_{k\ell} \quad \text{and} \quad \mathbf{z}_k{}^T\mathbf{z}_\ell = \sum_{i=1}^{a} m_i v_k(i) v_\ell(i) = \delta_{k\ell}.$$

Consequently, $\mathbf{z}_k, \mathbf{w}_k$ is a singular vector pair corresponding to the singular value s_k of the $a \times b$ matrix $\boldsymbol{D}_a^{1/2}\boldsymbol{P}\boldsymbol{D}_b^{1/2}$ ($k = 1, \ldots, r$). With the shrinking

$$\widetilde{\boldsymbol{D}}_a = \frac{1}{m}\boldsymbol{D}_a, \qquad \widetilde{\boldsymbol{D}}_b = \frac{1}{n}\boldsymbol{D}_b,$$

an equivalent form of (3.45) is yielded by

$$\widetilde{\boldsymbol{D}}_a^{1/2}\boldsymbol{P}\widetilde{\boldsymbol{D}}_b^{1/2}\mathbf{w} = \frac{s}{\sqrt{mn}}\mathbf{z},$$

that is the $a \times b$ matrix $\widetilde{\boldsymbol{D}}_a^{1/2}\boldsymbol{P}\widetilde{\boldsymbol{D}}_b^{1/2}$ has non-zero singular values $\frac{s_k}{\sqrt{mn}}$ with the same singular vector pairs $\mathbf{z}_k, \mathbf{w}_k$ ($k = 1, \ldots, r$). If the s_k's are not distinct numbers, the singular vector pairs corresponding to a multiple singular value are not unique, but still they can be obtained from the SVD of the shrunken matrix $\widetilde{\boldsymbol{D}}_a^{1/2}\boldsymbol{P}\widetilde{\boldsymbol{D}}_b^{1/2}$.

Now we want to establish relations between the singular values of $\boldsymbol{P}$ and $\widetilde{\boldsymbol{D}}_a^{1/2}\boldsymbol{P}\widetilde{\boldsymbol{D}}_b^{1/2}$. We will apply the minimax principle of A.3.16 of the Appendix. Since we are interested only in the first r singular values, where $r = \text{rank}(\boldsymbol{B}_{m\times n}) = \text{rank}(\widetilde{\boldsymbol{D}}_a^{1/2}\boldsymbol{P}\widetilde{\boldsymbol{D}}_b^{1/2})$, it suffices to consider vectors $\mathbf{x}$, for which $\widetilde{\boldsymbol{D}}_a^{1/2}\boldsymbol{P}\widetilde{\boldsymbol{D}}_b^{1/2}\mathbf{x} \neq \mathbf{0}$. Therefore, with $k \in \{1, \ldots, r\}$ and an arbitrary k-dimensional subspace $H \subset \mathbb{R}^b$ one can write

$$\begin{aligned}
\min_{\mathbf{x}\in H} \frac{\|\widetilde{\boldsymbol{D}}_a^{1/2}\boldsymbol{P}\widetilde{\boldsymbol{D}}_b^{1/2}\mathbf{x}\|}{\|\mathbf{x}\|} &= \min_{\mathbf{x}\in H} \frac{\|\widetilde{\boldsymbol{D}}_a^{1/2}\boldsymbol{P}\widetilde{\boldsymbol{D}}_b^{1/2}\mathbf{x}\|}{\|\boldsymbol{P}\widetilde{\boldsymbol{D}}_b^{1/2}\mathbf{x}\|} \cdot \frac{\|\boldsymbol{P}\widetilde{\boldsymbol{D}}_b^{1/2}\mathbf{x}\|}{\|\widetilde{\boldsymbol{D}}_b^{1/2}\mathbf{x}\|} \cdot \frac{\|\widetilde{\boldsymbol{D}}_b^{1/2}\mathbf{x}\|}{\|\mathbf{x}\|} \\
&\geq \min_a s_a(\widetilde{\boldsymbol{D}}_a^{1/2}) \cdot \min_{\mathbf{x}\in H} \frac{\|\boldsymbol{P}\widetilde{\boldsymbol{D}}_b^{1/2}\mathbf{x}\|}{\|\widetilde{\boldsymbol{D}}_b^{1/2}\mathbf{x}\|} \cdot \min_b s_b(\widetilde{\boldsymbol{D}}_b^{1/2}) \\
&\geq \sqrt{cd} \cdot \min_{\mathbf{x}\in H} \frac{\|\boldsymbol{P}\widetilde{\boldsymbol{D}}_b^{1/2}\mathbf{x}\|}{\|\widetilde{\boldsymbol{D}}_b^{1/2}\mathbf{x}\|},
\end{aligned}$$

with c, d of **GC1**. Now taking the maximum for all possible k-dimensional subspace H, we obtain that $s_k(\widetilde{\boldsymbol{D}}_a^{1/2}\boldsymbol{P}\widetilde{\boldsymbol{D}}_b^{1/2}) \geq \sqrt{cd} \cdot s_k(\boldsymbol{P}) > 0$. On the other hand,

$$s_k(\widetilde{\boldsymbol{D}}_a^{1/2}\boldsymbol{P}\widetilde{\boldsymbol{D}}_b^{1/2}) \leq \|\widetilde{\boldsymbol{D}}_a^{1/2}\boldsymbol{P}\widetilde{\boldsymbol{D}}_b^{1/2}\| \leq \|\widetilde{\boldsymbol{D}}_a^{1/2}\| \cdot \|\boldsymbol{P}\| \cdot \|\widetilde{\boldsymbol{D}}_b^{1/2}\| \leq \|\boldsymbol{P}\| \leq \sqrt{ab}.$$

These inequalities imply that $s_k(\widetilde{D}_a^{1/2} P \widetilde{D}_b^{1/2})$ is bounded from below and from above with a positive constant that does not depend on m and n, and because of $s_k(\widetilde{D}_m^{1/2} P \widetilde{D}_n^{1/2}) = \frac{s_k}{\sqrt{mn}}$, we obtain that $s_1, \ldots, s_r = \Theta(\sqrt{mn})$.

Theorem 3.2.6 *Let $A_{m\times n} = B_{m\times n} + W_{m\times n}$ be an $m \times n$ random matrix, where $B_{m\times n}$ is a blown-up matrix with positive singular values $s_1, \ldots, s_r$ and $W_{m\times n}$ is a Wigner-noise. Then, under* **GC1**, *the matrix $A_{m\times n}$ has r singular values $z_1, \ldots, z_r$, such that*

$$|z_i - s_i| = \mathcal{O}(\sqrt{m+n}), \quad i = 1, \ldots, r$$

and for the other singular values

$$z_j = \mathcal{O}(\sqrt{m+n}), \quad j = r+1, \ldots, \min\{m, n\}$$

holds, ***AS***.

Proof. The statement follows from the analog of the Weyl's perturbation theorem for the singular values of rectangular matrices (see Theorem A.3.20 of the Appendix): if $s_i(A_{m\times n})$ and $s_i(B_{m\times n})$ denote the ith largest singular values of the matrix in the argument then for the difference of the corresponding pairs

$$|s_i(A_{m\times n}) - s_i(B_{m\times n})| \le \|W_{m\times n}\|, \quad i = 1, \ldots, \min\{m, n\}.$$

As a consequence of (3.39), $\|W_{m\times n}\|$ is of order $\sqrt{m+n}$ **AS**, which finishes the proof.

Summarizing, with the notation

$$\varepsilon := \|W_{m\times n}\| = \mathcal{O}(\sqrt{m+n}) \quad \text{and} \quad \Delta := \min_{1\le i\le r} s_i(B_{m\times n}) = \Theta(\sqrt{mn}) \tag{3.46}$$

there is a spectral gap of size $\Delta - 2\varepsilon$ between the r largest and the other singular values of the perturbed matrix $A_{m\times n}$, and this gap is significantly larger than ε.

3.2.2 Clustering the rows and columns via singular vector pairs

Here perturbation results for the corresponding singular vector pairs are established. To this end, with the help of Theorem 3.2.6, we estimate the distances between the corresponding right- and left-hand side eigenspaces of the matrices $B_{m\times n}$ and $A_{m\times n} = B_{m\times n} + W_{m\times n}$.

With the notation of the Proof of Proposition 3.2.5, $\mathbf{v}_1, \ldots, \mathbf{v}_m \in \mathbb{R}^m$ and $\mathbf{u}_1, \ldots, \mathbf{u}_n \in \mathbb{R}^n$ are orthonormal left- and right-hand side singular vectors of $B_{m\times n}$. They are piecewise constant vectors over the row and column clusters determined by the blow-up; further, they satisfy

$$B_{m\times n}\mathbf{u}_i = s_i\mathbf{v}_i, \quad i = 1, \ldots, r \quad \text{and} \quad B_{m\times n}\mathbf{u}_j = 0, \quad j = r+1, \ldots, n.$$

Let us also denote the unit-norm, pairwise orthogonal left- and right-hand side singular vectors corresponding to the r outstanding singular values $z_1, \ldots, z_r$ of $A_{m\times n}$ by

$\mathbf{y}_1, \dots, \mathbf{y}_r \in \mathbb{R}^m$ and $\mathbf{x}_1, \dots, \mathbf{x}_r \in \mathbb{R}^n$, respectively. Then, for them $\boldsymbol{A}_{m\times n}\mathbf{x}_i = z_i\mathbf{y}_i$ holds ($i = 1, \dots, r$). Let

$$F := \operatorname{Span}\{\mathbf{v}_1, \dots, \mathbf{v}_r\} \quad \text{and} \quad G := \operatorname{Span}\{\mathbf{u}_1, \dots, \mathbf{u}_r\}$$

denote the spanned linear subspaces of piecewise constant vectors in $\mathbb{R}^m$ and $\mathbb{R}^n$, respectively.

Proposition 3.2.7 *With the above notation, under* **GC1**, *the following estimate holds* **AS**:

$$\sum_{i=1}^{r} \operatorname{dist}^2(\mathbf{y}_i, F) \le r\frac{\varepsilon^2}{(\Delta - \varepsilon)^2} = \mathcal{O}\left(\frac{m+n}{mn}\right), \tag{3.47}$$

and analogously,

$$\sum_{i=1}^{r} \operatorname{dist}^2(\mathbf{x}_i, G) \le r\frac{\varepsilon^2}{(\Delta - \varepsilon)^2} = \mathcal{O}\left(\frac{m+n}{mn}\right). \tag{3.48}$$

Proof. Let us choose one of the right-hand side singular vectors $\mathbf{x}_1, \dots, \mathbf{x}_r$ of $\boldsymbol{A}_{m\times n} = \boldsymbol{B}_{m\times n} + \boldsymbol{W}_{m\times n}$ and denote it simply by $\mathbf{x}$ with corresponding singular value z. We will estimate the distance between $\mathbf{y} = \frac{1}{z}\boldsymbol{A}\mathbf{x}$ and F. For this purpose, we expand $\mathbf{x}$ and $\mathbf{y}$ in the orthonormal bases $\mathbf{u}_1, \dots, \mathbf{u}_n$ and $\mathbf{v}_1, \dots, \mathbf{v}_m$, respectively:

$$\mathbf{x} = \sum_{i=1}^{n} t_i\mathbf{u}_i \quad \text{and} \quad \mathbf{y} = \sum_{i=1}^{m} l_i\mathbf{v}_i.$$

Then on the one hand,

$$\boldsymbol{A}_{m\times n}\mathbf{x} = (\boldsymbol{B}_{m\times n} + \boldsymbol{W}_{m\times n})\mathbf{x} = \sum_{i=1}^{r} t_i s_i \mathbf{v}_i + \boldsymbol{W}_{m\times n}\mathbf{x} \tag{3.49}$$

and, on the other hand,

$$\boldsymbol{A}_{m\times n}\mathbf{x} = z\mathbf{y} = \sum_{i=1}^{m} z l_i \mathbf{v}_i. \tag{3.50}$$

Equating the right-hand sides of (3.49) and (3.50), we obtain

$$\sum_{i=1}^{r}(zl_i - t_i s_i)\mathbf{v}_i + \sum_{i=r+1}^{m} zl_i\mathbf{v}_i = \boldsymbol{W}_{m\times n}\mathbf{x}.$$

In view of the Pythagorean theorem,

$$\sum_{i=1}^{r}(zl_i - t_i s_i)^2 + z^2\sum_{i=r+1}^{m} l_i^2 = \|\boldsymbol{W}_{m\times n}\mathbf{x}\|^2 \le \varepsilon^2, \tag{3.51}$$

because $\|\mathbf{x}\| = 1$ and $\|\boldsymbol{W}\| = \varepsilon$.

As $z \geq \Delta - \varepsilon$ holds **AS** by Theorem 3.2.6,

$$\text{dist}^2(\mathbf{y}, F) = \sum_{i=r+1}^{m} l_i^2 \leq \frac{\varepsilon^2}{z^2} \leq \frac{\varepsilon^2}{(\Delta - \varepsilon)^2}.$$

The order of the above estimate follows from the order of ε and Δ of (3.46):

$$\text{dist}^2(\mathbf{y}, F) = \mathcal{O}\left(\frac{m+n}{mn}\right), \tag{3.52}$$

AS. Applying (3.52) for the left-hand side singular vectors $\mathbf{y}_1, \dots, \mathbf{y}_r$, by Definition 3.1.4,

$$\mathbb{P}\left(\exists m_{0i}, n_{0i} \in \mathbb{N} \text{ s.t. for } m \geq m_{0i}\, n \geq n_{0i}\colon \text{dist}^2(\mathbf{y}_i, F) \leq \frac{\varepsilon^2}{(\Delta - \varepsilon)^2}\right) = 1,$$

for $i = 1, \dots, r$. Hence,

$$\mathbb{P}\left(\exists m_0, n_0 \in \mathbb{N} \text{ s.t. for } m \geq m_0\, n \geq n_0\colon \text{dist}^2(\mathbf{y}_i, F) \leq \frac{\varepsilon^2}{(\Delta - \varepsilon)^2},\ i = 1, \dots, r\right) = 1,$$

consequently,

$$\mathbb{P}\left(\exists m_0, n_0 \in \mathbb{N} \text{ s. t. for } m \geq m_0\, n \geq n_0\colon \sum_{i=1}^{r} \text{dist}^2(\mathbf{y}_i, F) \leq r\frac{\varepsilon^2}{(\Delta - \varepsilon)^2}\right) = 1$$

also holds, which finishes the proof of the first statement.

The estimate for the squared distance between G and a right-hand side singular vector $\mathbf{x}$ of $\boldsymbol{A}_{m\times n}$ follows in the same way, starting with $\boldsymbol{A}_{m\times n}^T \mathbf{y} = z\mathbf{x}$ and using the fact that $\boldsymbol{A}_{m\times n}^T$ has the same singular values as $\boldsymbol{A}_{m\times n}$.

By Proposition 3.2.7, the individual distances between the original and the perturbed subspaces and also the sum of these distances tend to zero **AS** as $m, n \to \infty$ under **GC1**.

Now let $\boldsymbol{A}_{m\times n}$ be a microarray on m genes and n conditions, with a_{ij} denoting the expression level of gene i under condition j. We assume that $\boldsymbol{A}_{m\times n}$ is a noisy random matrix obtained by adding a Wigner-noise $\boldsymbol{W}_{m\times n}$ to the blown-up matrix $\boldsymbol{B}_{m\times n}$. Let us denote by $R_1, \dots, R_a$ the partition of the genes and by $C_1, \dots, C_b$ the partition of the conditions with respect to the blow-up (they can also be thought of as clusters of the genes and conditions).

Proposition 3.2.7 also implies the well-clustering property of the representatives of the genes and conditions in the following representation. Let $\boldsymbol{Y}$ be the $m \times r$ matrix containing the left-hand side singular vectors $\mathbf{y}_1, \dots, \mathbf{y}_r$ of $\boldsymbol{A}_{m\times n}$ in its columns. Likewise, let $\boldsymbol{X}$ be the $n \times r$ matrix containing the right-hand side singular vectors $\mathbf{x}_1, \dots, \mathbf{x}_r$ of $\boldsymbol{A}_{m\times n}$ in its columns. Let the r-dimensional representatives of the genes be the row vectors of $\boldsymbol{Y}$: $\mathbf{y}^1, \dots, \mathbf{y}^m \in \mathbb{R}^r$, while the r-dimensional representatives of the conditions be the row vectors of $\boldsymbol{X}$: $\mathbf{x}^1, \dots, \mathbf{x}^n \in \mathbb{R}^r$. Let $S_a^2(\boldsymbol{Y})$ denote the a-variance, introduced in (C.5) of the genes' representatives:

$$S_a^2(\boldsymbol{Y}) = \min_{\{R'_1,\dots,R'_a\}} \sum_{i=1}^{a} \sum_{j\in R'_i} \|\mathbf{y}^j - \bar{\mathbf{y}}^i\|^2, \quad \text{where} \quad \bar{\mathbf{y}}^i = \frac{1}{m_i} \sum_{j\in R'_i} \mathbf{y}^j,$$

while $S_b^2(\boldsymbol{X})$ denotes the b-variance of the conditions' representatives:

$$S_b^2(\boldsymbol{X}) = \min_{\{C'_1,\dots,C'_b\}} \sum_{i=1}^{b} \sum_{j\in C'_i} \|\mathbf{x}^j - \bar{\mathbf{x}}^i\|^2, \quad \text{where} \quad \bar{\mathbf{x}}^i = \frac{1}{n_i} \sum_{j\in C'_i} \mathbf{x}^j,$$

when the partitions $\{R'_1, \dots, R'_a\}$ and $\{C'_1, \dots, C'_b\}$ vary over all a- and b-partitions of the genes and conditions, respectively.

Theorem 3.2.8 *With the above notation, under* **GC1**, *for the a- and b-variances of the representation of the microarray* $\mathbf{A}_{m\times n}$ *the relations*

$$S_a^2(\boldsymbol{Y}) = \mathcal{O}\left(\frac{m+n}{mn}\right) \quad \text{and} \quad S_b^2(\boldsymbol{X}) = \mathcal{O}\left(\frac{m+n}{mn}\right)$$

hold ***AS***.

Proof. Since $S_a^2(\boldsymbol{Y}) \le \sum_{i=1}^{a} \sum_{j\in R_i} \|\mathbf{y}^j - \bar{\mathbf{y}}^i\|^2$ and $S_b^2(\boldsymbol{X}) \le \sum_{i=1}^{b} \sum_{j\in C_i} \|\mathbf{x}^j - \bar{\mathbf{x}}^i\|^2$, by the considerations of the Proof of Proposition 2.1.8, the right-hand sides are equal to the left-hand sides of (3.47) and (3.48), respectively. Therefore, they are also of order $\frac{m+n}{mn}$.

Hence, the addition of any kind of a Wigner-noise to a rectangular matrix that has a blown-up structure $\boldsymbol{B}_{m\times n}$, will not change the order of the outstanding singular values, and the block structure of $\boldsymbol{B}_{m\times n}$ can be reconstructed from the representatives of the row and column items of the noisy matrix $\boldsymbol{A}_{m\times n}$. So far, we have only used **GC1**, and no restriction for the relation between m and n has been made. For noisy correspondence matrices, **GC2** will also be used.

3.2.3 Perturbation results for correspondence matrices

Let $\boldsymbol{B}_{m\times n}$ be the blown-up matrix of the $a \times b$ pattern matrix $\boldsymbol{P}$ of positive entries, with blow-up sizes $m_1, \dots, m_a$ and $n_1, \dots, n_b$. We perform the correspondence transformation, described in Section 1.2, on $\boldsymbol{B}_{m\times n}$. We are interested in the order of the singular values of $\boldsymbol{A}_{m\times n} = \boldsymbol{B}_{m\times n} + \boldsymbol{W}_{m\times n}$ when the correspondence transformation is applied to it. To this end, we introduce the following notation

$$\boldsymbol{D}_{Brow} = \mathrm{diag}(d_{Brow,1}, \ldots, d_{Brow,m}) = \mathrm{diag}\left(\sum_{j=1}^{n} b_{1j}, \ldots, \sum_{j=1}^{n} b_{mj}\right)$$

$$\boldsymbol{D}_{Bcol} = \mathrm{diag}(d_{Bcol,1}, \ldots, d_{Bcol,n}) = \mathrm{diag}\left(\sum_{i=1}^{m} b_{i1}, \ldots, \sum_{i=1}^{m} b_{in}\right)$$

$$\boldsymbol{D}_{Arow} = \mathrm{diag}(d_{Arow,1}, \ldots, d_{Arow,m}) = \mathrm{diag}\left(\sum_{j=1}^{n} a_{1j}, \ldots, \sum_{j=1}^{n} a_{mj}\right)$$

$$\boldsymbol{D}_{Acol} = \mathrm{diag}(d_{Acol,1}, \ldots, d_{Acol,n}) = \mathrm{diag}\left(\sum_{i=1}^{m} a_{i1}, \ldots, \sum_{i=1}^{m} a_{in}\right)$$

for the diagonal matrices containing the row- and column-sums of $\boldsymbol{B}_{m\times n}$ and $\boldsymbol{A}_{m\times n}$ in their main diagonals, respectively. For notational convenience, we discard the subscript $m \times n$ and set

$$\boldsymbol{B}_{corr} = \boldsymbol{D}_{Brow}^{-1/2}\boldsymbol{B}\boldsymbol{D}_{Bcol}^{-1/2} \quad \text{and} \quad \boldsymbol{A}_{corr} = \boldsymbol{D}_{Arow}^{-1/2}\boldsymbol{A}\boldsymbol{D}_{Acol}^{-1/2}$$

for the transformed matrices while carrying out correspondence transformation on $\boldsymbol{B}$ and $\boldsymbol{A}$, respectively. It is shown in Section C.3 that the leading singular value of $\boldsymbol{B}_{corr}$ is equal to 1, and it is a single singular value if $\boldsymbol{B}$ is non-decomposable (see Definition A.3.28), which will be assumed from now on. Let $1 > s_0 = s_1 \geq \cdots \geq s_{r-1}$ denote non-zero singular values of $\boldsymbol{B}_{corr}$ with unit-norm singular vector pairs $\mathbf{v}_i$, $\mathbf{u}_i$ $(i = 0, \ldots, r-1)$, where $r = \mathrm{rank}(\boldsymbol{B}_{corr}) = \mathrm{rank}(\boldsymbol{B}) = \mathrm{rank}(\boldsymbol{P})$. With the transformations

$$\mathbf{v}_{corr,i} = \boldsymbol{D}_{Brow}^{-1/2}\mathbf{v}_i \quad \text{and} \quad \mathbf{u}_{corr,i} = \boldsymbol{D}_{Bcol}^{-1/2}\mathbf{u}_i \quad (i = 0, 1, \ldots, r-1)$$

the so-called correspondence vector pairs are obtained. The trivial (constantly 1) vector pair corresponds to $\mathbf{v}_0$, $\mathbf{u}_0$. If the coordinates $v_{corr,i}(j)$, $u_{corr,i}(\ell)$ of such a pair are considered as possible values of two discrete random variables ψ_i and ϕ_i (often called the ith correspondence factor pair) with the prescribed marginals, then, as in the canonical analysis, their correlation is s_i, and this is the largest possible correlation under the condition that they are uncorrelated to the first $i-1$ correspondence factors within their own sets $(i = 1, \ldots, r-1)$, see Section C.3 of the Appendix.

In Section 1.2 we proved that the row vectors of the matrices $\mathbf{v}_{corr,1}, \ldots, \mathbf{v}_{corr,k-1}$ and $\mathbf{u}_{corr,1}, \ldots, \mathbf{u}_{corr,k-1}$ are optimal $(k-1)$-dimensional representatives of the rows and columns of the contingency table (where the trivial 0-index coordinates can as well be included to obtain a k-dimensional representation). Here we will establish **AS** properties for the k-variances of these representatives when we also make use of the growth rate condition **GC2**. In the proofs, we mostly follow Bolla *et al.* (2010).

Proposition 3.2.9 *Given the $m \times n$ blown-up matrix $\boldsymbol{B}$, under* **GC1** *there exists a constant $\delta \in (0, 1)$, independent of m and n, such that all the non-zero singular values of $\boldsymbol{B}_{corr}$ are in the interval $[\delta, 1]$.*

Proof. It is easy to see that $\boldsymbol{B}_{corr}$ is the blown-up matrix of the $a \times b$ pattern matrix $\tilde{\boldsymbol{P}}$ with entries

$$\tilde{p}_{ij} = \frac{p_{ij}}{\sqrt{(\sum_{\ell=1}^{b} p_{i\ell} n_\ell)(\sum_{k=1}^{a} p_{kj} m_k)}}.$$

Following the considerations of the proof of Proposition 3.2.5, the blown-up matrix $\boldsymbol{B}_{corr}$ has exactly $r = \text{rank}(\boldsymbol{P}) = \text{rank}(\tilde{\boldsymbol{P}})$ non-zero singular values that are the singular values of the $a \times b$ matrix $\boldsymbol{P}' = \boldsymbol{D}_a^{1/2} \tilde{\boldsymbol{P}} \boldsymbol{D}_b^{1/2}$ (where $\boldsymbol{D}_a$ and $\boldsymbol{D}_b$ are defined in (3.42)) with entries

$$p'_{ij} = \frac{p_{ij}\sqrt{m_i}\sqrt{n_j}}{\sqrt{(\sum_{\ell=1}^{b} p_{i\ell} n_\ell)(\sum_{k=1}^{a} p_{kj} m_k)}} = \frac{p_{ij}}{\sqrt{(\sum_{\ell=1}^{b} p_{i\ell} \frac{n_\ell}{n_j})(\sum_{k=1}^{a} p_{kj} \frac{m_k}{m_i})}}.$$

Since the matrix $\boldsymbol{P}$ contains no zero entries, under **GC1**, the matrix $\boldsymbol{P}'$ varies on a compact set of $a \times b$ matrices determined by the inequalities (3.35). The range of the non-zero singular values depends continuously on the matrix that does not depend on m and n. Therefore, the minimum non-zero singular value does not depend on m and n. Because the largest singular value is 1, this finishes the proof.

Theorem 3.2.10 *Under* **GC1** *and* **GC2**, *there exists a positive constant δ, that does not depend on m and n, such that for every $0 < \tau < 1/2$ the following statement holds* **AS**: *the r largest singular values of $\boldsymbol{A}_{corr}$ are in the interval $[\delta - \max\{n^{-\tau}, m^{-\tau}\}, 1 + \max\{n^{-\tau}, m^{-\tau}\}]$, while all the others are at most $\max\{n^{-\tau}, m^{-\tau}\}$.*

Proof. First notice that

$$\boldsymbol{A}_{corr} = \boldsymbol{D}_{Arow}^{-1/2} \boldsymbol{A} \boldsymbol{D}_{Acol}^{-1/2} = \boldsymbol{D}_{Arow}^{-1/2} \boldsymbol{B} \boldsymbol{D}_{Acol}^{-1/2} + \boldsymbol{D}_{Arow}^{-1/2} \boldsymbol{W} \boldsymbol{D}_{Acol}^{-1/2}. \tag{3.53}$$

The entries of $\boldsymbol{D}_{Brow}$ and those of $\boldsymbol{D}_{Bcol}$ are of order $\Theta(n)$ and $\Theta(m)$, respectively. Now we prove that for every $i = 1, \dots, m$ and $j = 1, \dots, n$, $|d_{Arow,i} - d_{Brow,i}| < n \cdot n^{-\tau}$ and $|d_{Acol,j} - d_{Bcol,j}| < m \cdot m^{-\tau}$ hold, **AS**. To this end, we use the Chernoff's inequality for large deviations (see Lemma 3.1.16):

$$\begin{aligned}
&\mathbb{P}\left(|d_{Arow\,i} - d_{Brow\,i}| > n \cdot n^{-\tau}\right) = \mathbb{P}\left(\left|\sum_{j=1}^{n} w_{ij}\right| > n^{1-\tau}\right) \\
&< \exp\left\{-\frac{n^{2-2\tau}}{2(\text{Var}(\sum_{j=1}^{n} w_{ij}) + K n^{1-\tau}/3)}\right\} \le \exp\left\{-\frac{n^{2-2\tau}}{2(n\sigma^2 + K n^{1-\tau}/3)}\right\} \\
&= \exp\left\{-\frac{n^{1-2\tau}}{2(\sigma^2 + K n^{-\tau}/3)}\right\}, \quad i = 1, \dots, m,
\end{aligned}$$

where the constant K is the uniform bound for $|w_{ij}|$'s and σ^2 is the bound for their variances. In view of **GC2** the following estimate holds with some $C_0 > 0$ and $C \geq 1$ (constants of **GC2**) and large enough n:

$$\begin{aligned}\mathbb{P}\left(|d_{Arow,i} - d_{Brow,i}| > n^{1-\tau}\,\forall i \in \{1,\dots,m\}\right)\\ \leq m \cdot \exp\left\{-\frac{n^{1-2\tau}}{2(\sigma^2 + Kn^{-\tau}/3)}\right\} \leq C_0 \cdot n^C \cdot \exp\left\{-\frac{n^{1-2\tau}}{2(\sigma^2 + Kn^{-\tau}/3)}\right\} \quad (3.54)\\ = \exp\left\{\ln C_0 + C\ln n - \frac{n^{1-2\tau}}{2(\sigma^2 + Kn^{-\tau}/3)}\right\}.\end{aligned}$$

The estimation of probability

$$\mathbb{P}\left(|d_{Acol,j} - d_{Bcol,j}| > m^{1-\tau} \quad \forall j \in \{1,\dots,n\}\right)$$

can be treated analogously (with $D_0 > 0$ and $D \geq 1$ of **GC2**). The right-hand side of (3.54) forms a convergent series; therefore, by the Borel–Cantelli lemma,

$$\min_{i\in\{1,\dots,m\}} |d_{Arow,i}| = \Theta(n), \qquad \min_{j\in\{1,\dots,n\}} |d_{Acol,j}| = \Theta(m) \quad (3.55)$$

hold **AS**.

Now it is straightforward to bound the norm of the second term of (3.53) by

$$\|\boldsymbol{D}_{Arow}^{-1/2}\| \cdot \|\boldsymbol{W}\| \cdot \|\boldsymbol{D}_{Acol}^{-1/2}\|. \quad (3.56)$$

As by (3.39), $\|\boldsymbol{W}\| = \mathcal{O}(\sqrt{m+n})$ holds **AS**, the quantity (3.56) is at most of order $\sqrt{\frac{m+n}{mn}}$, **AS**. Hence, it is less than $\max\{n^{-\tau}, m^{-\tau}\}$, **AS**.

In order to estimate the norm of the first term of (3.53) let us write it in the form

$$\begin{aligned}\boldsymbol{D}_{Arow}^{-1/2}\boldsymbol{B}\boldsymbol{D}_{Acol}^{-1/2} &= \boldsymbol{D}_{Brow}^{-1/2}\boldsymbol{B}\boldsymbol{D}_{Bcol}^{-1/2} + \left[\boldsymbol{D}_{Arow}^{-1/2} - \boldsymbol{D}_{Brow}^{-1/2}\right]\boldsymbol{B}\boldsymbol{D}_{Bcol}^{-1/2}\\ &\quad + \boldsymbol{D}_{Arow}^{-1/2}\boldsymbol{B}\left[\boldsymbol{D}_{Acol}^{-1/2} - \boldsymbol{D}_{Bcol}^{-1/2}\right].\end{aligned} \quad (3.57)$$

The first term is just $\boldsymbol{B}_{corr}$, so due to Proposition 3.2.9, we should prove only that the norms of both remainder terms are less than $\max\{n^{-\tau}, m^{-\tau}\}$, **AS**. These two terms have a similar appearance, therefore it is enough to estimate one of them. For example, the second term can be bounded by

$$\|\boldsymbol{D}_{Arow}^{-1/2} - \boldsymbol{D}_{Brow}^{-1/2}\| \cdot \|\boldsymbol{B}\| \cdot \|\boldsymbol{D}_{Bcol}^{-1/2}\|. \quad (3.58)$$

The estimation of the first factor in (3.58) is as follows:

$$\begin{aligned}\|\boldsymbol{D}_{Arow}^{-1/2} - \boldsymbol{D}_{Brow}^{-1/2}\| &= \max_{i\in\{1,\dots,m\}} \left(\frac{1}{\sqrt{d_{Arow,i}}} - \frac{1}{\sqrt{d_{Brow,i}}}\right) \\ &= \max_{i\in\{1,\dots,m\}} \frac{|d_{Arow,i} - d_{Brow,i}|}{\sqrt{d_{Arow,i}\cdot d_{Brow,i}}(\sqrt{d_{Arow,i}} + \sqrt{d_{Brow,i}})} \\ &\le \max_{i\in\{1,\dots,m\}} \frac{|d_{Arow,i} - d_{Brow,i}|}{\sqrt{d_{Arow,i}\cdot d_{Brow,i}}} \cdot \max_{i\in\{1,\dots,m\}} \frac{1}{(\sqrt{d_{Arow,i}} + \sqrt{d_{Brow,i}})}.\end{aligned} \tag{3.59}$$

By relations (3.55), $\sqrt{d_{Arow,i}\cdot d_{Brow,i}} = \Theta(n)$ for any $i = 1, \dots, m$, and hence,

$$\frac{|d_{Arow,i} - d_{Brow,i}|}{\sqrt{d_{Arow,i}\cdot d_{Brow,i}}} \le n^{-\tau},$$

AS; further, $\max_{i\in\{1,\dots,m\}} \frac{1}{\sqrt{d_{Arow,i}}+\sqrt{d_{Brow,i}}} = \Theta(\frac{1}{\sqrt{n}})$, **AS**.

Therefore, the left-hand side of (3.59) can be estimated by $n^{-\tau-1/2}$ from above, **AS**. For the further factors in (3.58) we obtain $\|\boldsymbol{B}\| = \Theta(\sqrt{mn})$ (see Proposition 3.2.5), while $\|\boldsymbol{D}_{Bcol}^{-1/2}\| = \Theta(\frac{1}{\sqrt{m}})$, **AS**. These together imply that

$$n^{-\tau-1/2}\cdot n^{1/2}m^{1/2}\cdot m^{-1/2} \le n^{-\tau} \le \max\{n^{-\tau}, m^{-\tau}\}.$$

This finishes the estimation of the first term in (3.53), and by the Weyl's perturbation theorem (see Theorem A.3.20) the proof, too.

Let us remark the following.

- If in the definition of the Wigner-noise we used Gaussian distributed entries, the large deviation principle could be replaced by the simple estimation of the Gaussian probabilities with any $\kappa > 0$:

$$\mathbb{P}\left(\left|\frac{1}{n}\sum_{j=1}^{n} w_{ij}\right| > \kappa\right) < \min\left(1, \frac{4\sigma}{\kappa\sqrt{2\pi n}}\exp\left\{-\frac{n}{2\sigma^2}\kappa^2\right\}\right).$$

  Setting $\kappa = n^{-\tau}$ we get an estimate, analogous to (3.54).

- Note that if the requirements of (3.36) for the uniform bound of the entries of the Wigner-noise are met, then the noisy matrix $\boldsymbol{A}$ has nonnegative entries; consequently the singular values of $\boldsymbol{A}_{corr}$ will be, in fact, within the interval $[\delta - \max\{n^{-\tau}, m^{-\tau}\}, 1]$.

Now, we are ready to estimate the k-variances. Recall that the correspondence matrix of a blown-up matrix is also a blown-up matrix. Therefore, the non-zero singular values of $\boldsymbol{B}_{corr}$ are the numbers $1 = s_0 > s_1 \ge \cdots \ge s_{r-1} > 0$ with unit-norm singular vector pairs $\mathbf{v}_i$, $\mathbf{u}_i$ having piecewise constant structure ($i = 0, \ldots, r-1$). Set

$$F := \text{Span}\{\mathbf{v}_0, \ldots, \mathbf{v}_{r-1}\} \quad \text{and} \quad G := \text{Span}\{\mathbf{u}_0, \ldots, \mathbf{u}_{r-1}\}.$$

Let $0 < \tau < 1/2$ be arbitrary and $\epsilon := \max\{n^{-\tau}, m^{-\tau}\}$. Let us also denote the unit-norm, pairwise orthogonal left- and right-hand side singular vectors corresponding to the singular values $z_0, \ldots, z_{r-1} \in [\delta - \epsilon, 1 + \epsilon]$ of $\boldsymbol{A}_{corr}$ – guaranteed by Theorem 3.2.10 under **GC2** – by $\mathbf{y}_0, \ldots, \mathbf{y}_{r-1} \in \mathbb{R}^m$ and $\mathbf{x}_0, \ldots, \mathbf{x}_{r-1} \in \mathbb{R}^n$, respectively.

Proposition 3.2.11 *With the above notation, under* **GC1** *and* **GC2**, *the following estimate holds* **AS** *for the distance between* $\mathbf{y}_i$ *and* F:

$$\text{dist}(\mathbf{y}_i, F) \le \frac{\epsilon}{(\delta - \epsilon)} = \frac{1}{(\frac{\delta}{\epsilon} - 1)} \quad i = 0, \ldots, r-1 \tag{3.60}$$

and analogously, for the distance between $\mathbf{x}_i$ *and* G:

$$\text{dist}(\mathbf{x}_i, G) \le \frac{\epsilon}{(\delta - \epsilon)} = \frac{1}{(\frac{\delta}{\epsilon} - 1)} \quad i = 0, \ldots, r-1. \tag{3.61}$$

Proof. Follow the method of proving Proposition 3.2.7 – under **GC1** – with δ instead of Δ and ϵ instead of ε. Here **GC2** is necessary only for $\boldsymbol{A}_{corr}$ to have r structural singular values.

Note that the left-hand sides of (3.60) and (3.61) are **AS** of order $\max\{n^{-\tau}, m^{-\tau}\}$, tending to zero as $m, n \to \infty$ under **GC1** and **GC2**.

Proposition 3.2.11 implies the well-clustering property of the representatives of the two discrete variables by means of the noisy correspondence vector pairs

$$\mathbf{y}_{corr,i} := \boldsymbol{D}_{Arow}^{-1/2}\mathbf{y}_i, \quad \mathbf{x}_{corr,i} := \boldsymbol{D}_{Acol}^{-1/2}\mathbf{x}_i \quad i = 0, \ldots, r-1.$$

Let $\boldsymbol{Y}_{corr} = (\mathbf{y}_{corr,0}, \ldots, \mathbf{y}_{corr,r-1})$ and $\boldsymbol{X}_{corr} = (\mathbf{x}_{corr,0}, \ldots, \mathbf{x}_{corr,r-1})$ be the matrices containing the optimal representatives of the rows and columns of the noisy contingency table in their columns. Let $\mathbf{y}_{corr}^1, \ldots \mathbf{y}_{corr}^m \in \mathbb{R}^r$ and $\mathbf{x}_{corr}^1, \ldots \mathbf{x}_{corr}^n \in \mathbb{R}^r$ denote the row vectors of $\boldsymbol{Y}_{corr}$ and $\boldsymbol{X}_{corr}$, that is, the representatives of the genes and conditions, respectively. With respect to the marginal distributions, the weighted a- and b-variances of these representatives are defined (see C.6) by

$$\tilde{S}_a^2(\boldsymbol{Y}_{corr}) = \min_{\{R_1', \ldots, R_a'\}} \sum_{i=1}^{a} \sum_{j \in R_i'} d_{Arow,j} \|\mathbf{y}_{corr}^j - \bar{\mathbf{y}}_{corr}^i\|^2$$

and

$$\tilde{S}_b^2(\boldsymbol{X}_{corr}) = \min_{\{C_1', \ldots, C_b'\}} \sum_{i=1}^{b} \sum_{j \in C_i'} d_{Acol,j} \|\mathbf{x}_{corr}^j - \bar{\mathbf{x}}_{corr}^i\|^2,$$

where $\{R'_1, \ldots, R'_a\}$ and $\{C'_1, \ldots, C'_b\}$ are a- and b-partitions of the genes and conditions, respectively; further,

$$\bar{\mathbf{y}}^i_{corr} = \sum_{j \in A'_i} d_{Arow,j} \mathbf{y}^j_{corr} \quad \text{and} \quad \bar{\mathbf{x}}^i_{corr} = \sum_{j \in B'_i} d_{Acol,j} \mathbf{x}^j_{corr}.$$

Theorem 3.2.12 *With the above notation, under* **GC1** *and* **GC2**,

$$\tilde{S}^2_a(\boldsymbol{Y}_{corr}) \le \frac{r}{(\frac{\delta}{\epsilon} - 1)^2} \quad \text{and} \quad \tilde{S}^2_b(\boldsymbol{X}_{corr}) \le \frac{r}{(\frac{\delta}{\epsilon} - 1)^2}$$

hold **AS**, *where* $\epsilon = \max\{n^{-\tau}, m^{-\tau}\}$ *with every* $0 < \tau < 1/2$.

Proof. With the considerations of the Proof of Theorem 3.2.8,

$$\tilde{S}^2_a(\boldsymbol{Y}_{corr}) \le \sum_{i=1}^{a} \sum_{j \in A_i} d_{Arow,j} \|\mathbf{y}^j_{corr} - \bar{\mathbf{y}}^i_{corr}\|^2 = \sum_{i=1}^{r} \text{dist}^2(\mathbf{y}_i, F),$$

and

$$\tilde{S}^2_b(\boldsymbol{X}_{corr}) \le \sum_{i=1}^{b} \sum_{j \in B_i} d_{Acol,j} \|\mathbf{x}^{(j)}_{corr} - \bar{\mathbf{x}}^i_{corr}\|^2 = \sum_{i=1}^{r} \text{dist}^2(\mathbf{x}_i, G),$$

hence the result of Proposition 3.2.11 can be used.

Under **GC1** and **GC2**, with m, n large enough, Theorem 3.2.12 implies that after performing correspondence analysis on the $m \times n$ noisy matrix $\boldsymbol{A}$, the representation through the correspondence vectors corresponding to $\boldsymbol{A}_{corr}$ will also reveal the block structure behind $\boldsymbol{A}$.

3.2.4 Finding the blown-up skeleton

One might wonder where the singular values of an $m \times n$ matrix $\boldsymbol{A} = (a_{ij})$ are located if $a := \max_{i,j} |a_{ij}|$ is independent of m and n. On the one hand, the maximum singular value cannot exceed $\mathcal{O}(\sqrt{mn})$, as it is at most $\sqrt{\sum_{i=1}^m \sum_{j=1}^n a_{ij}^2}$. On the other hand, let $\boldsymbol{Q}$ be an $m \times n$ random matrix with entries a or $-a$ (independently of each other). Consider the spectral norm of all such matrices and take the minimum of them: $\min_{\boldsymbol{Q} \in \{-a,+a\}^{m \times n}} \|\boldsymbol{Q}\|$. This quantity measures the minimum linear structure that a matrix of the same size and magnitude as $\boldsymbol{A}$ can possess. As the Frobenius norm of $\boldsymbol{Q}$ is $a\sqrt{mn}$, in view of the inequalities (A.14) between the spectral and Frobenius norms, the above minimum is at least $\frac{a}{\sqrt{2}}\sqrt{m+n}$, which is exactly the order of the spectral norm of a Wigner-noise.

So an $m \times n$ random matrix (whose entries are independent and uniformly bounded) under very general conditions has at least one singular value of order greater than $\sqrt{m+n}$. Assume there are k such singular values and the representatives by means of the corresponding singular vector pairs can be well classified into k clusters in terms of the k-variances. Under these conditions we can reconstruct a blown-up structure behind our matrix.

Theorem 3.2.13 *Let $\boldsymbol{A}_{m\times n}$ be a sequence of $m \times n$ matrices, where m and n tend to infinity. Assume that $\boldsymbol{A}_{m\times n}$ has exactly k singular values of order greater than $\sqrt{m+n}$ (k is fixed). If there are integers $a \geq k$ and $b \geq k$ such that the a- and b-variances of the optimal row- and column-representatives are $\mathcal{O}(\frac{m+n}{mn})$, then there is an explicit construction for a blown-up matrix $\boldsymbol{B}_{m\times n}$ such that $\boldsymbol{A}_{m\times n} = \boldsymbol{B}_{m\times n} + \boldsymbol{E}_{m\times n}$, with $\|\boldsymbol{E}_{m\times n}\| = \mathcal{O}(\sqrt{m+n})$.*

Proof. In the sequel the subscripts m and n will be dropped, for notational convenience. We will speak in terms of microarrays (genes and conditions). Let $\mathbf{y}_1, \dots, \mathbf{y}_k \in \mathbb{R}^m$ and $\mathbf{x}_1, \dots, \mathbf{x}_k \in \mathbb{R}^n$ denote the left- and right-hand side unit-norm singular vectors corresponding to $z_1, \dots, z_k$, the singular values of $\boldsymbol{A}$ of order greater than $\sqrt{m+n}$. The k-dimensional representatives of the genes and conditions – that are row vectors of the $m \times k$ matrix $\boldsymbol{Y} = (\mathbf{y}_1, \dots, \mathbf{y}_k)$ and those of the $n \times k$ matrix $\boldsymbol{X} = (\mathbf{x}_1, \dots, \mathbf{x}_k)$, respectively – by the assumption of the theorem form a and b clusters in $\mathbb{R}^k$, respectively, with sum of inner variances $\mathcal{O}(\frac{m+n}{mn})$. Reorder the rows and columns of $\boldsymbol{A}$ according to their respective cluster memberships. Denote by $\mathbf{y}^1, \dots, \mathbf{y}^m \in \mathbb{R}^k$ and $\mathbf{x}^1, \dots, \mathbf{x}^n \in \mathbb{R}^k$ the Euclidean representatives of the genes and conditions (the rows of the reordered $\boldsymbol{Y}$ and $\boldsymbol{X}$), and let $\bar{\mathbf{y}}^1, \dots, \bar{\mathbf{y}}^a \in \mathbb{R}^k$ and $\bar{\mathbf{x}}^1, \dots, \bar{\mathbf{x}}^b \in \mathbb{R}^k$ denote the cluster centers, respectively. Now let us choose the following new representation of the genes and conditions. The genes' representatives are row vectors of the $m \times k$ matrix $\widetilde{\boldsymbol{Y}}$ such that the first m_1 rows of $\widetilde{\boldsymbol{Y}}$ are equal to $\bar{\mathbf{y}}^1$, the next m_2 rows to $\bar{\mathbf{y}}^2$, and so on; the last m_a rows of $\widetilde{\boldsymbol{Y}}$ are equal to $\bar{\mathbf{y}}^a$. Then likewise, the conditions' representatives are row vectors of the $n \times k$ matrix $\widetilde{\boldsymbol{X}}$ such that the first n_1 rows of $\widetilde{\boldsymbol{X}}$ are equal to $\bar{\mathbf{x}}^1$, the next n_2 rows to $\bar{\mathbf{x}}^2$, and so on; the last n_b rows of $\widetilde{\boldsymbol{X}}$ are equal to $\bar{\mathbf{x}}^b$.

By the considerations of Theorem 3.2.8 and the assumption for the clusters,

$$\sum_{i=1}^{k} \operatorname{dist}^2(\mathbf{y}_i, F) = S_a^2(\boldsymbol{Y}) = \mathcal{O}\left(\frac{m+n}{mn}\right) \tag{3.62}$$

and

$$\sum_{i=1}^{k} \operatorname{dist}^2(\mathbf{x}_i, G) = S_b^2(\boldsymbol{X}) = \mathcal{O}\left(\frac{m+n}{mn}\right) \tag{3.63}$$

hold respectively, where the k-dimensional subspace $F \subset \mathbb{R}^m$ is spanned by the column vectors of $\widetilde{\boldsymbol{Y}}$, and the k-dimensional subspace $G \subset \mathbb{R}^n$ is spanned by the column vectors of $\widetilde{\boldsymbol{X}}$. We follow the construction given in Lemma 3.1.18 of a set $\mathbf{v}_1, \dots, \mathbf{v}_k$

of orthonormal vectors within F and another set $\mathbf{u}_1, \dots, \mathbf{u}_k$ of orthonormal vectors within G such that

$$\sum_{i=1}^{k} \|\mathbf{y}_i - \mathbf{v}_i\|^2 = \min_{\mathbf{v}'_1,\dots,\mathbf{v}'_k} \sum_{i=1}^{k} \|\mathbf{y}_i - \mathbf{v}'_i\|^2 \le 2\sum_{i=1}^{k} \mathrm{dist}^2(\mathbf{y}_i, F) \tag{3.64}$$

and

$$\sum_{i=1}^{k} \|\mathbf{x}_i - \mathbf{u}_i\|^2 = \min_{\mathbf{u}'_1,\dots,\mathbf{u}'_k} \sum_{i=1}^{k} \|\mathbf{x}_i - \mathbf{u}'_i\|^2 \le 2\sum_{i=1}^{k} \mathrm{dist}^2(\mathbf{x}_i, G) \tag{3.65}$$

hold, where the minimum is taken over orthonormal sets of vectors $\mathbf{v}'_1, \dots, \mathbf{v}'_k \in F$ and $\mathbf{u}'_1, \dots, \mathbf{u}'_k \in G$, respectively. The construction of the vectors $\mathbf{v}_1, \dots, \mathbf{v}_k$ is as follows ($\mathbf{u}_1, \dots, \mathbf{u}_k$ can be constructed in the same way). Let $\mathbf{v}'_1, \dots, \mathbf{v}'_k \in F$ be an arbitrary orthonormal system (obtained, e.g., by the Schmidt orthogonalization method; note that in the Lemma 3.1.18 they were given at the beginning). Let $\boldsymbol{V}' = (\mathbf{v}'_1, \dots, \mathbf{v}'_k)$ be an $m \times k$ matrix and

$$\boldsymbol{Y}^T \boldsymbol{V}' = \boldsymbol{Q}\boldsymbol{S}\mathbf{Z}^T$$

be SVD, where the matrix $\boldsymbol{S}$ contains the singular values of the $k \times k$ matrix $\boldsymbol{Y}^T\boldsymbol{V}'$ in its main diagonal and zeros otherwise, while $\boldsymbol{Q}$ and $\mathbf{Z}$ are $k \times k$ orthogonal matrices (containing the corresponding unit-norm singular vector pairs in their columns). The orthogonal matrix $\boldsymbol{R} = \mathbf{Z}\boldsymbol{Q}^T$ will give the convenient orthogonal rotation of the vectors $\mathbf{v}'_1, \dots, \mathbf{v}'_k$. That is, the column vectors of the matrix $\boldsymbol{V} = \boldsymbol{V}'\boldsymbol{R}$ form also an orthonormal set that is the desired set $\mathbf{v}_1, \dots, \mathbf{v}_k$. Define the error terms $\mathbf{r}_i$ and $\mathbf{q}_i$, respectively:

$$\mathbf{r}_i = \mathbf{y}_i - \mathbf{v}_i \quad \text{and} \quad \mathbf{q}_i = \mathbf{x}_i - \mathbf{u}_i \qquad (i = 1, \dots, k).$$

In view of (3.62) – (3.65),

$$\sum_{i=1}^{k} \|\mathbf{r}_i\|^2 = \mathcal{O}\left(\frac{m+n}{mn}\right) \quad \text{and} \quad \sum_{i=1}^{k} \|\mathbf{q}_i\|^2 = \mathcal{O}\left(\frac{m+n}{mn}\right). \tag{3.66}$$

Consider the following decomposition:

$$\boldsymbol{A} = \sum_{i=1}^{k} z_i \mathbf{y}_i \mathbf{x}_i^T + \sum_{i=k+1}^{\min\{m,n\}} z_i \mathbf{y}_i \mathbf{x}_i^T.$$

The spectral norm of the second term is at most of order $\sqrt{m+n}$. Now consider the first term,

$$\sum_{i=1}^{k} z_i \mathbf{y}_i \mathbf{x}_i^T = \sum_{i=1}^{k} z_i (\mathbf{v}_i + \mathbf{r}_i)(\mathbf{u}_i^T + \mathbf{q}_i^T) = \\ = \sum_{i=1}^{k} z_i \mathbf{v}_i \mathbf{u}_i^T + \sum_{i=1}^{k} z_i \mathbf{v}_i \mathbf{q}_i^T + \sum_{i=1}^{k} z_i \mathbf{r}_i \mathbf{u}_i^T + \sum_{i=1}^{k} z_i \mathbf{r}_i \mathbf{q}_i^T. \tag{3.67}$$

Since $\mathbf{v}_1, \dots, \mathbf{v}_k$ and $\mathbf{u}_1, \dots, \mathbf{u}_k$ are unit vectors, the last three terms in (3.67) can be estimated by means of the relations

$$\|\mathbf{v}_i \mathbf{u}_i^T\| = \sqrt{\|\mathbf{v}_i \mathbf{u}_i^T \mathbf{u}_i \mathbf{v}_i^T\|} = 1 \qquad (i = 1, \dots, k),$$

$$\|\mathbf{v}_i \mathbf{q}_i^T\| = \sqrt{\|\mathbf{q}_i \mathbf{v}_i^T \mathbf{v}_i \mathbf{q}_i^T\|} = \|\mathbf{q}_i\| \qquad (i = 1, \dots, k),$$

$$\|\mathbf{r}_i \mathbf{u}_i^T\| = \sqrt{\|\mathbf{r}_i \mathbf{u}_i^T \mathbf{u}_i \mathbf{r}_i^T\|} = \|\mathbf{r}_i\| \qquad (i = 1, \dots, k),$$

$$\|\mathbf{r}_i \mathbf{q}_i^T\| = \sqrt{\|\mathbf{r}_i \mathbf{q}_i^T \mathbf{q}_i \mathbf{r}_i^T\|} = \|\mathbf{q}_i\| \cdot \|\mathbf{r}_i\| \qquad (i = 1, \dots, k).$$

Taking into consideration that z_i cannot exceed $\Theta(\sqrt{mn})$, while k is fixed and, due to (3.66), we get that the spectral norms of the last three terms in (3.67) – for their finitely many subterms the triangle inequality is applicable – are at most of order $\sqrt{m+n}$. Let $\boldsymbol{B}$ be the first term, that is,

$$\boldsymbol{B} = \sum_{i=1}^{k} z_i \mathbf{v}_i \, u_i^T.$$

Then $\|\boldsymbol{A} - \boldsymbol{B}\| = \mathcal{O}(\sqrt{m+n})$.

By their definition, the vectors $\mathbf{v}_1, \dots, \mathbf{v}_k$ and the vectors $\mathbf{u}_1, \dots, \mathbf{u}_k$ are in the subspaces F and G, respectively. Both spaces consist of piecewise constant vectors; thus the matrix $\boldsymbol{B}$ is a blown-up matrix containing $a \times b$ blocks. The noise matrix is

$$\boldsymbol{E} = \sum_{i=1}^{k} z_i \mathbf{v}_i \mathbf{q}_i^T + \sum_{i=1}^{k} z_i \mathbf{r}_i \mathbf{u}_i^T + \sum_{i=1}^{k} z_i \mathbf{r}_i \mathbf{q}_j^T + \sum_{i=k+1}^{\min\{m,n\}} z_i \mathbf{y}_i \mathbf{x}_i^T$$

which finishes the proof.

Then, provided the conditions of Theorem 3.2.13 hold, by the construction given in the proof above, an algorithm can be written that uses several SVD's and produces the blown-up matrix $\boldsymbol{B}$. This $\boldsymbol{B}$ can be considered as the best blown-up approximation of the microarray $\boldsymbol{A}$. At the same time clusters of the genes and conditions are also obtained. More precisely, first we conclude the clusters from the SVD of $\boldsymbol{A}$, rearrange the rows and columns of $\boldsymbol{A}$ accordingly, and afterwards we use the above construction. If we decide to perform correspondence analysis on $\boldsymbol{A}$, then by (3.53) and (3.57), $\boldsymbol{B}_{corr}$

will give a good approximation to $\boldsymbol{A}_{corr}$ and likewise, the correspondence vectors obtained by the SVD of $\boldsymbol{A}_{corr}$ will give representatives of the genes and conditions.

Clustering microarray data via the k-means algorithm is also discussed in Dhillon (2001); Martella and Vichi (2012). To find the SVD for large rectangular matrices, randomized algorithms are favored, for example Achlioptas and McSherry (2007). A randomized, so-called fast Monte Carlo algorithm for the SVD and its application for clustering large graphs via the k-means algorithm is presented in Drineas *et al.* (2004). In the case of random matrices with an underlying linear structure (outstanding singular values), the random noise of the algorithm is just added to the noise in our data, but their sum is also a Wigner-noise, so it does not change the effect of our algorithm in finding the clusters. Under the conditions of Theorem 3.2.13, the separated error matrix is comparable with the noise matrix, and this fact guarantees that the underlying block structure can be extracted.

3.3 Regular cluster pairs

Here by clustering we understand the partition of the vertices into subsets of similar vertices. We will generalize the Laplacian and modularity-based spectral clustering methods to recover so-called regular cluster pairs such that the information flow between the pairs and within the clusters is as homogeneous as possible. The notion of volume-regularity is also extended to contingency tables. For this purpose, we take into account both ends of the normalized Laplacian spectrum, that is, large absolute value, so-called structural eigenvalues of the normalized modularity matrix, or the largest singular values of the correspondence matrix in the case of rectangular arrays.

3.3.1 Normalized modularity and volume regularity of edge-weighted graphs

Let us start with the one-cluster case. With the normalized modularity matrix of (1.20), the well-known expander mixing lemma (for simple graphs see, e.g., Alon and Spencer (2000); Hoory *et al.* (2006)) is formulated for edge-weighted graphs in the following way (see Bolla (2011b)):

Theorem 3.3.1 (Expander mixing lemma for edge-weighted graphs) *Let $G = (V, W)$ be an edge-weighted graph and assume that* $\mathrm{Vol}(V) = 1$. *Then for all $X, Y \subset V$*

$$|w(X, Y) - \mathrm{Vol}(X)\mathrm{Vol}(Y)| \leq \|\boldsymbol{M}_D\| \cdot \sqrt{\mathrm{Vol}(X)\mathrm{Vol}(Y)},$$

where $\|\boldsymbol{M}_D\|$ is the spectral norm of the normalized modularity matrix of G.

Proof. Recall that the spectrum of the normalized adjacency matrix $\boldsymbol{D}^{-1/2}\boldsymbol{W}\boldsymbol{D}^{-1/2}$ differs from that of $\boldsymbol{M}_D$ only in the following: it contains the eigenvalue $\mu_0 = 1$ with corresponding eigenvector $\mathbf{u}_0 = \sqrt{\mathbf{d}}$ instead of the eigenvalue 0 of

$\boldsymbol{M}_D$ with the same eigenvector. If $G = (V, \boldsymbol{W})$ is connected ($\boldsymbol{W}$ is irreducible), then 1 is a single eigenvalue. Let

$$\boldsymbol{D}^{-1/2}\boldsymbol{W}\boldsymbol{D}^{-1/2} = \sum_{i=0}^{n-1} \mu_i \mathbf{u}_i \mathbf{u}_i^T$$

be SD. In accord with the above remarks, $\|\boldsymbol{M}_D\| = \max_{i\geq 1} |\mu_i|$.

Let $X \subset V$, $Y \subset V$ be arbitrary and denote by $\mathbf{1}_X \in \mathbb{R}^n$ the indicator vector of $X \subset V$. Further, put $\mathbf{x} := \boldsymbol{D}^{1/2}\mathbf{1}_X$ and $\mathbf{y} := \boldsymbol{D}^{1/2}\mathbf{1}_Y$. Let $\mathbf{x} = \sum_{i=0}^{n-1} a_i \mathbf{u}_i$ and $\mathbf{y} = \sum_{i=0}^{n-1} b_i \mathbf{u}_i$ be the expansions of $\mathbf{x}$ and $\mathbf{y}$ in the orthonormal basis $\mathbf{u}_0, \ldots, \mathbf{u}_{n-1}$ with coordinates $a_i = \mathbf{x}^T\mathbf{u}_i$ and $b_i = \mathbf{y}^T\mathbf{u}_i$, respectively. Observe that $a_0 = \text{Vol}(X)$, $b_0 = \text{Vol}(Y)$ and $\sum_{i=0}^{n-1} a_i^2 = \|\mathbf{x}\|^2 = \text{Vol}(X)$, $\sum_{i=0}^{n-1} b_i^2 = \|\mathbf{y}\|^2 = \text{Vol}(Y)$; further, $w(X, Y) = \mathbf{1}_X^T \boldsymbol{W} \mathbf{1}_Y = \mathbf{x}^T(\boldsymbol{D}^{-1/2}\boldsymbol{W}\boldsymbol{D}^{-1/2})\mathbf{y}$. Based on these observations,

$$\begin{aligned}
|w(X, Y) - \text{Vol}(X)\text{Vol}(Y)| &= \left|\sum_{i=1}^{n-1} \mu_i a_i b_i\right| \leq \max_{i\geq 1} |\mu_i| \cdot \left|\sum_{i=1}^{n-1} a_i b_i\right| \\
&\leq \|\boldsymbol{M}_D\| \cdot \sqrt{\sum_{i=1}^{n-1} a_i^2 \sum_{i=1}^{n-1} b_i^2} \\
&\leq \|\boldsymbol{M}_D\| \cdot \sqrt{\text{Vol}(X)(1 - \text{Vol}(X))\text{Vol}(Y)(1 - \text{Vol}(Y))} \\
&\leq \|\boldsymbol{M}_D\| \cdot \sqrt{\text{Vol}(X)\text{Vol}(Y)},
\end{aligned}$$

where we also used the triangle and the Cauchy-Schwarz inequalities.

In fact, we proved the stronger result

$$|w(X, Y) - \text{Vol}(X)\text{Vol}(Y)| \leq \|\boldsymbol{M}_D\| \sqrt{\text{Vol}(X)(1 - \text{Vol}(X))\text{Vol}(Y)(1 - \text{Vol}(Y))}.$$

We remark that Theorem 3.3.1 carries the following implications and equivalences concerning spectral gap, discrepancy and other notions related to expanding and quasirandom properties.

- Since the *spectral gap* of G is $1 - \|\boldsymbol{M}_D\|$, in view of the expander mixing lemma, a 'large' spectral gap is an indication that the weighted cut between any two subsets of the graph is near to what is expected in a random graph, the vertices of which are connected independently, with probability proportional to their generalized degrees.

- The notion of discrepancy is just introduced to measure the deviation from this random attachment of the vertices. The discrepancy of the edge-weighted graph $G = (V, \boldsymbol{W})$ with $\text{Vol}(V) = 1$ is the smallest $\alpha > 0$ such that for all $X, Y \subset V$

$$|w(X, Y) - \text{Vol}(X)\text{Vol}(Y)| \leq \alpha\sqrt{\text{Vol}(X)\text{Vol}(Y)}.$$

  In view of this, the result of Theorem 3.3.1 can be interpreted as follows: α-eigenvalue separation causes α-discrepancy, where the eigenvalue separation is the spectral norm of the normalized modularity matrix, which is the smaller the bigger the separation between the largest normalized adjacency eigenvalue (the 1) and the absolute values of its other eigenvalues is.

- The notion of discrepancy together with the expander mixing lemma was first used for simple (sometimes regular) graphs, see for example Alon and Spencer (2000); Hoory *et al.* (2006), and extended to Hermitian matrices in Bollobás and Nikiforov (2004). Chung and Graham (2008) discuss 'small' discrepancy as a so-called quasirandom property that characterizes random-looking graphs with given degree sequences. The multiclass extension of quasirandomness is discussed in Section 4.5, where the notion of generalized quasirandom graphs is defined based on Lovász and T.-Sós (2008).

- Chung, Graham and Wilson (1989) use the term quasirandom for simple graphs that satisfy any of some equivalent properties, where these properties are closly related to the properties of expander graphs, including the 'large' spectral gap. For a sampler of these quasirandom properties see also Lovász (2008). The above proof of Theorem 3.3.1 follows that of Chung and Graham (2008) for simple graphs: they prove that the absolute value of the largest (except the trivial 1) normalized adjacency eigenvalue bounds the constant α of the discrepancy from above. They also prove that in the case of dense enough graphs (the minimum degree is cn for some constant c and number of vertices n) the converse implication is also true. At the same time, for sparse graphs, Bollobás and Nikiforov (2004) describe a general construction when the converse implication is not true.

- However, Bilu and Linial (2006) prove the following converse to the expander mixing lemma for a d-regular simple graph on n vertices. Assume that for any disjoint vertex-subsets S, T: $|e(S,T) - \frac{|S||T|d}{n}| \le \alpha\sqrt{|S||T|}$. Then all but the largest adjacency eigenvalue of G are bounded, in absolute value, by $\mathcal{O}(\alpha(1 + \log \frac{d}{\alpha}))$. Note that for a d-regular graph the adjacency eigenvalues are d times larger than the normalized adjacency ones, and the same holds for the relation between the α of the above statement and that of the discrepancy's definition when the total volume is normalized to 1.

- Alon *et al.* (2010) relax the notion of eigenvalue separation to essential eigenvalue separation (by introducing a parameter for it and requiring the separation only for the eigenvalues of a relatively large part of the graph). Then they prove bilateral relations between the constants of this kind of eigenvalue separation and discrepancy.

Now the question naturally arises, what if the gap is not at the ends of the spectrum? We want to partition the vertices into clusters so that a relation similar to the above discrepancy property for the edge-densities between the cluster pairs would hold. For this purpose, we use a slightly modified version of the volume regularity's notion introduced in Alon *et al.* (2010). The authors of the aforementioned paper also give an algorithm that computes a regular partition of a given (possibly sparse) graph in polynomial time giving some kind of construction for the regularity lemma of Szemerédi (1976), cited in Theorem 3.1.20.

Definition 3.3.2 *Let $G = (V, \mathbf{W})$ be an edge-weighted graph with* $\mathrm{Vol}(V) = 1$. *The disjoint pair $A, B \subseteq V$ is α-volume regular if for all $X \subset A$, $Y \subset B$ we have*

$$|w(X, Y) - \rho(A, B)\mathrm{Vol}(X)\mathrm{Vol}(Y)| \le \alpha\sqrt{\mathrm{Vol}(A)\mathrm{Vol}(B)},$$

where $\rho(A, B) = \frac{w(A,B)}{\mathrm{Vol}(A)\mathrm{Vol}(B)}$ is the relative inter-cluster density of (A, B).

In the above definition of volume regularity, the X, Y pairs are also disjoint, and a 'small' α indicates that the (A, B) pair is like a bipartite expander, see for example Alon (1986) and Chung (1997).

In Bolla (2011a) we proved the following statement for the $k = 2$ case: if one eigenvalue jumps out of the bulk of the normalized modularity spectrum, then clustering the coordinates of the corresponding transformed eigenvector into 2 parts (by minimizing the 2-variance of its coordinates) will result in an α-volume regular partition of the vertices, where α depends on the gap between the outstanding eigenvalue and the other ones.

In the next theorem, for a general k, we will estimate the constant of volume regularity in terms of the absolute value of the kth largest absolute value eigenvalue and the k-variance of the optimal vertex representatives constructed by the eigenvectors corresponding to the $k-1$ largest absolute value, so-called *structural*, eigenvalues of the normalized modularity matrix.

Theorem 3.3.3 *Let $G = (V, \mathbf{W})$ be an edge-weighted graph on n vertices, with generalized degrees $d_1, \dots, d_n$ and degree-matrix $\mathbf{D}$. Assume that G is connected,* $\mathrm{Vol}(V) = 1$, *and there are no dominant vertices: $d_i = \Theta(1/n)$, $i = 1, \dots, n$ as $n \to \infty$. Let the eigenvalues of the normalized modularity matrix, of G, enumerated in decreasing absolute values, be*

$$1 \ge |\mu_1| \ge \cdots \ge |\mu_{k-1}| > \varepsilon \ge |\mu_k| \ge \cdots \ge |\mu_n| = 0.$$

The partition $(V_1, \dots, V_k)$ of V is defined so that it minimizes the weighted k-variance $\tilde{S}_k^2(\mathbf{X}^*)$ of the optimum vertex representatives obtained as row vectors of the $n \times (k-1)$ matrix $\mathbf{X}^*$ of column vectors $\mathbf{D}^{-1/2}\mathbf{u}_i$, where $\mathbf{u}_i$ is the unit-norm eigenvector corresponding to μ_i $(i = 1, \dots, k-1)$. Assume that there is a constant $0 < K \le \frac{1}{k}$ such that $|V_i| \ge Kn$, $i = 1, \dots, k$. With the notation $s = \sqrt{\tilde{S}_k^2(\mathbf{X}^*)}$, the (V_i, V_j) pairs are $\mathcal{O}(\sqrt{2k}s + \varepsilon)$-volume regular $(i \ne j)$ and for the clusters V_i $(i = 1, \dots, k)$ the following holds: for all $X, Y \subset V_i$,

$$|w(X, Y) - \rho(V_i)\mathrm{Vol}(X)\mathrm{Vol}(Y)| = \mathcal{O}(\sqrt{2k}s + \varepsilon)\mathrm{Vol}(V_i),$$

where $\rho(V_i) = \frac{w(V_i,V_i)}{\mathrm{Vol}^2(V_i)}$ is the relative intra-cluster density of V_i.

Note that in Subsection 2.4.2 we indexed the eigenvalues of $\boldsymbol{M}_D$ in non-increasing order and denoted them by β's. The set of all β_i's is the same as that of all μ_is. Nonetheless, we need a different notation for the eigenvalues indexed in order of their absolute values. Recall that 1 cannot be an eigenvalue of $\boldsymbol{M}_D$ if G is connected. Consequently, $|\mu_1| = 1$ can be an eigenvalue of $\boldsymbol{M}_D$ if and only if $\mu_1 = -1$, that is, if G is bipartite. Therefore, provided the conditions of the above theorem hold with $k = 2$ and $|\mu_1| = 1$, our graph is a bipartite expander discussed in Alon (1986). If the conditions hold with $k \geq 2$ and $\mu_1 = \cdots = \mu_{k-1} \in [-1, -\delta]$ and $|\mu_i| \leq \varepsilon < \delta$ $(i \geq k)$, then we have a k-partite expander. Especially, the modularity spectrum of the complete k-partite graph has $k - 1$ eigenvalues in the $[-1, -\delta]$ interval (with some $\delta > 0$), and all the other eigenvalues are 0 in accord with Table 3.1.

For the proof we need the definition of the cut-norm and the relation between it and the spectral norm (see also Frieze and Kannan (1999)).

Definition 3.3.4 *The cut-norm of the real matrix* $\boldsymbol{A}$ *with row-set Row and column-set Col is*

$$\|\boldsymbol{A}\|_{\square} = \max_{R \subset Row,\, C \subset Col} \left| \sum_{i \in R} \sum_{j \in C} a_{ij} \right| .$$

Lemma 3.3.5 *For the* $m \times n$ *real matrix* $\boldsymbol{A}$,

$$\|\boldsymbol{A}\|_{\square} \leq \sqrt{mn}\|\boldsymbol{A}\|,$$

where the right hand side contains the spectral norm, that is, the largest singular value of $\boldsymbol{A}$.

Proof.

$$\begin{aligned}\|\boldsymbol{A}\|_{\square} &= \max_{\mathbf{x} \in \{0,1\}^m,\, \mathbf{y} \in \{0,1\}^n} |\mathbf{x}^T \boldsymbol{A} \mathbf{y}| = \max_{\mathbf{x} \in \{0,1\}^m,\, \mathbf{y} \in \{0,1\}^n} \left| \left(\frac{\mathbf{x}}{\|\mathbf{x}\|}\right)^T \boldsymbol{A} \left(\frac{\mathbf{y}}{\|\mathbf{y}\|}\right) \right| \cdot \|\mathbf{x}\| \cdot \|\mathbf{y}\|| \\ &\leq \sqrt{mn} \max_{\|\mathbf{x}\|=1,\, \|\mathbf{y}\|=1} |\mathbf{x}^T \boldsymbol{A} \mathbf{y}| = \sqrt{mn}\|\boldsymbol{A}\|,\end{aligned}$$

since for $\mathbf{x} \in \{0, 1\}^m$, $\|\mathbf{x}\| \leq \sqrt{m}$, and for $\mathbf{y} \in \{0, 1\}^n$, $\|\mathbf{y}\| \leq \sqrt{n}$.

The definition of the cut-norm and the result of the above lemma naturally extends to symmetric matrices with $m = n$.

Proof. [**Theorem** 3.3.3] The optimum $(k - 1)$-dimensional representatives of the vertices, in view of Theorem 1.3.3, are row vectors of the matrix $\boldsymbol{X}^* = (\mathbf{x}_1^*, \ldots, \mathbf{x}_{k-1}^*)$, where $\mathbf{x}_i^* = \boldsymbol{D}^{-1/2}\mathbf{u}_i$ $(i = 1, \ldots, k - 1)$. The representatives can as well be regarded as k-dimensional ones, as inserting the vector $\mathbf{x}_0^* = \boldsymbol{D}^{-1/2}\mathbf{u}_0 = \mathbf{1}$ will not change the weighted k-variance $s^2 = \tilde{S}_k^2(\boldsymbol{X}^*)$. Assume that this minimum weighted k-variance is attained on the k-partition $(V_1, \dots, V_k)$ of the vertices. By the considerations of the Proof of Proposition 2.1.8, it follows that

$$s^2 = \sum_{i=0}^{k-1} \text{dist}^2(\mathbf{u}_i, F), \tag{3.68}$$

where $F = \text{Span}\{\boldsymbol{D}^{1/2}\mathbf{z}_1, \dots, \boldsymbol{D}^{1/2}\mathbf{z}_k\}$ with the so-called normalized partition vectors $\mathbf{z}_1, \dots, \mathbf{z}_k$ of coordinates $z_{ji} = \frac{1}{\sqrt{\text{Vol}(V_i)}}$ if $j \in V_i$ and 0, otherwise $(i = 1, \dots, k)$, see (2.25). Note that the vectors $\boldsymbol{D}^{1/2}\mathbf{z}_1, \dots, \boldsymbol{D}^{1/2}\mathbf{z}_k$ form an orthonormal system in $\mathbb{R}^n$. In view of Lemma 3.1.18, we can find another orthonormal system $\mathbf{v}_0, \dots, \mathbf{v}_{k-1} \in F$ such that

$$s^2 \le \sum_{i=0}^{k-1} \|\mathbf{u}_i - \mathbf{v}_i\|^2 \le 2s^2 \tag{3.69}$$

($\mathbf{v}_0 = \mathbf{u}_0$, since $\mathbf{u}_0 \in F$).

Recall that the spectrum of $\boldsymbol{D}^{-1/2}\boldsymbol{W}\boldsymbol{D}^{-1/2}$ only differs from that of $\boldsymbol{M}_D$ in the following: it contains the eigenvalue $\mu_0 = 1$ instead of the eigenvalue 0 of $\boldsymbol{M}_D$ with the same corresponding eigenvector $\sqrt{\mathbf{d}}$. Now we approximate the matrix $\boldsymbol{D}^{-1/2}\boldsymbol{W}\boldsymbol{D}^{-1/2} = \sum_{i=0}^{n-1} \mu_i \mathbf{u}_i \mathbf{u}_i^T$ by the rank k matrix $\sum_{i=0}^{k-1} \mu_i \mathbf{v}_i \mathbf{v}_i^T$ with the following accuracy (in spectral norm):

$$\left\| \sum_{i=0}^{n-1} \mu_i \mathbf{u}_i \mathbf{u}_i^T - \sum_{i=0}^{k-1} \mu_i \mathbf{v}_i \mathbf{v}_i^T \right\| \le \sum_{i=0}^{k-1} |\mu_i| \cdot \|\mathbf{u}_i \mathbf{u}_i^T - \mathbf{v}_i \mathbf{v}_i^T\| + \left\| \sum_{i=k}^{n-1} \mu_i \mathbf{u}_i \mathbf{u}_i^T \right\|, \tag{3.70}$$

which can be estimated from above with

$$\sum_{i=0}^{k-1} \sin \alpha_i + \varepsilon \le \sum_{i=0}^{k-1} \|\mathbf{u}_i - \mathbf{v}_i\| + \varepsilon \le \sqrt{2k}s + \varepsilon$$

where α_i is the angle between $\mathbf{u}_i$ and $\mathbf{v}_i$, and for it, $\sin \frac{\alpha_i}{2} = \frac{1}{2}\|\mathbf{u}_i - \mathbf{v}_i\|$ holds, $i = 0, \dots, k-1$.

Based on these considerations and relation between the cut-norm and the spectral norm (see Definition 3.3.4 and Lemma 3.3.5), the densities to be estimated in the defining formula of volume regularity can be written in terms of piecewise constant vectors in the following way. The vectors $\mathbf{y}_i := \boldsymbol{D}^{-1/2}\mathbf{v}_i$ are piecewise constants on the partition $(V_1, \dots, V_k)$, $i = 0, \dots, k-1$. The matrix $\sum_{i=0}^{k-1} \mu_i \mathbf{y}_i \mathbf{y}_i^T$ is therefore a symmetric block-matrix on $k \times k$ blocks corresponding to the above partition of the vertices. Let $\hat{w}_{ab}$ denote its entries in the (a, b) block $(a, b = 1, \dots, k)$. Using (3.70), the rank k approximation of the matrix $\boldsymbol{W}$ is performed with the following accuracy of the perturbation $\boldsymbol{E}$:

$$\|\boldsymbol{E}\| = \left\| \boldsymbol{W} - \boldsymbol{D}\left(\sum_{i=0}^{k-1} \mu_i \mathbf{y}_i \mathbf{y}_i^T\right) \boldsymbol{D} \right\| = \left\| \boldsymbol{D}^{1/2} \left(\boldsymbol{D}^{-1/2}\boldsymbol{W}\boldsymbol{D}^{-1/2} - \sum_{i=0}^{k-1} \mu_i \mathbf{v}_i \mathbf{v}_i^T \right) \boldsymbol{D}^{1/2} \right\|.$$

Therefore, the entries of $\boldsymbol{W}$ – for $i \in V_a$, $j \in V_b$ – can be decomposed as $w_{ij} = d_i d_j \hat{w}_{ab} + \eta_{ij}$, where the cut-norm of the $n \times n$ symmetric error matrix $\boldsymbol{E} = (\eta_{ij})$ restricted to $V_a \times V_b$ (otherwise it contains entries which are all zeros) and denoted by $\boldsymbol{E}_{ab}$, is estimated as follows:

$$\begin{aligned}
\|\boldsymbol{E}_{ab}\|_\square &\leq n\|\boldsymbol{E}_{ab}\| \leq n \cdot \|\boldsymbol{D}_a^{1/2}\| \cdot (\sqrt{2}ks + \varepsilon) \cdot \|\boldsymbol{D}_b^{1/2}\| \\
&\leq n \cdot \sqrt{c_1 \frac{\mathrm{Vol}(V_a)}{|V_a|}} \cdot \sqrt{c_1 \frac{\mathrm{Vol}(V_b)}{|V_b|}} \cdot (\sqrt{2}ks + \varepsilon) \\
&= c_1 \cdot \sqrt{\frac{n}{|V_a|}} \cdot \sqrt{\frac{n}{|V_b|}} \cdot \sqrt{\mathrm{Vol}(V_a)}\sqrt{\mathrm{Vol}(V_b)}(\sqrt{2}ks + \varepsilon) \\
&\leq c_1 \cdot \frac{1}{K} \sqrt{\mathrm{Vol}(V_a)}\sqrt{\mathrm{Vol}(V_b)}(\sqrt{2}ks + \varepsilon) \\
&= c\sqrt{\mathrm{Vol}(V_a)}\sqrt{\mathrm{Vol}(V_b)}(\sqrt{2}ks + \varepsilon).
\end{aligned}$$

Here the diagonal matrix $\boldsymbol{D}_a$ contains the diagonal part of $\boldsymbol{D}$ restricted to V_a, otherwise zeros, and the constant c does not depend on n. Consequently, for $a, b = 1, \ldots, k$ and $X \subset V_a$, $Y \subset V_b$:

$$\begin{aligned}
&|w(X, Y) - \rho(V_a, V_b)\mathrm{Vol}(X)\mathrm{Vol}(Y)| = \\
&\left| \sum_{i \in X} \sum_{j \in Y} (d_i d_j \hat{w}_{ab} + \eta_{ij}) - \frac{\mathrm{Vol}(X)\mathrm{Vol}(Y)}{\mathrm{Vol}(V_a)\mathrm{Vol}(V_b)} \sum_{i \in V_a} \sum_{j \in V_b} (d_i d_j \hat{w}_{ab} + \eta_{ij}) \right| = \\
&\left| \sum_{i \in X} \sum_{j \in Y} \eta_{ij} - \frac{\mathrm{Vol}(X)\mathrm{Vol}(Y)}{\mathrm{Vol}(V_a)\mathrm{Vol}(V_b)} \sum_{i \in V_a} \sum_{j \in V_b} \eta_{ij} \right| \leq 2c(\sqrt{2}ks + \varepsilon)\sqrt{\mathrm{Vol}(V_a)\mathrm{Vol}(V_b)},
\end{aligned}$$

that gives the required statement both in the $a \neq b$ and $a = b$ case.

The above theorem only has relevance if there is a noticeable spectral gap between $|\mu_{k-1}|$ and $|\mu_k|$. Note that in the $k = 1$ special two-cluster case, due to Theorem 2.2.3, the 2-variance of the optimal one-dimensional representatives can be directly estimated from above by the gap between the two largest absolute value eigenvalues of $\boldsymbol{M}_D$, and hence, the statement of Theorem 3.3.3 simplifies as follows. The optimal pair (V_1, V_2) based on minimizing the weighted 2-variance of the coordinates of $\mathbf{u}_1$ is $\mathcal{O}(\sqrt{\frac{1-\delta}{1-\varepsilon}})$-volume regular, where $\delta = |\mu_1|$ and $\varepsilon = |\mu_2|$. With other methods, the same estimate is obtained in Bolla (2011b). In fact, Theorem 2.2.3 is formulated for the case when the two largest absolute value eigenvalues of the normalized modularity matrix are positive, though it can be adopted to the other situations too; see, for example, the Dual Cheeger inequality (Proposition 2.3.5).

For a general k, we can make the following considerations. Assume that the normalized modularity spectrum (with decreasing absolute values) of $G_n = (V, \boldsymbol{W})$ satisfies

$$1 \geq |\mu_1| \geq \cdots \geq |\mu_{k-1}| \geq \delta > \varepsilon \geq |\mu_k| \geq \cdots \geq |\mu_n| = 0.$$

Our purpose is to estimate s with the gap $\theta := \delta - \varepsilon$. We will use the notation of the Proof of Theorem 3.3.3 and apply Theorem A.3.22 for the perturbation of spectral subspaces of the symmetric matrices

$$\boldsymbol{A} = \sum_{i=0}^{n-1} \mu_i \mathbf{u}_i \mathbf{u}_i^T \quad \text{and} \quad \boldsymbol{B} = \sum_{i=0}^{k-1} \mu_i \mathbf{v}_i \mathbf{v}_i^T$$

in the following situation. The subsets $S_1 = \{\mu_k, \dots, \mu_{n-1}\}$ and $S_2 = \{\mu_0, \dots, \mu_{k-1}\}$ of the eigenvalues of $\boldsymbol{D}^{-1/2}\boldsymbol{W}\boldsymbol{D}^{-1/2}$ are separated by an annulus, where $\text{dist}(S_1, S_2) = \theta > 0$. Denote by $\boldsymbol{P}_A(S_1)$ and $\boldsymbol{P}_B(S_2)$ the projections onto the spectral subspaces of $\boldsymbol{A}$ and $\boldsymbol{B}$ spanned by the eigenvectors corresponding to the eigenvalues in S_1 and S_2, respectively:

$$\boldsymbol{P}_A(S_1) = \sum_{j=k}^{n-1} \mathbf{u}_j \mathbf{u}_j^T, \quad \boldsymbol{P}_B(S_2) = \sum_{i=0}^{k-1} \mathbf{v}_i \mathbf{v}_i^T.$$

Then Theorem A.3.22 implies that

$$\|\boldsymbol{P}_A(S_1)\boldsymbol{P}_B(S_2)\|_2 \le \frac{1}{\theta}\|\boldsymbol{P}_A(S_1)(\boldsymbol{A}-\boldsymbol{B})\boldsymbol{P}_B(S_2)\|_2, \tag{3.71}$$

where $\|.\|_2$ denotes the Frobenius norm.

Since $\boldsymbol{P}_A^{\perp}(S_1)$ and $\boldsymbol{P}_B(S_2)$ project onto subspaces of the same dimension, the non-zero singular values of $\boldsymbol{P}_A(S_1)\boldsymbol{P}_B(S_2)$ are the sines of the the canonical (principal) angles α_i's between the subspaces $\text{Span}\{\mathbf{u}_0, \mathbf{u}_1, \dots, \mathbf{u}_{k-1}\}$ and $\text{Span}\{\mathbf{v}_0, \mathbf{v}_1, \dots, \mathbf{v}_{k-1}\}$, see Stewart and Sun (1990) for details. Further, $\|\boldsymbol{P}_A(S_1)\boldsymbol{P}_B(S_2)\|_2$ is the distance between the two above subspaces, where $\sin\alpha_0 = 0$.

On the left hand side, $\|\boldsymbol{P}_A(S_1)\boldsymbol{P}_B(S_2)\|_2 = \sqrt{\sum_{i=0}^{k-1}\sin^2\alpha_i}$, and in view of $\|\mathbf{u}_i - \mathbf{v}_i\| = 2\sin\frac{\alpha_i}{2}$ and (3.69), this is between $\frac{\sqrt{3}}{2}s$ and s. On the right hand side,

$$\begin{aligned}
&\boldsymbol{P}_A(S_1)\boldsymbol{A}\boldsymbol{P}_B(S_2) - \boldsymbol{P}_A(S_1)\boldsymbol{B}\boldsymbol{P}_B(S_2)\\
&= (\boldsymbol{P}_A(S_1)\boldsymbol{A})\boldsymbol{P}_B(S_2) - \boldsymbol{P}_A(S_1)(\boldsymbol{P}_B(S_2)\boldsymbol{B})\\
&= \sum_{i=0}^{k-1}\sum_{j=k}^{n-1}(\mu_j - \mu_i)(\mathbf{u}_j^T\mathbf{v}_i)\mathbf{u}_j\mathbf{v}_i^T
\end{aligned}$$

where the Frobenius norm of the rank one matrices $\mathbf{u}_j\mathbf{v}_i^T$ is 1, and the inner product $\mathbf{u}_j^T\mathbf{v}_i$ is the 'smaller', the $\mathbf{v}_i$ is the 'closer' to $\mathbf{u}_i$. Therefore, by the Inequality (3.71), s is the 'smaller', θ is the 'larger' and the $|\mu_j - \mu_i|$ differences for $i = 0, \dots, k-1$; $j = k, \dots, n-1$ are 'closer' to θ. If $|\mu_k| = \varepsilon$ is 'small', then $|\mu_1|, \dots, |\mu_{k-1}|$ should be 'close' to each other ($\mu_0 = 1$ does not play an important role because of $\mathbf{u}_0 = \mathbf{v}_0$).

Theorem 3.3.3 implies that for a general edge-weighted graph, the existence of $k-1$ structural eigenvalues of the normalized modularity matrix, separated from 0, is indication of a k-cluster structure such that the cluster-pairs are volume regular with constant depending on the inner eigengap and the above k-variance.

The clusters themselves can be recovered by applying the k-means algorithm for the vertex representatives.

Note that for simple graphs Rohe *et al.* (2011) also make use of the large absolute value eigenvalues of the normalized adjacency matrix – which differs from the normalized modularity matrix just in the trivial factor – but no regular cluster pairs are investigated. Also, in their latent space model, the above authors prove the convergence of the eigenvectors of the empirical normalized Laplacian to the eigenvectors of the population one as $n \to \infty$ in such a way that the minimum expected degree grows fast enough (the graph is dense enough) and the eigengap that separates the large absolute value eigenvalues from the rest of the spectrum does not shrink too quickly. However, they do not discuss the relation between the eigengap and the regularity of the cluster pairs. Based on the normalized Laplacian eigenvalues and eigenvectors, the consistency of spectral clustering is also proved in von Luxburg *et al.* (2008) when the number of data points, selected randomly from a population distribution, tends to infinity. Their model builds a graph on the data points based on their metric distances, for example by kernelized methods, see Section 1.5.

Recently, new techniques have been developed, making use of simplicial complexes, see for example Gundert and Wagner (2012); Meshulam and Wallach (2009); Parzanchevski *et al.* (2012). We can imagine that k-way cuts in the statements of this section can better be treated with the help of $(k-1)$-dimensional, k-partite complexes.

3.3.2 Correspondence matrices and volume regularity of contingency tables

Let us start again with the one-cluster case. Let $\boldsymbol{C}$ be $m \times n$ contingency table and let $\boldsymbol{C}_{corr}$ be the correspondence matrix of Section 1.2 belonging to it. Theorem 3.3.1 naturally extends to this situation, see for example Butler (2006), though he does not use the terminology of correspondence matrices.

Theorem 3.3.6 (Expander mixing lemma for contingency tables) *Let $\boldsymbol{C}$ be a non-decomposable contingency table on row set Row and column set Col, and of total volume 1 (that is, the sum of its nonnegative entries is 1). Then for all $R \subset Row$ and $C \subset Col$*

$$|c(R, C) - \mathrm{Vol}(R)\mathrm{Vol}(C)| \le s_1\sqrt{\mathrm{Vol}(R)\mathrm{Vol}(C)},$$

where s_1 is the the second largest singular value of the normalized contingency table $\boldsymbol{C}_{corr}$ and $c(R, C)$ is defined in Section 2.5.

Proof. We repeat the idea of the proof of Theorem 3.3.1. Let

$$\boldsymbol{C}_{corr} = \boldsymbol{D}_{row}^{-1/2}\boldsymbol{C}\boldsymbol{D}_{col}^{-1/2} = \sum_{i=0}^{r-1} s_i \mathbf{v}_i \mathbf{u}_i^T \tag{3.72}$$

be SVD, $1 = s_0 > s_1, \cdots > s_{r-1}$, where r is the rank of the contingency table. In accord with Section 1.2, $\mathbf{v}_0 = \sqrt{\mathbf{d}_{row}}$, $\mathbf{u}_0 = \sqrt{\mathbf{d}_{col}}$, and $s_1 < 1$, thanks to $\boldsymbol{C}$ non-decomposable.

Let $R \subset Row$ and $C \subset Col$ be arbitrary, and with the indicator vectors corresponding to them, put $\mathbf{x} := \boldsymbol{D}_{row}^{1/2}\mathbf{1}_R$ and $\mathbf{y} := \boldsymbol{D}_{col}^{1/2}\mathbf{1}_C$. Let $\mathbf{x} = \sum_{i=0}^{r-1} a_i \mathbf{v}_i$ and $\mathbf{y} = \sum_{i=0}^{r-1} b_i \mathbf{u}_i$ be the expansions of $\mathbf{x}$ and $\mathbf{y}$ in the orthonormal systems $\mathbf{v}_0, \ldots, \mathbf{v}_{r-1}$ $\mathbf{u}_0, \ldots, \mathbf{u}_{r-1}$ with coordinates $a_i = \mathbf{x}^T\mathbf{v}_i$ and $b_i = \mathbf{y}^T\mathbf{u}_i$, respectively. Observe that $a_0 = \mathrm{Vol}(R)$, $b_0 = \mathrm{Vol}(C)$ and $\sum_{i=0}^{r-1} a_i^2 = \|\mathbf{x}\|^2 = \mathrm{Vol}(R)$, $\sum_{i=0}^{r-1} b_i^2 = \|\mathbf{y}\|^2 = \mathrm{Vol}(C)$. Based on these observations,

$$
\begin{aligned}
|c(R,C) - \mathrm{Vol}(R)\mathrm{Vol}(C)| &= \left|\sum_{i=1}^{r-1} s_i a_i b_i\right| \le s_1 \cdot \left|\sum_{i=1}^{r-1} a_i b_i\right| \\
&\le s_1 \cdot \sqrt{\sum_{i=1}^{r-1} a_i^2 \sum_{i=1}^{r-1} b_i^2} \\
&\le s_1 \cdot \sqrt{\mathrm{Vol}(R)(1-\mathrm{Vol}(R))\mathrm{Vol}(C)(1-\mathrm{Vol}(C))} \\
&\le s_1 \cdot \sqrt{\mathrm{Vol}(R)\mathrm{Vol}(C)},
\end{aligned}
$$

where we also used the triangle and the Cauchy–Schwarz inequalities.

Here we also proved the stronger result

$$|c(R,C) - \mathrm{Vol}(R)\mathrm{Vol}(C)| \le s_1\sqrt{\mathrm{Vol}(R)(1-\mathrm{Vol}(R))\mathrm{Vol}(C)(1-\mathrm{Vol}(C))}.$$

Since the spectral gap of $\boldsymbol{C}_{corr}$ is $1 - s_1$, in view of the above expander mixing lemma, 'large' spectral gap is an indication that the weighted cut between any row and column subset of the contingency table is is near to what is expected in a random table. The following notion of discrepancy just measures the deviation from this random situation. The discrepancy (see Butler (2006)) of the contingency table $\boldsymbol{C}$ of total volume 1 is the smallest $\alpha > 0$ such that for all $R \subset Row$ and $C \subset Col$

$$|c(R,C) - \mathrm{Vol}(R)\mathrm{Vol}(C)| \le \alpha\sqrt{\mathrm{Vol}(R)\mathrm{Vol}(C)}.$$

In view of this, the result of Theorem 3.3.6 can be interpreted as follows: α singular value separation causes α-discrepancy, where the singular value separation is the second largest singular value of the normalized contingency table, which is the smaller the bigger the separation between the largest singular value (the 1) of the normalized contingency table and the other singular values is. Based on the ideas of Bilu and Linial (2006); Bollobás and Nikiforov (2004), Butler (2006) proves the converse of the expander mixing lemma for contingency tables, namely that

$$s_1 \le 150\alpha(1 - 8\log\alpha).$$

Now we extend this notion to volume regular pairs.

Definition 3.3.7 *The row–column cluster pair $R \subset Row$, $C \subset Col$ of the contingency table $\mathbf{C}$ of total volume 1 is α-volume regular if for all $X \subset R$ and $Y \subset C$ the relation*

$$|c(X, Y) - \rho(R, C)\mathrm{Vol}(X)\mathrm{Vol}(Y)| \leq \alpha\sqrt{\mathrm{Vol}(R)\mathrm{Vol}(C)} \tag{3.73}$$

holds, where $\rho(R, C) = \frac{c(R,C)}{\mathrm{Vol}(R)\mathrm{Vol}(C)}$ is the relative inter-cluster density of the row–column pair R, C.

Theorem 3.3.8 *Let $\mathbf{C}$ be a non-decomposable contingency table of m rows and n columns, with row- and column sums $d_{row,1}, \ldots, d_{row,m}$ and $d_{col,1}, \ldots, d_{col,n}$, respectively. Assume that $\sum_{i=1}^{n} \sum_{j=1}^{m} c_{ij} = 1$ and there are no dominant rows and columns: $d_{row,i} = \Theta(\frac{1}{m})$, $i = 1, \ldots, m$ and $d_{col,j} = \Theta(\frac{1}{n})$, $j = 1, \ldots, n$ as $m, n \to \infty$. Let the singular values of $\mathbf{C}_{corr}$ be*

$$1 = s_0 > s_1 \geq \cdots \geq s_{k-1} > \varepsilon \geq s_i, \quad i \geq k.$$

The partition $(R_1, \ldots, R_k)$ of Row and $(C_1, \ldots, C_k)$ of Col are defined so that they minimize the weighted k-variances $\tilde{S}_k^2(\mathbf{X})$ and $\tilde{S}_k^2(\mathbf{Y})$ of the optimal row and column representatives defined in Section 1.2. Assume that there are constants $0 < K_1, K_2 \leq \frac{1}{k}$ such that $|R_i| \geq K_1 n$ and $|C_i| \geq K_2 m$ $(i = 1, \ldots, k)$, respectively. Then the R_i, C_j pairs are $\mathcal{O}(\sqrt{2k}(\tilde{S}_k(\mathbf{X})\tilde{S}_k(\mathbf{Y})) + \varepsilon)$-volume regular $(i, j = 1, \ldots, k)$.

Proof. Recall that provided $\mathbf{C}$ is non-decomposable, the largest singular value $s_0 = 1$ of $\mathbf{C}_{corr}$ – defined in (3.72) – is single with corresponding singular vector pair $\mathbf{v}_0 = \mathbf{D}_{row}^{1/2}\mathbf{1}_m$ and $\mathbf{u}_0 = \mathbf{D}_{col}^{1/2}\mathbf{1}_n$, respectively. The optimal k-dimensional representatives of the rows and columns are row vectors of the matrices $\mathbf{X} = (\mathbf{x}_0, \ldots, \mathbf{x}_{k-1})$ and $\mathbf{Y} = (\mathbf{y}_0, \ldots, \mathbf{y}_{k-1})$, where $\mathbf{x}_i = \mathbf{D}_{row}^{-1/2}\mathbf{v}_i$ and $\mathbf{y}_i = \mathbf{D}_{col}^{-1/2}\mathbf{u}_i$, respectively $(i = 0, \ldots, k-1)$. (Note that the first columns of equal coordinates can as well be omitted.) Assume that the minimum weighted k-variance is attained on the k-partition $(R_1, \ldots, R_k)$ of the rows and $(C_1, \ldots, C_k)$ of the columns, respectively. By the argument of the Proof of Proposition 2.1.8, it follows that

$$\tilde{S}_k^2(\mathbf{X}) = \sum_{i=0}^{k-1} \mathrm{dist}^2(\mathbf{v}_i, F), \quad \tilde{S}_k^2(\mathbf{Y}) = \sum_{i=0}^{k-1} \mathrm{dist}^2(\mathbf{u}_i, G),$$

where $F = \mathrm{Span}\{\mathbf{D}_{row}^{1/2}\mathbf{w}_1, \ldots, \mathbf{D}_{row}^{1/2}\mathbf{w}_k\}$ and $G = \mathrm{Span}\{\mathbf{D}_{col}^{1/2}\mathbf{z}_1, \ldots, \mathbf{D}_{col}^{1/2}\mathbf{z}_k\}$ with the so-called normalized row partition vectors $\mathbf{w}_1, \ldots, \mathbf{w}_k$ of coordinates $w_{ji} = \frac{1}{\sqrt{\mathrm{Vol}(R_i)}}$ if $j \in R_i$ and 0, otherwise; and column partition vectors $\mathbf{z}_1, \ldots, \mathbf{z}_k$ of coordinates $z_{ji} = \frac{1}{\sqrt{\mathrm{Vol}(C_i)}}$ if $j \in C_i$ and 0, otherwise $(i = 1, \ldots, k)$. Note that the vectors $\mathbf{D}_{row}^{1/2}\mathbf{w}_1, \ldots, \mathbf{D}_{row}^{1/2}\mathbf{w}_k$ and $\mathbf{D}_{col}^{1/2}\mathbf{z}_1, \ldots, \mathbf{D}_{col}^{1/2}\mathbf{z}_k$ form orthonormal systems in $\mathbb{R}^n$ and

$\mathbb{R}^m$, respectively (but they are, usually, not complete). By Lemma 3.1.18, we can find orthonormal systems $\tilde{\mathbf{v}}_0, \dots, \tilde{\mathbf{v}}_{k-1} \in F$ and $\tilde{\mathbf{u}}_0, \dots, \tilde{\mathbf{u}}_{k-1} \in G$ such that

$$\tilde{S}_k^2(\boldsymbol{X}) \leq \sum_{i=0}^{k-1} \|\mathbf{v}_i - \tilde{\mathbf{v}}_i\|^2 \leq 2\tilde{S}_k^2(\boldsymbol{X}), \quad \tilde{S}_k^2(\boldsymbol{Y}) \leq \sum_{i=0}^{k-1} \|\mathbf{u}_i - \tilde{\mathbf{u}}_i\|^2 \leq 2\tilde{S}_k^2(\boldsymbol{Y}).$$

Let $\boldsymbol{C}_{corr} = \sum_{i=0}^{r-1} s_i \mathbf{v}_i \mathbf{u}_i^T$ be SVD, where $r = \text{rank}(\boldsymbol{C}) = \text{rank}(\boldsymbol{C}_{corr})$. We approximate $\boldsymbol{C}_{corr}$ by the rank k matrix $\sum_{i=0}^{k-1} s_i \tilde{\mathbf{v}}_i \tilde{\mathbf{u}}_i^T$ with the following accuracy (in spectral norm):

$$\left\| \sum_{i=0}^{r-1} s_i \mathbf{v}_i \mathbf{u}_i^T - \sum_{i=0}^{k-1} s_i \tilde{\mathbf{v}}_i \tilde{\mathbf{u}}_i^T \right\| \leq \sum_{i=0}^{k-1} s_i \left\| \mathbf{v}_i \mathbf{u}_i^T - \tilde{\mathbf{v}}_i \tilde{\mathbf{u}}_i^T \right\| + \left\| \sum_{i=k}^{r-1} s_i \mathbf{v}_i \mathbf{u}_i^T \right\| \tag{3.74}$$

where the spectral norm of the last term is at most ε, and the the individual terms of the first one are estimated from above in the following way.

$$\begin{aligned}
s_i \|\mathbf{v}_i \mathbf{u}_i^T - \tilde{\mathbf{v}}_i \tilde{\mathbf{u}}_i^T\| &\leq \|(\mathbf{v}_i \mathbf{u}_i^T - \tilde{\mathbf{v}}_i \mathbf{u}_i^T) + (\tilde{\mathbf{v}}_i \mathbf{u}_i^T - \tilde{\mathbf{v}}_i \tilde{\mathbf{u}}_i^T)\| \\
&\leq \|(\mathbf{v}_i - \tilde{\mathbf{v}}_i)\mathbf{u}_i^T\| + \|\tilde{\mathbf{v}}_i(\mathbf{u}_i - \tilde{\mathbf{u}}_i)^T\| \\
&= \sqrt{\|(\mathbf{v}_i - \tilde{\mathbf{v}}_i)\mathbf{u}_i^T \mathbf{u}_i (\mathbf{v}_i - \tilde{\mathbf{v}}_i)^T\|} + \sqrt{\|(\mathbf{u}_i - \tilde{\mathbf{u}}_i)\tilde{\mathbf{v}}_i^T \tilde{\mathbf{v}}_i (\mathbf{u}_i - \tilde{\mathbf{u}}_i)^T\|} \\
&= \sqrt{(\mathbf{v}_i - \tilde{\mathbf{v}}_i)^T (\mathbf{v}_i - \tilde{\mathbf{v}}_i)} + \sqrt{(\mathbf{u}_i - \tilde{\mathbf{u}}_i)^T (\mathbf{u}_i - \tilde{\mathbf{u}}_i)} \\
&= \|\mathbf{v}_i - \tilde{\mathbf{v}}_i\| + \|\mathbf{u}_i - \tilde{\mathbf{u}}_i\|
\end{aligned}$$

where we exploited that the spectral norm (i.e., the largest singular value) of an $n \times m$ matrix $\boldsymbol{A}$ is equal to either the squareroot of the largest eigenvalue of the matrix $\boldsymbol{A}\boldsymbol{A}^T$ or equivalently, that of $\boldsymbol{A}^T\boldsymbol{A}$. In the above calculations all of these matrices are of rank 1, hence, the largest eigenvalue of the symmetric, positive semidefinite matrix under the squareroot is the only non-zero eigenvalue of it, therefore, it is equal to its trace; finally, we used the commutativity of the trace, and in the last line we have the usual vector norm.

Therefore, the first term in (3.74) can be estimated from above by

$$\begin{aligned}
\sum_{i=0}^{k-1} \|\mathbf{v}_i \mathbf{u}_i^T - \tilde{\mathbf{v}}_i \tilde{\mathbf{u}}_i^T\| &\leq \sqrt{k} \sqrt{\sum_{i=0}^{k-1} \|\mathbf{v}_i - \tilde{\mathbf{v}}_i\|^2} + \sqrt{k} \sqrt{\sum_{i=0}^{k-1} \|\mathbf{u}_i - \tilde{\mathbf{u}}_i\|^2} \\
&\leq \sqrt{k} \left(\sqrt{2\tilde{S}_k^2(\boldsymbol{X})} + \sqrt{2\tilde{S}_k^2(\boldsymbol{Y})} \right) = \sqrt{2k}(\tilde{S}_k(\boldsymbol{X}) + \tilde{S}_k(\boldsymbol{Y})).
\end{aligned}$$

Based on these considerations and the relation between the cut-norm and the spectral norm (see Definition 3.3.4 and Lemma 3.3.5), the densities to be estimated in the defining formula (3.73) of volume regularity can be written in terms of stepwise constant vectors in the following way. The vectors $\hat{\mathbf{v}}_i := \boldsymbol{D}_{row}^{-1/2} \tilde{\mathbf{v}}_i$ are stepwise constants on the partition $(R_1, \dots, R_k)$ of the rows, whereas the vectors $\hat{\mathbf{u}}_i := \boldsymbol{D}_{col}^{-1/2}\tilde{\mathbf{u}}_i$ are stepwise constants on the partition $(C_1, \dots, C_k)$ of the columns, $i = 0, \dots, k-1$. The matrix

$$\sum_{i=0}^{k-1} s_i \hat{\mathbf{v}}_i \hat{\mathbf{u}}_i^T$$

is therefore an $n \times m$ block-matrix on $k \times k$ blocks corresponding to the above partition of the rows and columns. Let $\hat{c}_{ab}$ denote its entries in the ab block $(a, b = 1, \dots, k)$. Using (3.74), the rank k approximation of the matrix $\boldsymbol{C}$ is performed with the following accuracy of the perturbation $\boldsymbol{E}$ in spectral norm:

$$\|\boldsymbol{E}\| = \left\| \boldsymbol{C} - \boldsymbol{D}_{row} \left(\sum_{i=0}^{k-1} s_i \hat{\mathbf{v}}_i \hat{\mathbf{u}}_i^T \right) \boldsymbol{D}_{col} \right\| = \left\| \boldsymbol{D}_{row}^{1/2} \left(\boldsymbol{C}_{corr} - \sum_{i=0}^{k-1} s_i \tilde{\mathbf{v}}_i \tilde{\mathbf{u}}_i^T \right) \boldsymbol{D}_{col}^{1/2} \right\|.$$

Therefore, the entries of $\boldsymbol{C}$ – for $i \in R_a$, $j \in C_b$ – can be decomposed as

$$c_{ij} = d_{row,i} d_{col,j} \hat{c}_{ab} + \eta_{ij}$$

where the cut-norm of the $n \times m$ error matrix $\boldsymbol{E} = (\eta_{ij})$ is restricted to $R_a \times C_b$ (otherwise it contains entries which are all zeros) and denoted by $\boldsymbol{E}_{ab}$, is estimated as follows. Making use of Lemma 3.3.5,

$$\begin{aligned}
\|\boldsymbol{E}_{ab}\|_\square &\le \sqrt{mn}\|\boldsymbol{E}_{ab}\| \le \sqrt{nm} \cdot \|\boldsymbol{D}_{row,a}^{1/2}\| \cdot (\sqrt{2k}(\tilde{S}_k(\boldsymbol{X}) + \tilde{S}_k(\boldsymbol{Y})) + \varepsilon) \cdot \|\boldsymbol{D}_{col,b}^{1/2}\| \\
&\le \sqrt{nm}\sqrt{c_1 \frac{\mathrm{Vol}(R_a)}{|R_a|}} \cdot \sqrt{c_2 \frac{\mathrm{Vol}(C_b)}{|C_b|}} (\sqrt{2k}(\tilde{S}_k(\boldsymbol{X}) + \tilde{S}_k(\boldsymbol{Y})) + \varepsilon) \\
&= \sqrt{c_1 c_2} \cdot \sqrt{\frac{n}{|R_a|}} \cdot \sqrt{\frac{m}{|C_b|}} \cdot \sqrt{\mathrm{Vol}(R_a)}\sqrt{\mathrm{Vol}(C_b)}(\sqrt{2k}(\tilde{S}_k(\boldsymbol{X}) + \tilde{S}_k(\boldsymbol{Y})) + \varepsilon) \\
&\le \sqrt{\frac{c_1 c_2}{K_1 K_2}} \sqrt{\mathrm{Vol}(R_a)}\sqrt{\mathrm{Vol}(C_b)}(\sqrt{2k}s + \varepsilon) \\
&= c\sqrt{\mathrm{Vol}(R_a)}\sqrt{\mathrm{Vol}(C_b)}(\sqrt{2k}(\tilde{S}_k(\boldsymbol{X}) + \tilde{S}_k(\boldsymbol{Y})) + \varepsilon)
\end{aligned}$$

where the $n \times n$ diagonal matrix $\boldsymbol{D}_{row,a}$ inherits $\boldsymbol{D}_{row}$'s diagonal entries over R_a, whereas the $m \times m$ diagonal matrix $\boldsymbol{D}_{col,b}$ inherits $\boldsymbol{D}_{col}$'s diagonal entries over C_b, otherwise they are zeros. Further, the constants c_1, c_2 are due to the fact that there are no dominant rows and columns, while K_1, K_2 are derived from the cluster size balancing conditions. Hence,

$$\|\boldsymbol{E}_{ab}\|_\square \le c\sqrt{\mathrm{Vol}(R_a)}\sqrt{\mathrm{Vol}(C_b)}(\sqrt{2k}(\tilde{S}_k(\boldsymbol{X}) + \tilde{S}_k(\boldsymbol{Y})) + \varepsilon)$$

where the constant c does not depend on n and m. Consequently, for $a, b = 1, \dots, k$ and $X \subset R_a$, $Y \subset C_b$,

$$|c(X, Y) - \rho(R_a, C_b)\mathrm{Vol}(X)\mathrm{Vol}(Y)| =$$

$$\left| \sum_{i \in X} \sum_{j \in Y} (d_{row,i} d_{col,j} \hat{c}_{ab} + \eta_{ij}) - \frac{\mathrm{Vol}(X)\mathrm{Vol}(Y)}{\mathrm{Vol}(R_a)\mathrm{Vol}(C_b)} \sum_{i \in R_a} \sum_{j \in C_b} (d_{row,i} d_{col,j} \hat{c}_{ab} + \eta_{ij}) \right| =$$

$$\left| \sum_{i \in X} \sum_{j \in Y} \eta_{ij} - \frac{\mathrm{Vol}(X)\mathrm{Vol}(Y)}{\mathrm{Vol}(R_a)\mathrm{Vol}(C_b)} \sum_{i \in R_a} \sum_{j \in C_b} \eta_{ij} \right| \leq 2\|\boldsymbol{E}_{ab}\|_{\square}$$

$$\leq 2c(\sqrt{2k}(\tilde{S}_k(\boldsymbol{X}) + \tilde{S}_k(\boldsymbol{Y})) + \varepsilon)\sqrt{\mathrm{Vol}(R_a)\mathrm{Vol}(C_b)}$$

that gives the required statement for $a, b = 1, \dots, k$.

If we use the Definition 3.3.7 of α-volume regularity for the row–column cluster pairs R_a, C_b $(a, b = 1, \dots, k)$, then we may say that the *k-way discrepancy* of the underlying contingency table is the minimum α for which all the row–column cluster pairs are α-volume regular. With this nomenclature, Theorem 3.3.8 states that the k-way discrepancy of a contingency table can be estimated from above by the the kth largest singular value of the correspondence matrix and the k-variance of the clusters obtained by the left and right singular vectors corresponding to the $k-1$ largest singular values of this matrix.

We remark the following. When we perform correspondence analysis on a large $m \times n$ contingency table and consider the rank k approximation of it, the entries of this matrix will not necessarily be positive at all. Nonetheless, the entries $\hat{c}_{ij}$'s of the block-matrix constructed in the proof of Theorem 3.3.8 will already be positive provided the weighted k-variances $\tilde{S}_k^2(\boldsymbol{X})$ and $\tilde{S}_k^2(\boldsymbol{Y})$ are 'small' enough. Let us discuss this issue more precisely.

In accord with the notation used in the proof, denote by ab in the lower index the matrix restricted to the $R_a \times C_b$ block (otherwise it has zero entries). Then for the squared Frobenius norm of the rank k approximation of $\boldsymbol{D}_{row}^{-1}\boldsymbol{C}\boldsymbol{D}_{col}^{-1}$, restricted to the ab block, we have that

$$\left\| \boldsymbol{D}_{row,a}^{-1} \boldsymbol{C}_{ab} \boldsymbol{D}_{col,b}^{-1} - \left(\sum_{i=0}^{k-1} s_i \hat{\mathbf{v}}_i \hat{\mathbf{u}}_i^T \right)_{ab} \right\|_2^2 = \sum_{i \in R_a} \sum_{j \in C_b} \left(\frac{c_{ij}}{d_{row,i} d_{col,j}} - \hat{c}_{ab} \right)^2$$
$$= \sum_{i \in R_a} \sum_{j \in C_b} \left(\frac{c_{ij}}{d_{row,i} d_{col,j}} - \bar{c}_{ab} \right)^2 + |R_a||C_b|(\bar{c}_{ab} - \hat{c}_{ab})^2 \tag{3.75}$$

where we used the Steiner equality with the average $\bar{c}_{ab}$ of the entries of $\boldsymbol{D}_{row}^{-1}\boldsymbol{C}\boldsymbol{D}_{col}^{-1}$ in the ab block. Now we estimate the above Frobenius norm by a constant multiple of the spectral norm, where for the spectral norm

$$\left\| \boldsymbol{D}_{row,a}^{-1} \boldsymbol{C}_{ab} \boldsymbol{D}_{col,b}^{-1} - \left(\sum_{i=0}^{k-1} s_i \hat{\mathbf{v}}_i \hat{\mathbf{u}}_i^T \right)_{ab} \right\| = \left\| \boldsymbol{D}_{row,a}^{-1/2} \left(\boldsymbol{C}_{corr} - \sum_{i=0}^{k-1} s_i \tilde{\mathbf{v}}_i \tilde{\mathbf{u}}_i^T \right)_{ab} \boldsymbol{D}_{col,b}^{-1/2} \right\|$$

$$\leq \max_{i \in R_a} \frac{1}{\sqrt{d_{row,i}}} \cdot \max_{j \in C_b} \frac{1}{\sqrt{d_{col,j}}} \cdot [\sqrt{2k}(\tilde{S}_k(\boldsymbol{X}) + \tilde{S}_k(\boldsymbol{Y})) + \varepsilon]$$

holds. Therefore,

$$\left\| \boldsymbol{D}_{row,a}^{-1} \boldsymbol{C}_{ab} \boldsymbol{D}_{col,b}^{-1} - \left(\sum_{i=0}^{k-1} s_i \hat{\mathbf{v}}_i \hat{\mathbf{u}}_i^T \right)_{ab} \right\|_2^2$$

$$\leq \min\{|R_a|, |C_b|\} \cdot \max_{i \in R_a} \frac{1}{d_{row,i}} \cdot \max_{j \in C_b} \frac{1}{d_{col,j}} \cdot [\sqrt{2k}(\tilde{S}_k(\boldsymbol{X}) + \tilde{S}_k(\boldsymbol{Y})) + \varepsilon]^2.$$

Consequently, in view of (3.75),

$$(\bar{c}_{ab} - \hat{c}_{ab})^2 \leq \frac{1}{\max\{|R_a|, |C_b|\}} \cdot \max_{i \in R_a} \frac{1}{d_{row,i}} \cdot \max_{j \in C_b} \frac{1}{d_{col,j}} \cdot [\sqrt{2k}(\tilde{S}_k(\boldsymbol{X}) + \tilde{S}_k(\boldsymbol{Y})) + \varepsilon]^2.$$

Then using the conditions on the block sizes and the row- and column-sums of Theorem 3.3.8, provided

$$\sqrt{2k}(\tilde{S}_k(\boldsymbol{X}) + \tilde{S}_k(\boldsymbol{Y})) + \varepsilon = \mathcal{O}\left(\frac{1}{\min\{m, n\}^{\frac{1}{2}+\tau}} \right)$$

holds with some 'small' $\tau > 0$, the relation $\bar{c}_{ab} - \hat{c}_{ab} \to 0$ also holds as $n, m \to \infty$. Therefore, both $\hat{c}_{ab}$ and $\hat{c}_{ab} d_{row,i} d_{col,j}$ will be positive if m and n are large enough, over blocks that are not constantly zero in the original table.

3.3.3 Directed graphs

We can consider quadratic, but not symmetric contingency tables with zero diagonal as edge-weight matrices of directed graphs. The $n \times n$ edge-weight matrix $\boldsymbol{W}$ of a directed graph has zero diagonal, but is usually not symmetric: w_{ij} is the weight of the $i \to j$ edge $(i, j = 1, \dots, n;\ i \neq j)$. Equivalently, for the pair $i < j$ of vertices, w_{ij} is the weight of the $i \to j$, whereas w_{ji} is that of the $j \to i$ edge. In this setup, the generalized in- and out-degrees are

$$d_{in,j} = \sum_{i=1}^{n} w_{ij} \quad (j = 1, \dots, n) \quad \text{and} \quad d_{out,i} = \sum_{j=1}^{n} w_{ij} \quad (i = 1, \dots, n);$$

further, $\boldsymbol{D}_{in} = \operatorname{diag}(d_{in,1}, \dots, \mathbf{d}_{in,n})$ and $\boldsymbol{D}_{out} = \operatorname{diag}(d_{out,1}, \dots, \mathbf{d}_{out,n})$ are the in- and out-degree matrices. Assume that there are no sources and sinks (i.e., no identically zero rows or columns), and that $\boldsymbol{W}$ is non-decomposable. Then the correspondence matrix belonging to $\boldsymbol{W}$ is

$$\boldsymbol{W}_{corr} = \boldsymbol{D}_{out}^{-1/2} \boldsymbol{W} \boldsymbol{D}_{in}^{-1/2},$$

and its SVD is used to minimize the normalized bicut of $\boldsymbol{W}$ as a contingency table, see Section 2.5. In this way, for given k, we obtain regular in- and out-cluster pairs in the following sense.

The V_{in}, V_{out} in- and out-vertex cluster pair of the directed graph (with sum of the weights of directed edges 1) is α-volume regular if for all $X \subset V_{out}$ and $Y \subset V_{in}$ the relation

$$|w(X,Y) - \rho(V_{out}, V_{in})\mathrm{Vol}_{out}(X)\mathrm{Vol}_{in}(Y)| \leq \alpha\sqrt{\mathrm{Vol}_{out}(V_{out})\mathrm{Vol}_{in}(V_{in})}$$

holds, where the *directed cut* $w(X,Y)$ is the sum the weights of the $X \to Y$ edges, $\mathrm{Vol}_{out}(X) = \sum_{i \in X} d_{out,i}$, $\mathrm{Vol}_{in}(Y) = \sum_{j \in Y} d_{in,j}$, and $\rho(V_{out}, V_{in}) = \frac{w(V_{out}, V_{in})}{\mathrm{Vol}_{out}(V_{out})\mathrm{Vol}_{in}(V_{in})}$ is the relative inter-cluster density of the out–in cluster pair V_{out}, V_{in}. The clustering $(V_{in,1}, \ldots, V_{in,k})$ and $(V_{out,1}, \ldots, V_{out,k})$ of the columns and rows – guaranteed by Theorem 3.3.8 – corresponds to in- and out-clusters of the same vertex set such that the directed information flow $V_{out,a} \to V_{in,b}$ is as homogeneous as possible for all $a, b = 1, \ldots, k$ pairs.

References

Achlioptas D and McSherry F 2007 Fast computation of low-rank matrix approximations. *J. ACM* **54** (2), Article 9.

Almaas E, Kovács B, Vicsek T, Oltvai ZN and Barabási A-L 2004 Global organization of the metabolic Fluxes in the bacterium escherichia coli. *Nature* **427**, 839–843.

Alon N 1986 Eigenvalues and expanders. *Combinatorica* **6**, 83–96.

Alon N, Duke RA, Lefmann H, Rödl V and Yuster R 1994 The algotithmic aspects of the Regularity Lemma. *J. Algorithms* **16**, 80–109.

Alon N and Spencer JH 2000 *The Probabilistic Method*. John Wiley & Sons, Ltd.

Alon N, Coja-Oghlan A, Han H, Kang M, Rödl V and Schacht M 2010 Quasi-randomness and algorithmic regularity for graphs with general degree distributions. *Siam J. Comput.* **39** (6), 2336–2362.

Barabási A-L and Albert R 1999 Emergence of scaling in random networks. *Science* **286**, 509–512.

Bickel PJ and Chen A 2009 A nonparametric view of network models and Newman-Girvan and other modularities. *Proc. Natl. Acad. Sci. USA* **106** (50), 21068–21073.

Bilu Y and Linial N 2006 Lifts, discrepancy and nearly optimal spectral gap. *Combinatorica* **26** (5), 495–519.

Bolla M 2004 Distribution of the eigenvalues of random block-matrices *Linear Algebra Appl.* **377**, 219–240.

Bolla M 2005 Recognizing linear structure in noisy matrices. *Linear Algebra Appl.* **402**, 228–244.

Bolla M 2008 Noisy random graphs and their Laplacians. *Discret. Math.* **308**, 4221–4230.

Bolla M 2011a Beyond the expanders. *Int. J. Comb.*, 787596.

Bolla M 2011b Spectra and structure of weighted graphs. *Electronic Notes in Discrete Mathematics* **38**, 149–154.

Bolla M, Friedl K and Krámli A 2010 Singular value decomposition of large random matrices (for two-way classification of microarrays). *J. Multivariate Anal.* **101**, 434–446.

Bollobás B 2001 *Random Graphs*, 2nd edn Cambridge University Press, Cambridge.

Bollobás B, Riordan O, Spencer J and Tusnády G 2001 The degree sequence of a scale-free random graph process. *Random Struct. Algorithms* **18** (1), 279–290.

Bollobás B and Riordan O 2003 Robustness and vulnerability of scale-free random graphs. *Internet Math.* **1** (1), 1–35.

Bollobás B and Nikiforov V 2004 Hermitian matrices and graphs: singular values and discrepancy. *Discret. Math.* **285**, 17–32.

Bollobás B, Janson S and Riordan O 2007 The phase transition in inhomogeneous random graphs. *Random Struct. Algorithms* **31**, 3–122.

Butler S 2006 Using discrepancy to control singular values for nonnegative matrices. *Linear Algebra Appl.* **419**, 486–493.

Chaudhuri K, Chung F and Tsiatas A 2012 Spectral clustering of graphs with general degrees in the extended planted partition model, in *JMLR Workshop and Conference Proceedings. Proc. 25th Annual Conference on Learning Theory (COLT 2012)* (eds Mannor S, Srebro N and Williamson RC), Vol. 23, Edinburgh, Scotland, pp. 35.1–35.23.

Choi DS, Wolfe PJ and Airoldi EM 2012 Stochastic blockmodels with growing number of classes. *Biometrika* **99** (2), 273–284.

Chung F 1997 *Spectral Graph Theory*, CBMS Regional Conference Series in Mathematics **92**, American Mathematical Society, Providence RI.

Chung F and Graham R 2008 Quasi-random graphs with given degree sequences, *Random Struct. Algorithms* **12**, 1–19.

Chung F, Graham RL and Wilson RK 1989 Quasi-random graphs, *Combinatorica* **9**, 345–362.

Chung F, Lu L and Vu V 2003 Eigenvalues of random power law graphs. *Ann. Comb.* **7**, 21–33.

Coja-Oghlan A 2010 Graph partitioning via adaptive spectral techniques. *Combin. Probab. Comput.* **19** (2), 227–284.

Coja-Oghlan A and Lanka A 2009a The spectral gap of random graphs with given expected degrees. *Electron. J. Comb.* **16** (1), R138.

Coja-Oghlan A and Lanka A 2009b Finding planted partitions in random graphs with general degree distributions. *J. Discret. Math.* **23** (4), 1682–1714.

Dhillon IS 2001 Co-clustering documents and words using bipartite spectral graph partitioning, in *Proc. ACM Int. Conf. Knowledge Disc. Data Mining (KDD 2001)* (eds Provost FJ and Srikant R), Association for Computer Machinery, New York, pp. 269–274.

Drineas P, Frieze A, Kannan R *et al.* 2004 Clustering large graphs via the singular value decomposition. *Mach. Learn.* **56**, 9–33.

Erdős P and Rényi A 1960 On the evolution of random graphs. *Publ. Math. Inst. Hung. Acad. Sci.* **5**, 17–61.

Farkas IJ, Derényi I, Barabási, A-L. and Vicsek T 2001 Spectra of 'Real-World' Graphs: Beyond the Semi-Circle Law. *Phys. Rev. E* **64**, 026704.

Frieze A and Kannan R 1999 Quick approximation to matrices and applications. *Combinatorica* **19** (2), 175–220.

Füredi Z and Komlós J 1981 The eigenvalues of random symmetric matrices. *Combinatorica* **1**, 233–241.

Golub GH and van Loan CF 1996 *Matrix Computations*, 3rd edn. Johns Hopkins University Press, Baltimore, MD.

Granovetter MS 1973 The strength of weak ties. *Amer. J. Sociology* **78**, 1360–1380.

Gundert A and Wagner U 2012 On Laplacians of random complexes, in *Proc. 28th Annual ACM Symposium on Computational Geometry, SoCG* (eds Dey TK and Whitesides S), pp. 151–160.

Hoeffding W 1963 Probability inequalities for sums of bounded random variables. *J. Am. Stat. Assoc.* **58** (301), 13–30.

Holland P, Laskey KB and Leinhardt S 1983 Stochastic blockmodels: some first steps. *Social Networks* **5**, 109–137.

Hoory S, Linial N and Widgerson A 2006 Expander graphs and their applications. *Bull. Amer. Math. Soc. (N. S.)* **43** (4), 439–561.

Juhász F 1996 On the structural eigenvalues of block random matrices. *Linear Algebra Appl.* **246**, 225–231.

Juhász F and Mályusz K 1980 Problems of cluster analysis from the viewpoint of numerical analysis, in *Numerical Methods, Coll. Math. Soc. J. Bolyai* (ed. Rózsa P), North-Holland, Amsterdam, vol 22, pp. 405–415.

Kannan R, Salmasian H and Vempala S 2005 The spectral method for general mixture models, in *Proc. 18th Annual Conference on Learning Theory* (eds Auer P and Meir R), Bertinoro, Italy, pp. 444–457.

Karrer B and Newman MEJ 2011 Stochastic blockmodels and community structure in networks. *Phys. Rev. E* **83**, 016107.

Karrer B Levina E and Newman MEJ 2008 Robustness of community structure in networks. *Phys. Rev. E* **77**, 046119.

Kluger Y, Basri R, Chang JT and Gerstein M 2003 Spectral biclustering of microarray data: coclustering genes and conditions. *Genome Research* **13**, 703–716.

Komlós J, Shokoufanden A, Simonovits M and Szemerédi E 2002 Szemerédi's regularity lemma and its applications in graph theory, in *Lecture Notes in Computer Science*, Springer, Berlin, vol 2292, pp. 84–112.

Leskovec J, Lang KI, Dasgupta A and Mahoney MW 2008 Statistical properties of community structure in large social and information networks, in *Proc. 17th International Conference on World Wide Web, Bejing, China*, ACM New York NY, pp. 695–704.

Leskovec J, Lang KI, Dasgupta A and Mahoney MW 2009 Community structure in large networks: natural cluster sizes and the absence of well-defined clusters. *Internet Math.* **6** (1), 29–123.

Lovász L 2008 Very large graphs, in *Current Developments in Mathematics* (eds Jerison D, Mazur B, Mrowka T *et al.*), International Press, Somerville, MA, pp. 67–128.

Lovász L and T.-Sós V 2008 Generalized quasirandom graphs. *J. Comb. Theory B* **98**, 146–163.

Martella F and Vichi M Clustering microarray data using model-based double K-means. *J. Appl. Stat.*, **39** (9), 1853–1869.

McSherry F 2001 Spectral partitioning of random graphs, in *Proc/42nd Annual Symposium on Foundations of Computer Science (FOCS 2001)*, Las Vegas, Nevada, pp. 529–537.

Meshulam R and Wallach N 2009 Homological connectivity of random k-dimensional complexes. *Random Struct. Algorithms* **34** (3), 408–417.

Móri TF 2005 The maximum degree of the Barabási–Albert random tree. *Combin. Probab. Comput.* **14** (3), 339–348.

Parzanchevski O, Rosenthal R and Tessler J 2012 Isoperimetric inequalities in simplicial complexes. arXiv:math.CO/1207.0638.

Rohe K, Chatterjee S and Yu B 2011 Spectral clustering and the high-dimensional stochastic blockmodel. *Ann. Stat.* **39** (4), 1878–1915.

Simonovits M and T.-Sós V 1991 Szemerédi's partition and quasi-randomness. *Random Struct. Algorithms* **2**, 1–10.

Söderberg B 2003 Properties of random graphs with hidden color. *Phys. Rev. E* **68**, 026107.

Stewart GW and Sun J-g. 1990 *Matrix Pertubation Theory*, Academic Press.

Szemerédi E 1978 Regular partitions of graphs, in *Colloque Inter. CNRS. No. 260, Problémes Combinatoires et Théorie Graphes* (eds Bermond J-C, Fournier J-C, Las Vergnas M and Sotteau D), CNRS, Paris, pp. 399–401.

Tao T 2006 Szemerédi's regularity lemma revisited. *Contrib. Discret. Math.* **1**, 8–28.

von Luxburg U Belkin M and Bousquet O 2008 Consistency of spectral clustering. *Ann. Stat.* **36**, 555–586.

Wasserman S and Faust K 1994 *Social Network Analysis: Methods and Applications*, Cambridge Univ. Press, Cambridge.

4

Testable graph and contingency table parameters

In this chapter, the theory of convergent graph sequences, elaborated by graph theorists (see e.g., Borgs *et al.* (2006, 2008); Elek (2008); Lovász and Szegedy (2006)), will be used for vertex- and edge-weighted graphs and contingency tables. Roughly speaking, graphs and contingency tables of a convergent sequence become more and more similar to each other in small details, which fact is made exact in terms of the convergence of homomorphism densities when 'small' simple graphs or binary tables are mapped into the 'large' networks of the sequence. The convergence can as well be formulated in terms of the cut-distance, and limit objects are defined. This cut-distance also makes it possible to classify graphs and contingency tables, or to assign them to given prototypes. These issues will be further discussed in Chapter 5.

Testable parameters are, in fact, nonparametric statistics defined on graphs and contingency tables that can be consistently estimated based on a smaller sample, selected randomly from the underlying huge network. Real-world graphs or rectangular arrays are sometimes considered as samples from a large network, and we want to conclude the network's parameters from the same parameters of its smaller parts. The theory guarantees that this can be done if the investigated parameter is testable. We will prove that certain balanced minimum multiway cut densities are indeed testable.

4.1 Convergent graph sequences

Let $G = G_n$ be a weighted graph on the vertex set $V(G) = \{1, \dots, n\} = [n]$ and edge set $E(G)$. Both the edges and vertices have weights: the edge-weights are pairwise similarities $\beta_{ij} = \beta_{ji} \in [0, 1]$, $i, j \in [n]$ between the vertices, while the vertex-weights

Spectral Clustering and Biclustering: Learning Large Graphs and Contingency Tables, First Edition.
Marianna Bolla.

$\alpha_i > 0$ ($i \in [n]$) indicate relative significance of the vertices. For example, in a social network, the edge-weights are pairwise associations between people, while the vertex-weights can be individual abilities in the same context in which relations between them exist; for example in the actors' network, relative frequencies of costarring of actors are the pairwise similarities, whereas individual abilities of the actors are their weights. In strategic interaction networks, the edge-weights represent the mutual effect of the pairs of individuals on each other's strategies, while vertex-weights are actions of the individuals that they would follow themselves, without being aware of actions of their neighbors (neighbors are persons to whom an individual is connected with positive weights), see for example Ballester *et al.* (2006).

It is important that the edge-weights are nonnegative ($\beta_{ij} = 0$ means that vertices i and j are not connected at all), the normalization into the [0,1] interval is for the sake of treating them later as probabilities for random sampling. A simple graph has edge-weights 0 or 1, and vertex-weights all 1. Note that the edge-weights resemble the w_{ij}'s used in Chapter 1, but here loops may also be present, namely, if $\beta_{ii} > 0$ for some i. If all the edge-weights are positive, our graph is called *soft-core*. Soft-core graphs are dense, and among the simple graphs they are the complete graphs.

Let $\mathcal{G}$ denote the set of all such edge- and vertex-weighted graphs. The *volume* of $G \in \mathcal{G}$ is defined by $\alpha_G = \sum_{i=1}^{n} \alpha_i$, while that of the vertex-subset S by $\alpha_S = \sum_{i \in S} \alpha_i$. Note that this notion of volume coincides with that of Chapter 2 only if the vertex-weights are the generalized degrees. Further,

$$e_G(S, T) = \sum_{s \in S} \sum_{t \in T} \alpha_s \alpha_t \beta_{st}$$

denotes the *vertex- and edge-weighted cut* between the (not necessarily disjoint) vertex-subsets S and T, which is equal to $w(S, T)$ of Chapter 2 only if the vertex-weights are all 1. However, for brevity, we will also call $e_G(S, T)$ weighted cut throughout this chapter.

Borgs *et al.* (2008) define the *homomorphism density* between the simple graph F (on vertex set $V(F) = [k]$) and the above weighted graph G as

$$t(F, G) = \frac{1}{(\alpha_G)^k} \sum_{\Phi: V(F) \to V(G)} \prod_{i=1}^{k} \alpha_{\Phi(i)} \prod_{ij \in E(F)} \beta_{\Phi(i)\Phi(j)}.$$

Note that under homomorphism, an edge-preserving map is understood in the following sense. If F is a simple and G is an edge-weighted graph, then $\Phi : V(F) \to V(G)$ is a homomorphism when for every $ij \in E(F)$, $\beta_{\Phi(i)\Phi(j)} > 0$. Therefore, in the above summation, a zero term will correspond to a Φ which is not a homomorphism. For 'large' n, Borgs *et al.* (2008) relate this quantity to the probability that the following sampling from G results in F: k vertices of G are selected with replacement out of the n ones, with respective probabilities α_i/α_G ($i = 1, \dots, n$); given the vertex-subset $\{\Phi(1), \dots, \Phi(k)\}$, the edges come into existence conditionally independently, with probabilities of the edge-weights. Such a *random simple graph is denoted by* $\xi(k, G)$. In the sequel only the $k \ll n$ case – when $t(F, G)$ is very close to to the probability that the above sampling from G results in F – makes sense, and this is the situation we need: k is kept fixed, while n tends to infinity. (For a more precise formulation with induced and injective homomorphisms, see Borgs *et al.* (2008)).

Definition 4.1.1 *The weighted graph sequence* (G_n) *is (left-)convergent if the sequence* $t(F, G_n)$ *converges for any simple graph* F *as* $n \to \infty$.

As other kinds of convergence are not discussed here, in the sequel, the word left will be omitted, and we simply use convergence. Note that Borgs *et al.* (2012) define other kinds of graph convergence too, together with equivalences and implications between them.

Lovász and Szegedy (2006) also construct the limit object that is a symmetric, bounded, measurable function $W : [0, 1] \times [0, 1] \to \mathbb{R}$, called *graphon*. Let $\mathcal{W}$ denote the set of these functions. The interval [0,1] corresponds to the vertices and the values $W(x, y) = W(y, x)$ to the edge-weights. In view of the conditions imposed on the edge-weights, the range is also the [0,1] interval. The set of symmetric, measurable functions $W : [0, 1] \times [0, 1] \to [0, 1]$ is denoted by $\mathcal{W}_{[0,1]}$. The stepfunction graphon $W_G \in \mathcal{W}_{[0,1]}$ is assigned to the weighted graph $G \in \mathcal{G}$ in the following way: the sides of the unit square are divided into intervals $I_1, \dots, I_n$ of lengths $\alpha_1/\alpha_G, \dots, \alpha_n/\alpha_G$, and over the rectangle $I_i \times I_j$ the stepfunction takes on the value β_{ij}.

The so-called *cut distance* between the graphons W and U is

$$\delta_\square(W, U) = \inf_\nu \|W - U^\nu\|_\square \tag{4.1}$$

where the *cut-norm* of the graphon $W \in \mathcal{W}$ is defined by

$$\|W\|_\square = \sup_{S,T \subset [0,1]} \left| \iint_{S \times T} W(x, y)\, dx\, dy \right| ,$$

and the infimum in (4.1) is taken over all measure-preserving bijections $\nu : [0, 1] \to [0, 1]$, while U^ν denotes the transformed U after performing the same measure-preserving bijection ν on both sides of the unit square. (You can also think of ν such that for a uniformly distributed random variable ξ over $(0, 1)$, $\nu(\xi)$ has the same distribution.) An equivalence relation is defined over the set of graphons: two graphons belong to the same class if they can be transformed into each other by a measure-preserving bijection, that is, their $\delta_\square$ distance is zero. In the sequel, we consider graphons modulo measure preserving maps, and under graphon we understand the whole equivalence class. By Theorem 5.1 of Lovász and Szegedy (2007), the classes of $\mathcal{W}_{[0,1]}$ form a compact metric space with the $\delta_\square$ metric. Based on this fact, the authors give an analytic proof for the weak version of the Szemerédi's regularity lemma (for the strong version see Theorem 3.1.20, and for the weak one the forthcoming Lemma 4.3.1).

We will intensively use the following reversible relation between convergent weighted graph sequences and graphons (see Corollary 3.9 of Borgs *et al.* (2008)).

Fact 4.1.2 *For any convergent sequence (G_n) of weighted graphs with uniformly bounded edge-weights there exists a graphon such that $\delta_\square(W_{G_n}, W) \to 0$. Conversely, any graphon W can be obtained as the limit of a sequence of weighted graphs with uniformly bounded edge-weights. The limit of a convergent graph sequence is essentially unique: if $G_n \to W$, then also $G_n \to W'$ for precisely those graphons W' for which $\delta_\square(W, W') = 0$.*

Authors of Borgs *et al.* (2008, 2011) define the $\delta_\square$ distance of two weighted graphs and that of a graphon and a graph in the following way. For the weighted graphs G, G', and for the graphon W

$$\delta_\square(G, G') = \delta_\square(W_G, W_{G'}) \quad \text{and} \quad \delta_\square(W, G) = \delta_\square(W, W_G).$$

Theorem 2.6 of Borgs *et al.* (2008) states that a sequence of weighted graphs with uniformly bounded edge-weights is convergent if and only if it is a Cauchy sequence in the metric $\delta_\square$.

A simple graph on k vertices can be sampled based on W in the following way: k uniform random numbers, $X_1, \ldots, X_k$ are generated on [0,1] independently. Then we connect the vertices corresponding to X_i and X_j with probability $W(X_i, X_j)$. For the so obtained simple graph $\xi(k, W)$ the following large deviation result is proved (see Theorem 4.7 of Borgs *et al.* (2008)).

Fact 4.1.3 *Let k be a positive integer and $W \in \mathcal{W}_{[0,1]}$ be a graphon. Then with probability at least $1 - e^{-k^2/(2\log_2 k)}$, we have*

$$\delta_\square(W, \xi(k, W)) \leq \frac{10}{\sqrt{\log_2 k}}. \tag{4.2}$$

Fixing k, Inequality (4.2) holds uniformly for any graphon $W \in \mathcal{W}_{[0,1]}$, especially for W_G. Further, the sampling from W_G is identical to the previously defined sampling with replacement from G, that is, $\xi(k, G)$ and $\xi(k, W_G)$ are identically distributed. In fact, this argument is relevant in the $k \leq |V(G)|$ case.

4.2 Testability of weighted graph parameters

A function $f : \mathcal{G} \to \mathbb{R}$ is called a *graph parameter* if it is invariant under isomorphism. In fact, a graph parameter is a statistic evaluated on the graph, and hence, we are interested in weighted graph parameters that are not sensitive to minor changes in the weights of the graph. By the definition of Borgs *et al.* (2008), the simple graph parameter f is testable if for every $\varepsilon > 0$ there is a positive integer k such that for every simple graph G on at least k vertices,

$$\boldsymbol{P}(|f(G) - f(\xi(k, G))| > \varepsilon) \leq \varepsilon,$$

where $\xi(k, G)$ is a random simple graph on k vertices selected randomly from G as described above. Then they prove equivalent statements of testability for simple graphs.

These results remain valid if we consider weighted graph sequences (G_n) with *no dominant vertex-weights*, that is, $\max_i \frac{\alpha_i(G_n)}{\alpha_{G_n}} \to 0$ as $n \to \infty$. To use this condition imposed on the vertex-weights, Bolla *et al.* (2012) slightly modified the definition of a testable graph parameter for weighted graphs.

Definition 4.2.1 *A weighted graph parameter f is testable if for every $\varepsilon > 0$ there is a positive integer k such that if $G \in \mathcal{G}$ satisfies*

$$\max_i \frac{\alpha_i(G)}{\alpha_G} \le \frac{1}{k}, \tag{4.3}$$

then

$$\boldsymbol{P}(|f(G) - f(\xi(k, G))| > \varepsilon) \le \varepsilon,$$

where $\xi(k, G)$ is a random simple graph on k vertices selected randomly from G as described in Section 4.1.

Note that for simple G, Inequality (4.3) implies that $|V(G)| \ge k$, and we get back the definition applicable to simple graphs.

By the above definition, a testable graph parameter can be consistently estimated based on a fairly large sample. As the randomization depends only on the $\alpha_i(G)/\alpha_G$ ratios, it is not able to distinguish between weighted graphs whose vertex-weights differ only in a constant factor. Thus, a testable weighted graph parameter is invariant under scaling the vertex-weights.

As a straightforward generalization of Theorem 6.1 in Borgs *et al.* (2008), Bolla *et al.* (2012) introduced the following equivalent statements of the testability for weighted graphs. We state this theorem without proof as it uses the ideas of the proof in Borgs *et al.* (2008), where some details for such a generalization are also elaborated.

Theorem 4.2.2 *For the weighted graph parameter f the following are equivalent:*

(a) f is testable.

(b) For every $\varepsilon > 0$ there is a positive integer k such that for every weighted graph $G \in \mathcal{G}$ satisfying the node-condition $\max_i \alpha_i(G)/\alpha_G \le 1/k$,

$$|f(G) - \boldsymbol{E}(f(\xi(k, G)))| \le \varepsilon.$$

(c) For every convergent weighted graph sequence (G_n) with $\max_i \alpha_i(G_n)/\alpha_{G_n} \to 0$, *$f(G_n)$ is also convergent ($n \to \infty$).*

(d) f can (essentially uniquely) be extended to graphons such that the graphon functional $\tilde{f}$ is continuous in the cut-norm and $\tilde{f}(W_{G_n}) - f(G_n) \to 0$, whenever $\max_i \alpha_i(G_n)/\alpha_{G_n} \to 0$ *($n \to \infty$).*

(e) *For every* $\varepsilon > 0$ *there is an* $\varepsilon_0 > 0$ *real and an* $n_0 > 0$ *integer such that if* G_1, G_2 *are weighted graphs satisfying* $\max_i \alpha_i(G_1)/\alpha_{G_1} \le 1/n_0$, $\max_i \alpha_i(G_2)/\alpha_{G_2} \le 1/n_0$, *and* $\delta_\square(G_1, G_2) < \varepsilon_0$, *then* $|f(G_1) - f(G_2)| < \varepsilon$ *also holds.*

This theorem indicates that a testable parameter depends continuously on the whole graph, and hence, it is not sensitive to minor changes in the edge-weights. Some of these equivalences will be used in the proofs of the next section.

4.3 Testability of minimum balanced multiway cuts

The testability of the maximum cut density is stated in Borgs *et al.* (2006) based on earlier algorithmic results of Arora *et al.* (1995); Frieze and Kannan (1999); Goldreich *et al.* (1998). Inspired by Chapter 2, we are rather interested in the minimum cut density, which is somewhat different. We will show that it trivially tends to zero as the number of the graph's vertices tends to infinity, whereas the normalized version of it (cuts are penalized by the volumes of the clusters they connect) is not testable. For example, if a single vertex is loosely connected to a dense part, the minimum cut density of the whole graph is 'small', however, randomizing a smaller sample, with high probability, it comes from the dense part with a 'large' minimum normalized cut density. Nonetheless, if we impose conditions on the cluster volumes in anticipation, the so obtained balanced minimum cut densities are testable. Balanced multiway cuts are frequently looked for in contemporary cluster analysis when we want to find groups of a large network's vertices with sparse inter-cluster connections, where the clusters do not differ significantly in sizes.

For the proofs of the testability results we use Theorem 4.2.2 and some notion of statistical physics in the same way as in Borgs *et al.* (2012).

Let $G \in \mathcal{G}$ be a weighted graph on n vertices with vertex-weights $\alpha_1, \dots, \alpha_n$ and edge-weights β_{ij}'s. Let $q \le n$ be a fixed positive integer, and $\mathcal{P}_q$ denote the set of q-partitions $P = (V_1, \dots, V_q)$ of the vertex set V. The non-empty, disjoint vertex-subsets sometimes are referred to as clusters or states. The *factor graph* or *q-quotient* of G with respect to the q-partition P is denoted by G/P and it is defined as the weighted graph on q vertices with vertex- and edge-weights

$$\alpha_i(G/P) = \frac{\alpha_{V_i}}{\alpha_G} \quad (i \in [q]) \quad \text{and} \quad \beta_{ij}(G/P) = \frac{e_G(V_i, V_j)}{\alpha_{V_i}\alpha_{V_j}} \quad (i, j \in [q]),$$

respectively.

In terms of the factor graph, the following weak version of the Szemerédi's regularity lemma (see Theorem 3.1.20) is stated in Lemma 2.4 of Borgs *et al.* (2008).

Lemma 4.3.1 (Weak regularity lemma) *For every* $\varepsilon > 0$, *every weighted graph* G *has a partition* P *into at most* $4^{\frac{1}{\varepsilon^2}}$ *clusters such that*

$$\delta_\square(G, G/P) \le \varepsilon \|G\|_2$$

where

$$\|G\|_2 = \left(\sum_{i,j} \frac{\alpha_i \alpha_j}{\alpha_G^2} \beta_{ij}^2 \right)^{1/2} .$$

For the proof, see Lovász and Szegedy (2007). Moreover, Lovász (2008) gives an algorithm to compute a weak Szemerédi partition in a huge graph. The way of presenting the output of the algorithm for a large graph was formerly proposed by Frieze and Kannan (1999).

Let $\hat{\mathcal{S}}_q(G)$ denote the set of all q-quotients of G. The *Hausdorff distance* between $\hat{\mathcal{S}}_q(G)$ and $\hat{\mathcal{S}}_q(G')$ is defined by

$$\begin{aligned} &d^{\mathrm{Hf}}(\hat{\mathcal{S}}_q(G), \hat{\mathcal{S}}_q(G')) \\ &= \max\{ \sup_{H \in \hat{\mathcal{S}}_q(G)} \inf_{H' \in \hat{\mathcal{S}}_q(G')} d_1(H, H'), \sup_{H' \in \hat{\mathcal{S}}_q(G')} \inf_{H \in \hat{\mathcal{S}}_q(G)} d_1(H, H')\} \end{aligned}$$

where

$$\begin{aligned} d_1(H, H') = &\sum_{i,j \in [q]} \left| \frac{\alpha_i(H)\alpha_j(H)\beta_{ij}(H)}{\alpha_H^2} - \frac{\alpha_i(H')\alpha_j(H')\beta_{ij}(H')}{\alpha_{H'}^2} \right| \\ &+ \sum_{i \in [q]} \left| \frac{\alpha_i(H)}{\alpha_H} - \frac{\alpha_i(H')}{\alpha_{H'}} \right| \end{aligned}$$

is the l^1-distance between two weighted graphs H and H' on the same number of vertices. (If especially, H and H' are factor graphs, then $\alpha_H = \alpha_{H'} = 1$.)

Given the real symmetric $q \times q$ matrix $\boldsymbol{J}$ and the vector $\mathbf{h} \in \mathbb{R}^q$, the partitions $P \in \mathcal{P}_q$ also define a spin system on the weighted graph G. The so-called *ground state energy* (Hamiltonian) of such a spin configuration is

$$\hat{\mathcal{E}}_q(G, \boldsymbol{J}, \mathbf{h}) = -\max_{P \in \mathcal{P}_q} \left(\sum_{i \in [q]} \alpha_i(G/P) h_i + \sum_{i,j \in [q]} \alpha_i(G/P)\alpha_j(G/P)\beta_{ij}(G/P) J_{ij} \right)$$

where $\boldsymbol{J}$ is the so-called *coupling-constant matrix* with J_{ij} representing the strength of interaction between states i and j, and $\mathbf{h}$ is the magnetic field. They carry physical meaning. We will use only special $\boldsymbol{J}$ and $\mathbf{h}$.

Sometimes, we need balanced q-partitions to regulate the proportion of the cluster volumes. A slight balancing between the cluster volumes is achieved by fixing a positive real number c ($c \le 1/q$). Let $\mathcal{P}_q^c$ denote the set of q-partitions of V such that $\frac{\alpha_{V_i}}{\alpha_G} \ge c \quad (i \in [q])$, or equivalently, $c \le \frac{\alpha_{V_i}}{\alpha_{V_j}} \le \frac{1}{c}$ $(i \ne j)$. A more accurate balancing is defined by fixing a probability vector $\mathbf{a} = (a_1, \dots, a_q)$ with components forming a probability distribution over $[q]$: $a_i > 0$ $(i \in [q])$, $\sum_{i=1}^q a_i = 1$. Let $\mathcal{P}_q^{\mathbf{a}}$ denote the set of q-partitions of V such that $\left(\frac{\alpha_{V_1}}{\alpha_G}, \dots, \frac{\alpha_{V_q}}{\alpha_G}\right)$ is approximately $\mathbf{a}$-distributed, that is $\left|\frac{\alpha_{V_i}}{\alpha_G} - a_i\right| \le \frac{\alpha_{\max}(G)}{\alpha_G}$ $(i = 1, \dots, q)$. Observe that the above difference tends to 0 as $|V(G)| \to \infty$ for weighted graphs with no dominant vertex-weights.

The *microcanonical ground state energy* of G given $\mathbf{a}$ and $\boldsymbol{J}$ ($\mathbf{h} = \mathbf{0}$) is

$$\hat{\mathcal{E}}_q^{\mathbf{a}}(G, \boldsymbol{J}) = -\max_{P \in \mathcal{P}_q^{\mathbf{a}}} \sum_{i,j \in [q]} \alpha_i(G/P)\alpha_j(G/P)\beta_{ij}(G/P)J_{ij}.$$

Theorem 2.14 and 2.15 of Borgs *et al.* (2012) state the following important facts.

Fact 4.3.2 *The convergence of the weighted graph sequence (G_n) with no dominant vertex-weights is equivalent to the convergence of its microcanonical ground state energies for any q, $\mathbf{a}$, and $\boldsymbol{J}$. Also, it is equivalent to the convergence of its q-quotients in Hausdorff distance for any q.*

Fact 4.3.3 *Under the same conditions, the convergence of the above (G_n) implies the convergence of its ground state energies for any q, $\boldsymbol{J}$, and $\mathbf{h}$.*

Using these facts, we investigate the testability of some special multiway cut densities defined in the forthcoming definitions.

Definition 4.3.4 *The minimum q-way cut density of G is*

$$f_q(G) = \min_{P \in \mathcal{P}_q} \frac{1}{\alpha_G^2} \sum_{i=1}^{q-1} \sum_{j=i+1}^{q} e_G(V_i, V_j),$$

the minimum c-balanced q-way cut density of G is

$$f_q^c(G) = \min_{P \in \mathcal{P}_q^c} \frac{1}{\alpha_G^2} \sum_{i=1}^{q-1} \sum_{j=i+1}^{q} e_G(V_i, V_j),$$

and the minimum $\mathbf{a}$-balanced q-way cut density of G is

$$f_q^{\mathbf{a}}(G) = \min_{P \in \mathcal{P}_q^{\mathbf{a}}} \frac{1}{\alpha_G^2} \sum_{i=1}^{q-1} \sum_{j=i+1}^{q} e_G(V_i, V_j).$$

Occasionally, we want to penalize cluster volumes that significantly differ. We therefore introduce the notions of minimum normalized cut densities.

Definition 4.3.5 *The minimum normalized q-way cut density of G is*

$$\mu_q(G) = \min_{P \in \mathcal{P}_q} \sum_{i=1}^{q-1} \sum_{j=i+1}^{q} \frac{1}{\alpha_{V_i} \cdot \alpha_{V_j}} \cdot e_G(V_i, V_j),$$

the minimum normalized c-balanced q-way cut density of G is

$$\mu_q^c(G) = \min_{P \in \mathcal{P}_q^c} \sum_{i=1}^{q-1} \sum_{j=i+1}^{q} \frac{1}{\alpha_{V_i} \cdot \alpha_{V_j}} \cdot e_G(V_i, V_j),$$

and the minimum normalized **a***-balanced q-way cut density of G is*

$$\mu_q^{\mathbf{a}}(G) = \min_{P \in \mathcal{P}_q^c} \sum_{i=1}^{q-1} \sum_{j=i+1}^{q} \frac{1}{\alpha_{V_i} \cdot \alpha_{V_j}} \cdot e_G(V_i, V_j).$$

Proposition 4.3.6 *$f_q(G)$ is testable for any $q \leq |V(G)|$.*

Proof. Observe that $f_q(G)$ is a special ground state energy: $f_q(G) = \hat{\mathcal{E}}_q(G, \boldsymbol{J}, \mathbf{0})$, where the magnetic field is $\mathbf{0}$ and the $q \times q$ symmetric matrix $\boldsymbol{J}$ is the following: $J_{ii} = 0$ $(i \in [q])$, further $J_{ij} = -1/2$ $(i \neq j)$. By Fact 4.3.3 and the equivalent statement (c) of Theorem 4.2.2, the minimum q-way cut density is testable for any q.

However, this statement is not of much use, since $f_q(G_n) \to 0$ as $n \to \infty$, in the lack of dominant vertex-weights. Indeed, the minimum q-way cut density is trivially estimated from above by

$$f_q(G_n) \leq (q-1)\frac{\alpha_{max}(G_n)}{\alpha_{G_n}} + \binom{q-1}{2}\left(\frac{\alpha_{max}(G_n)}{\alpha_{G_n}}\right)^2$$

that tends to 0 provided $\alpha_{\max}(G_n)/\alpha_{G_n} \to 0$ as $n \to \infty$.

Proposition 4.3.7 *$f_q^{\mathbf{a}}(G)$ is testable for any $q \leq |V(G)|$ and probability vector* **a** *over $[q]$.*

Proof. Choose $\boldsymbol{J}$ as in the proof of Proposition 4.3.6. In this way, $f_q^{\mathbf{a}}(G)$ is a special microcanonical ground state energy:

$$f_q^{\mathbf{a}}(G) = \hat{\mathcal{E}}_q^{\mathbf{a}}(G, \boldsymbol{J}). \tag{4.4}$$

Hence, by Fact 4.3.2, the convergence of (G_n) is equivalent to the convergence of $f_q^{\mathbf{a}}(G_n)$ for any q and any distribution **a** over $[q]$. Therefore, by the equivalent statement (c) of Theorem 4.2.2, the testability of the minimum **a**-balanced q-way cut density also follows.

Proposition 4.3.7 and Fact 4.3.2 together imply the following less obvious statement.

Proposition 4.3.8 *$f_q^c(G)$ is testable for any $q \leq |V(G)|$ and $c \leq 1/q$.*

Proof. Theorem 4.7 and Theorem 5.5 of Borgs *et al.* (2012) imply that for any two weighted graphs G, G'

$$\left|\hat{\mathcal{E}}_q^{\mathbf{a}}(G, \boldsymbol{J}) - \hat{\mathcal{E}}_q^{\mathbf{a}}(G', \boldsymbol{J})\right| \leq (3/2 + \kappa) \cdot d^{\mathrm{Hf}}(\hat{S}_q(G), \hat{S}_q(G')), \tag{4.5}$$

where $\kappa = o(\min\{|V(G)|, |V(G')|\})$ is a negligible small constant, provided the number of vertices of G and G' is sufficiently large. By Fact 4.3.2, we know that if (G_n) converges, its q-quotients also converge in Hausdorff distance, consequently, form a Cauchy sequence. This means that for any $\varepsilon > 0$ there is an N_0 such that for $n, m > N_0$: $d^{\mathrm{Hf}}(\hat{S}_q(G_n), \hat{S}_q(G_m)) < \varepsilon$. We want to prove that for $n, m > N_0$: $|f_q^c(G_n) - f_q^c(G_m)| < 2\varepsilon$. On the contrary, assume that there are $n, m > N_0$ such that $|f_q^c(G_n) - f_q^c(G_m)| \geq 2\varepsilon$. Say, $f_q^c(G_n) \geq f_q^c(G_m)$. Let $A := \{\mathbf{a} : a_i \geq c\,,\ i = 1, \ldots, q\}$ is the subset of special c-balanced distributions over $[q]$. On the one hand,

$$f_q^c(G_m) = \min_{\mathbf{a} \in A} f_q^{\mathbf{a}}(G_m) = f_q^{\mathbf{a}^*}(G_m)$$

for some $\mathbf{a}^* \in A$. On the other hand, by (4.4) and (4.5), $f_q^{\mathbf{a}^*}(G_n) - f_q^{\mathbf{a}^*}(G_m) \leq (\frac{3}{2} + \kappa)\varepsilon$, that together with the indirect assumption implies that $f_q^c(G_n) - f_q^{\mathbf{a}^*}(G_n) \geq (\frac{1}{2} - \kappa)\varepsilon > 0$ for this $\mathbf{a}^* \in A$. But this contradicts to the fact that $f_q^c(G_n)$ is the minimum of $f_q^{\mathbf{a}}(G_n)$'s over A. Thus, $f_q^c(G_n)$ is also a Cauchy sequence, and being a real sequence, is also convergent.

Now consider the normalized density $\mu_q(G) = \min_{P \in \mathcal{P}_q} \sum_{i=1}^{q-1} \sum_{j=i+1}^{q} \beta_{ij}(G/P)$. It is not testable as the following example shows: let $q = 2$ and G_n be a simple graph on n vertices such that about $\sqrt{n}$ vertices are connected with a single edge to the remaining vertices that form a complete graph. Then $\mu_2(G_n) \to 0$, but randomizing a sufficiently large part of the graph, with high probability, it will be a subgraph of the complete graph, whose minimum normalized 2-way cut density is of constant order. In the $q = 2$, $\alpha_i = 1$ $(\forall i)$ special case, $\mu_2(G)$ of a regular graph G is its normalized 2-way cut; consequently, the normalized cut is not testable either.

However, balanced versions of the minimum normalized q-way cut density are testable.

Proposition 4.3.9 *$\mu_q^{\mathbf{a}}(G)$ is testable for any $q \leq |V(G)|$ and probability vector* $\mathbf{a}$ *over $[q]$.*

Proof. By the definition of Hausdorff distance, the convergence of q-quotients guarantees the convergence of

$$\mu_q^{\mathbf{a}}(G) = \min_{P \in \mathcal{P}_q^{\mathbf{a}}} \sum_{i=1}^{q-1} \sum_{j=i+1}^{q} \beta_{ij}(G/P) \tag{4.6}$$

for any $\mathbf{a}$ and q in the following way. Let $\hat{\mathcal{S}}_q^{\mathbf{a}}(G)$ denote the set of factor graphs of G with respect to partitions in $\mathcal{P}_q^{\mathbf{a}}$. As a consequence of Lemma 4.5 and Theorem 5.4

of Borgs *et al.* (2012), for any two weighted graphs G, G'

$$\max_{\mathbf{a}} d^{Hf}(\hat{\mathcal{S}}_q^{\mathbf{a}}(G), \hat{\mathcal{S}}_q^{\mathbf{a}}(G')) \leq (3+\kappa) \cdot d^{Hf}(\hat{\mathcal{S}}_q(G), \hat{\mathcal{S}}_q(G')), \tag{4.7}$$

where $\kappa = o(\min\{|V(G)|, |V(G')|\})$.

By Fact 4.3.2, for a convergent graph-sequence (G_n), the sequence $\hat{\mathcal{S}}_q(G_n)$ converges, and by the inequality (4.7), $\hat{\mathcal{S}}_q^{\mathbf{a}}(G_n)$ also converges in Hausdorff distance for any distribution $\mathbf{a}$ over $[q]$. As they form a Cauchy sequence, $\forall \varepsilon\ \exists N_0$ such that for $n, m > N_0$

$$d^{Hf}(\hat{\mathcal{S}}_q^{\mathbf{a}}(G_n), \hat{\mathcal{S}}_q^{\mathbf{a}}(G_m)) < \varepsilon$$

uniformly for any $\mathbf{a}$. In view of the Hausdorff distance's definition, this means that for any q-quotient $H \in \hat{\mathcal{S}}_q^{\mathbf{a}}(G_n)$ there exists (at least one) q-quotient $H' \in \hat{\mathcal{S}}_q^{\mathbf{a}}(G_m)$, and vice versa, for any $H' \in \hat{\mathcal{S}}_q^{\mathbf{a}}(G_m)$ there exists (at least one) $H \in \hat{\mathcal{S}}_q^{\mathbf{a}}(G_n)$ such that $d_1(H, H') < \varepsilon$. Therefore, the maximum distance between the elements of the above pairs is less than ε. (Note that the symmetry in the definition of the Hausdorff distance is important: the pairing exhausts the sets even if they have different cardinalities.)

Using the fact that the vertex-weights of such a pair H and H' are almost the same (the coordinates of the vector $\mathbf{a}$), by the notation $a = \min_{i\in[q]} a_i$, the following argument is valid for n, m large enough:

$$\begin{aligned} 2a^2 \sum_{i\neq j} |\beta_{ij}(H) - \beta_{ij}(H')| &\leq \sum_{i,j=1}^{q} a^2 |\beta_{ij}(H) - \beta_{ij}(H')| \\ &\leq \sum_{i,j=1}^{q} |\alpha_i(H)\alpha_j(H)\beta_{ij}(H) - \alpha_i(H')\alpha_j(H')\beta_{ij}(H')| = d_1(H, H') < \varepsilon. \end{aligned}$$

Therefore

$$\left| \sum_{i=1}^{q-1} \sum_{j=i+1}^{q} \beta_{ij}(H) - \sum_{i=1}^{q-1} \sum_{j=i+1}^{q} \beta_{ij}(H') \right| < \frac{\varepsilon}{2a^2} := \varepsilon',$$

and because $\sum_{i=1}^{q-1}\sum_{j=i+1}^{q} \beta_{ij}(H)$ and $\sum_{i=1}^{q-1}\sum_{j=i+1}^{q} \beta_{ij}(H')$ are individual terms behind the minimum in (4.6), the above inequality holds for their minima over $\mathcal{P}_q^{\mathbf{a}}$ as well:

$$|\mu_q^{\mathbf{a}}(G_n) - \mu_q^{\mathbf{a}}(G_m)| < \varepsilon'. \tag{4.8}$$

Consequently, the sequence $\mu_q^{\mathbf{a}}(G_n)$ is a Cauchy sequence, and being a real sequence, is also convergent. Thus, $\mu_q^{\mathbf{a}}$ is testable.

Proposition 4.3.10 *$\mu_q^c(G)$ is testable for any $q \leq |V(G)|$ and $c \leq 1/q$.*

Proof. The proof is analogous to that of Proposition 4.3.8 using equation (4.8) instead of equation (4.5). By the pairing argument of the proof of Proposition 4.3.9, the real sequence $\mu_q^c(G_n)$ is a Cauchy sequence, and therefore, convergent. This immediately implies the testability of μ_q^c.

4.4 Balanced cuts and fuzzy clustering

In cluster analysis of large data sets, the testable parameters $f_q^c(G)$ and $\mu_q^c(G)$ have the greatest relevance as they require only a slight balancing between the clusters. Now, they are continuously extended to graphons by an explicit construction.

Proposition 4.4.1 *Let us define the graphon functional $\tilde{f}_q^c$ in the following way:*

$$\tilde{f}_q^c(W) := \inf_{Q\in\mathcal{Q}_q^c} \sum_{i=1}^{q-1} \sum_{j=i+1}^{q} \iint_{S_i\times S_j} W(x,y)\,dx\,dy = \inf_{Q\in\mathcal{Q}_q^c} \tilde{f}_q(W; S_1,\dots,S_q) \quad (4.9)$$

where the infimum is taken over all the c-balanced Lebesgue-measurable partitions $Q = (S_1,\dots,S_q)$ of [0,1]. For these, $\sum_{i=1}^q \lambda(S_i) = 1$ and $\lambda(S_i) \geq c \quad (i \in [q])$, where λ denotes the Lebesgue-measure, and $\mathcal{Q}_q^c$ the set of c-balanced q-partitions of [0,1]. We state that $\tilde{f}_q^c$ is the extension of f_q^c in the following sense: if (G_n) is a convergent weighted graph sequence with no dominant vertex-weights and edge-weights in the [0,1] interval, then denoting by W the essentially unique limit graphon of the sequence, $f_q^c(G_n) \to \tilde{f}_q^c(W)$ as $n \to \infty$.

Proof. First we show that $\tilde{f}_q^c$ is continuous in the cut-distance. As $\tilde{f}_q^c(W) = \tilde{f}_q^c(W^\nu)$, where $\nu : [0,1] \to [0,1]$ is a measure preserving bijection, it suffices to prove that to any ε we can find ε' such that for any two graphons W, U with $\|W - U\|_\square < \varepsilon'$, the relation $|\tilde{f}_q^c(W) - \tilde{f}_q^c(U)| < \varepsilon$ also holds. Indeed, by the definition of the cut-norm, for any Lebesgue-measurable q-partition $(S_1,\dots,S_q)$ of [0,1], the relation

$$\left| \iint_{S_i\times S_j} (W(x,y) - U(x,y))\,dx\,dy \right| \leq \varepsilon' \qquad (i \neq j)$$

holds. Summing up for the $i \neq j$ pairs

$$\left| \sum_{i=1}^{q-1} \sum_{j=i+1}^{q} \iint_{S_i\times S_j} W(x,y)\,dx\,dy - \sum_{i=1}^{q-1} \sum_{j=i+1}^{q} \iint_{S_i\times S_j} U(x,y)\,dx\,dy \right| \leq \binom{q}{2}\varepsilon'.$$

Therefore,

$$\begin{aligned} &\inf_{(S_1,\dots,S_q)\in\mathcal{Q}_q^c} \sum_{i=1}^{q-1} \sum_{j=i+1}^{q} \iint_{S_i\times S_j} W(x,y)\,dx\,dy \\ \geq &\inf_{(S_1,\dots,S_q)\in\mathcal{Q}_q^c} \sum_{i=1}^{q-1} \sum_{j=i+1}^{q} \iint_{S_i\times S_j} U(x,y)\,dx\,dy - \binom{q}{2}\varepsilon' \end{aligned}$$

and vice versa,

$$\inf_{(S_1,\dots,S_q)\in\mathcal{Q}_q^c} \sum_{i=1}^{q-1}\sum_{j=i+1}^{q} \iint_{S_i\times S_j} U(x,y)\,dx\,dy$$

$$\geq \inf_{(S_1,\dots,S_q)\in\mathcal{Q}_q^c} \sum_{i=1}^{q-1}\sum_{j=i+1}^{q} \iint_{S_i\times S_j} W(x,y)\,dx\,dy - \binom{q}{2}\varepsilon'.$$

Consequently the absolute difference of the two infima is bounded from above by $\binom{q}{2}\varepsilon'$. Thus, $\varepsilon' = \varepsilon/\binom{q}{2}$ will do.

Let (G_n) be a convergent weighted graph sequence with no dominant vertex-weights and edge-weights in the [0,1] interval. By Fact 4.1.2, there is an essentially unique graphon W such that $G_n \to W$, that is, $\delta_\square(W_{G_n}, W) \to 0$ as $n \to \infty$. By the continuity of $\tilde{f}_q^c$,

$$\tilde{f}_q^c(W_{G_n}) \to \tilde{f}_q^c(W), \qquad n \to \infty. \tag{4.10}$$

Assume that

$$\tilde{f}_q^c(W_{G_n}) = \tilde{f}_q(W_{G_n}; S_1^*, \dots, S_q^*),$$

that is the infimum in (4.9) is attained at the c-balanced Lebesgue-measurable q-partition $(S_1^*, \dots, S_q^*)$ of [0,1].

Let G_{nq}^* be the q-fold blow-up of G_n with respect to $(S_1^*, \dots, S_q^*)$. It is a weighted graph on at most nq vertices defined in the following way. Let $I_1, \dots, I_n$ be consecutive intervals of [0,1] such that $\lambda(I_j) = \alpha_j(G_n)$, $j = 1, \dots, n$. The weight of the vertex labeled by ju of G_{nq}^* is $\lambda(I_j \cap S_u^*)$, $u \in [q]$, $j \in [n]$, while the edge-weights are $\beta_{ju,iv}(G_{nq}^*) = \beta_{ji}(G_n)$. Trivially, the graphons W_{G_n} and $W_{G_{nq}^*}$ essentially define the same stepfunction, hence $\tilde{f}_q^c(W_{G_n}) = \tilde{f}_q^c(W_{G_{nq}^*})$. Therefore, by (4.10),

$$\tilde{f}_q^c(W_{G_{nq}^*}) \to \tilde{f}_q^c(W), \qquad n \to \infty. \tag{4.11}$$

As $\delta_\square(G_n, G_{nq}^*) = \delta_\square(W_{G_n}, W_{G_{nq}^*}) = 0$, by part (e) of Theorem 4.2.2 it follows that

$$|f_q^c(G_{nq}^*) - f_q^c(G_n)| \to 0, \qquad n \to \infty. \tag{4.12}$$

Finally, by the construction of G_{nq}^*, $\tilde{f}_q^c(W_{G_{nq}^*}) = f_q^c(G_{nq}^*)$, and hence,

$$|f_q^c(G_n) - \tilde{f}_q^c(W)| \leq |f_q^c(G_n) - f_q^c(G_{nq}^*)| + |\tilde{f}_q^c(W_{G_{nq}^*}) - \tilde{f}_q^c(W)|$$

which, in view of (4.11) and (4.12), implies the required statement.

The above continuous extension of $f_q^c(G)$ to graphons is essentially unique, and part (d) of Theorem 4.2.2 implies that for a weighted graph sequence (G_n) with $\max_i \frac{\alpha_i(G_n)}{\alpha_{G_n}} \to 0$, the limit relation $\tilde{f}_q^c(W_{G_n}) - f_q^c(G_n) \to 0$ also holds as $n \to \infty$. This gives rise to an approximation of the minimum c-balanced q-way cut density of a weighted graph on a large number of vertices with no dominant vertex-weights by the extended c-balanced q-way cut density of the stepfunction graphon assigned to the graph. In this way, the discrete optimization problem can be formulated as a quadratic programming task with linear equality and inequality constraints, and fuzzy clusters are obtained.

To this end, let us investigate a fixed weighted graph G on n vertices (n is large). To simplify notation we will drop the subscript n, and G in the arguments of the vertex- and edge-weights. As $f_q^c(G)$ is invariant under scaling the vertex-weights, we can assume that $\alpha_G = \sum_{i=1}^n \alpha_i = 1$. As $\beta_{ij} \in [0, 1]$, W_G is uniformly bounded by 1. Recall that $W_G(x, y) = \beta_{ij}$, if $x \in I_i$, $y \in I_j$, where $\lambda(I_j) = \alpha_j$ $(j = 1, \dots, n)$ and $I_1, \dots, I_n$ are consecutive intervals which form a partition of [0,1].

For fixed q and $c \le 1/q$, $f_q(G; V_1, \dots, V_q) = \frac{1}{\alpha_G^2} \sum_{i=1}^{q-1} \sum_{j=i+1}^{q} e_G(V_i, V_j)$ is a function taking on discrete values over c-balanced q-partitions $P = (V_1, \dots, V_q) \in \mathcal{P}_q^c$ of the vertices of G. As $n \to \infty$, this function approaches $\tilde{f}_q(W_G; S_1, \dots, S_q)$ that is already a continuous function over c-balanced q-partitions $(S_1, \dots, S_q) \in \mathcal{Q}_q^c$ of [0,1]. In fact, this continuous function is a multilinear function of the variable

$$\mathbf{x} = (x_{11}, \dots, x_{1n}, x_{21}, \dots, x_{2n}, \dots, x_{q1}, \dots, x_{qn})^T \in \mathbb{R}^{nq}$$

where the coordinate indexed by ij is

$$x_{ij} = \lambda(S_i \cap I_j), \quad j = 1, \dots, n; \quad i = 1, \dots q.$$

Hence,

$$\tilde{f}_q(W_G; S_1, \dots, S_q) = \tilde{f}_q(\mathbf{x}) = \sum_{i=1}^{q-1} \sum_{i'=i+1}^{q} \sum_{j=1}^{n} \sum_{j'=1}^{n} x_{ij} x_{i'j'} \beta_{jj'} = \frac{1}{2} \mathbf{x}^T (\boldsymbol{A} \otimes \boldsymbol{B}) \mathbf{x},$$

where – denoting by $\mathbf{1}_{q\times q}$ and $\boldsymbol{I}_{q\times q}$ the $q \times q$ all 1's and the identity matrix, respectively – the eigenvalues of the $q \times q$ symmetric matrix $\boldsymbol{A} = \mathbf{1}_{q\times q} - \boldsymbol{I}_{q\times q}$ are the number $q - 1$ and -1 with multiplicity $q - 1$, while those of the $n \times n$ symmetric matrix $\boldsymbol{B} = (\beta_{ij})$ are $\lambda_1 \ge \dots \ge \lambda_n$. The eigenvalues of the Kronecker-product $\boldsymbol{A} \otimes \boldsymbol{B}$ are the numbers $(q - 1)\lambda_i$ $(i = 1, \dots, n)$ and $-\lambda_i$ with multiplicity $q - 1$ $(i = 1, \dots, n)$. Therefore, the above quadratic form is indefinite.

Hence, we have the following *quadratic programming* task:

$$\begin{aligned} &\text{minimize} \quad \tilde{f}_q(\mathbf{x}) = \frac{1}{2} \mathbf{x}^T (\boldsymbol{A} \otimes \boldsymbol{B}) \mathbf{x} \\ &\text{subject to} \quad \mathbf{x} \ge 0; \quad \sum_{i=1}^{q} x_{ij} = \alpha_j \quad (j \in [n]); \quad \sum_{j=1}^{n} x_{ij} \ge c \quad (i \in [q]). \end{aligned} \tag{4.13}$$

The feasible region is the closed convex polytope of (4.13), and it is, in fact, in an $n(q - 1)$-dimensional hyperplane of $\mathbb{R}^{nq}$. The gradient of the objective function $\nabla \tilde{f}_q(\mathbf{x}) = (\boldsymbol{A} \otimes \boldsymbol{B})\mathbf{x}$ cannot be $\mathbf{0}$ in the feasible region, provided the weight matrix $\boldsymbol{B}$, or equivalently, $\boldsymbol{A} \otimes \boldsymbol{B}$ is not singular.

The argmin $\mathbf{x}^*$ of the quadratic programming task (4.13) is one of the Kuhn–Tucker points (giving relative minima of the indefinite quadratic form over the feasible region), that can be found by numerical algorithms, see for example Bazaraa and Shetty (1979). In this way, for large n, we also obtain *fuzzy clustering* of the vertices, in that $x_{ij}^*/\lambda(S_i)$ is the intensity of vertex j belonging to cluster i, see Drineas *et al.* (2004). If we still want to find disjoint clusters, the index i giving the largest proportion can be considered as the cluster membership of vertex j.

We conjecture that most of the vertices j can be uniquely assigned to a cluster i for which $x_{ij}^* = \alpha_j$, and there will be $\Theta(q)$ vertices with $0 < x_{ij}^* < \alpha_j$ $(i = 1, \ldots, q)$. Indeed, the minimum is more likely to be attained at a low-dimensional face, as here a lot of inequality constraints of (4.13) are satisfied by equalities that causes many coordinates of $\mathbf{x}^*$ to be zero. On higher dimensional faces the small number of equalities may come from the last q ones.

More generally, the above problem can be solved with other equality or inequality constraints making it possible to solve individual fuzzy clustering problems by quadratic programming. Similarly, the testable weighted graph parameter $\mu_q^c(G)$ can be extended to graphons and used for fuzzy clustering.

4.5 Noisy graph sequences

Now, we use the above theory for perturbations, showing that special noisy weighted graph sequences converge in the sense of Section 4.1. If not stated otherwise, the vertex-weights are equal (say, equal to 1), and a weighted graph G on n vertices is determined by its $n \times n$ symmetric weight matrix $\boldsymbol{A}$. Let $G_{\boldsymbol{A}}$ denote the weighted graph with unit vertex-weights and edge-weights that are entries of $\boldsymbol{A}$. We will use the notion of a Wigner-noise (Definition 3.1.1) and blown-up matrix (Definition 3.1.2).

Let us fix the $q \times q$ symmetric pattern matrix $\boldsymbol{P}$ of entries $0 < p_{ij} < 1$, and blow it up to an $n \times n$ blown-up matrix $\boldsymbol{B}_n$ of blow-up sizes $n_1, \ldots, n_q$ (note that $\boldsymbol{B}_n$ is a soft-core graph). Consider the noisy matrix $\boldsymbol{A}_n = \boldsymbol{B}_n + \boldsymbol{W}_n$ as $n_1, \ldots, n_q \to \infty$ at the same rate, where $\boldsymbol{W}_n$ is an $n \times n$ Wigner-noise. While perturbing $\boldsymbol{B}_n$ by $\boldsymbol{W}_n$, assume that for the uniform bound of the entries of $\boldsymbol{W}_n$ the condition

$$K \le \min\{\min_{i,j\in[q]} p_{ij}\,,\ 1 - \max_{i,j\in[q]} p_{ij}\} \tag{4.14}$$

is satisfied. In this way, the entries of $\boldsymbol{A}_n$ are in the [0,1] interval, and hence, $G_{\boldsymbol{A}_n} \in \mathcal{G}$. We remark that $G_{\boldsymbol{W}_n} \notin \mathcal{G}$, but $W_{G_{\boldsymbol{W}_n}} \in \mathcal{W}$ and the theory of bounded graphons applies to it. In Section 3.1 we showed that by adding an appropriate Wigner-noise to $\boldsymbol{B}_n$, we can achieve that $\boldsymbol{A}_n$ becomes a 0-1 matrix: its entries are equal to 1 with probability p_{ij} and 0 otherwise within the block of size $n_i \times n_j$ (after rearranging its rows and columns). In this case, the corresponding noisy graph $G_{\boldsymbol{A}_n}$ is a generalized random graph.

By routine large deviation techniques we are able to prove that the cut-norm of the stepfunction graphon assigned to a Wigner-noise tends to zero with probability 1 as $n \to \infty$.

Proposition 4.5.1 *For any sequence $\mathbf{W}_n$ of Wigner-noises*

$$\lim_{n\to\infty} \|W_{G_{W_n}}\|_\square = 0$$

almost surely.

The main idea of the proof is that the definition of the cut-norm of a stepfunction graphon and formulas (7.2), (7.3) of Borgs *et al.* (2008) yield

$$\|W_{G_{W_n}}\|_\square = \frac{1}{n^2} \max_{U,T\subset[n]} \left| \sum_{i\in U} \sum_{j\in T} w_{ij} \right| \le 6 \max_{U\subset[n]} \frac{1}{n^2} \left| \sum_{i\in U} \sum_{j\in[n]\setminus U} w_{ij} \right|,$$

where the entries behind the latter double summation are independent random variables. Hence, the Azuma's inequality (see Chung and Lu (2005)) is applicable, and the statement follows by the Borel–Cantelli lemma.

Let $\mathbf{A}_n := \mathbf{B}_n + \mathbf{W}_n$ and $n_1, \dots, n_q \to \infty$ in such a way that $\lim_{n\to\infty} \frac{n_i}{n} = r_i$ $(i = 1, \dots, q)$, $n = \sum_{i=1}^q n_i$; further, for the uniform bound K of the entries of the matrix $\mathbf{W}_n$ the condition (4.14) is assumed. Under these conditions, Proposition 4.5.1 implies that the noisy graph sequence $(G_{\mathbf{A}_n}) \subset \mathcal{G}$ converges almost surely in the $\delta_\square$ metric. It is easy to see that the almost sure limit is the stepfunction W_H, where the vertex- and edge-weights of the weighted graph H are

$$\alpha_i(H) = r_i \quad (i \in [q]), \qquad \beta_{ij}(H) = p_{ij} \quad (i, j \in [q]).$$

By adding a special Wigner-noise, the noisy graph sequence $(G_{\mathbf{A}_n})$ becomes a generalized random graph sequence on the model graph H, under which the following is understood. Let $V = [n]$ be the desired vertex-set. A vertex is put into the vertex-subset V_i with probability $\alpha_i(H)$; then vertices of V_i and V_j are connected with probability $\beta_{ij}(H)$, $i, j = 1, \dots, q$ (see also Section 3.1).

The deterministic counterparts of the generalized random graphs are the *generalized quasirandom graphs* introduced by Lovász and T.-Sós (2008) in the following way. We have a model graph graph H on q vertices with vertex-weights $r_1, \dots, r_q$ and edge-weights $p_{ij} = p_{ji}$, $i, j = 1, \dots, q$. Then (G_n) is H-quasirandom if $G_n \to H$ as $n \to \infty$ in the sense of Definition 4.1.1. Lovász and T.-Sós (2008) also prove that the vertex set V of a generalized quasirandom graph G_n can be partitioned into clusters $V_1, \dots, V_q$ in such a way that $\frac{|V_i|}{|V|} \to r_i$ $(i = 1, \dots, q)$ and the subgraph of G_n induced by V_i is the general term of a quasirandom graph sequence with edge-density tending to p_{ii} $(i = 1, \dots, q)$, whereas the bipartite subgraph between V_i and V_j is the general term of a quasirandom bipartite graph sequence with edge-density tending to p_{ij} $(i \ne j)$ as $n \to \infty$. Consequently, for any fixed finite graph F, the number of copies of F is asymptotically the same in the above generalized random and generalized quasirandom graphs on the same model graph and number of vertices.

4.6 Convergence of the spectra and spectral subspaces

In Section 4.3 we proved the testability of some normalized balanced multiway cut densities such that we imposed balancing conditions on the cluster volumes. Under similar conditions, for fixed number of clusters k, the unnormalized and normalized multiway modularities are also testable, provided our edge-weighted graph has no dominant vertices. The proofs rely on statistical physics notions of Borgs *et al.* (2008), utilizing the fact that the graph convergence implies the convergence of the ground state energy. Reichardt and Bornholdt (2007) showed that the Newman-Girvan modularity is an energy function (Hamiltonian), and hence, testability of the maximum normalized modularities, under appropriate balancing conditions, can be shown analogously. Here we rather discuss the testability of spectra and k-variances, because in spectral clustering methods these provide us with polynomial time algorithms, though only approximate solutions are expected via the spectral relaxation.

In Theorem 6.6 of Borgs *et al.* (2012), the authors prove that the normalized spectrum of a convergent graph sequence converges in the following sense. Let W be a graphon and (G_n) be a sequence of weighted graphs – with uniformly bounded edge-weights – tending to W. (For simplicity, we assume that $|V(G_n)| = n$). Let $|\lambda_{n,1}| \ge |\lambda_{n,2}| \ge \cdots \ge |\lambda_{n,n}|$ be the adjacency eigenvalues of G_n indexed by their decreasing absolute values, and let $\mu_{n,i} = \lambda_{n,i}/n$ $(i = 1, \dots, n)$ be the normalized eigenvalues. Further, let T_W be the $L^2[0,1] \to L^2[0,1]$ integral operator corresponding to W:

$$(T_W f)(x) = \int_0^1 W(x, y) f(y)\, dy.$$

It is well-known that this operator is self-adjoint and compact, and hence, it has a discrete real spectrum, whose only possible point of accumulation is the 0 (see Appendix A.2). Let $\mu_i(W)$ denote the ith largest absolute value eigenvalue of T_W. Then for every $i \ge 1$, $\mu_{n,i} \to \mu_i(W)$ as $n \to \infty$. In fact, the authors prove a bit more (see Theorem 6.7 of Borgs *et al.* (2012)): if a sequence W_n of uniformly bounded graphons converges to a graphon W, then for every $i \ge 1$, $\mu_i(W_n) \to \mu_i(W)$ as $n \to \infty$. Note that the spectrum of W_G is the normalized spectrum of G, together with countably infinitely many 0's. Therefore, the convergence of the spectrum of (G_n) is the consequence of that of (W_{G_n}).

We will prove that in the absence of dominant vertices, the normalized modularity spectrum is testable. To this end, both the modularity matrix and the graphon are related to kernels of special integral operators, described in Section 1.4. We recall the most important facts herein. Let (ξ, ξ') be a pair of identically distributed real-valued random variables defined over the product space $\mathcal{X} \times \mathcal{X}$ having a symmetric joint distribution $\mathbb{W}$ with equal margins $\mathbb{P}$. Suppose that the dependence between ξ and ξ' is regular, that is, their joint distribution $\mathbb{W}$ is absolutely continuous with respect to the product measure $\mathbb{P} \times \mathbb{P}$, and let w denote its Radon–Nikodym derivative. Let $H = L^2(\xi)$ and $H' = L^2(\xi')$ be the Hilbert spaces of random variables which are functions of ξ and ξ' and have zero expectation and finite variance with respect to $\mathbb{P}$. Observe that H and H' are isomorphic Hilbert spaces with the covariance as inner product; further, they are embedded as subspaces into the L^2-space defined similarly over the product space. (Here H and H' are also isomorphic in the sense that for any $\psi \in H$ there exists a $\psi' \in H'$ and vice versa, such that ψ and ψ' are identically distributed.)

Consider the linear operator taking conditional expectation between H' and H with respect to the joint distribution. It is an integral operator and will be denoted by $P_{\mathbb{W}} : H' \to H$ as it is a projection restricted to H' and projects onto H. To $\psi' \in H'$ the operator $P_{\mathbb{W}}$ assigns $\psi \in H$ such that $\psi = \mathbb{E}_{\mathbb{W}}(\psi' \mid \xi)$, that is,

$$\psi(x) = \int_{\mathcal{Y}} w(x, y)\psi'(y)\,\mathbb{P}(dy), \quad x \in \mathcal{X}.$$

If

$$\int_{\mathcal{X}}\int_{\mathcal{X}} w^2(x, y)\mathbb{P}(dx)\mathbb{P}(dy) < \infty,$$

then $P_{\mathbb{W}}$ is a Hilbert–Schmidt operator, therefore it is compact and has SD

$$P_{\mathbb{W}} = \sum_{i=1}^{\infty} \lambda_i \langle ., \psi_i' \rangle_{H'} \psi_i$$

where for the eigenvalues $|\lambda_i| \leq 1$ holds and the eigenvalue–eigenfunction equation looks like

$$P_{\mathbb{W}}\psi_i' = \lambda_i \psi_i \quad (i = 1, 2, \ldots)$$

where ψ_i and ψ_i' are identically distributed, whereas their joint distribution is $\mathbb{W}$. It is easy to see that $P_{\mathbb{W}}$ is self-adjoint and it takes the constantly 1 random variable of H' into the constantly 1 random variable of H; however, the $\psi_0 = 1$, $\psi_0' = 1$ pair is not regarded as a function pair with eigenvalue $\lambda_0 = 1$, since they have no zero expectation. More precisely, the kernel is reduced to $w(x, y) - 1$.

Theorem 4.6.1 *Let $G_n = (V_n, \mathbf{W}_n)$ be the general entry of a convergent sequence of connected edge-weighted graphs whose edge-weights are in [0,1] and the vertex-weights are the generalized vertex-degrees. Assume that there are no dominant vertices. Let W denote the limit graphon of the sequence (G_n), and let*

$$1 \geq |\mu_{n,1}| \geq |\mu_{n,2}| \geq \cdots \geq |\mu_{n,n}| = 0$$

be the normalized modularity spectrum of G_n (the eigenvalues are indexed by their decreasing absolute values). Further, let $\mu_i(P_{\mathbb{W}})$ be the ith largest absolute value eigenvalue of the integral operator $P_{\mathbb{W}} : L^2(\xi') \to L^2(\xi)$ taking conditional expectation with respect to the joint measure $\mathbb{W}$ embodied by the normalized limit graphon W, and ξ, ξ' are identically distributed random variables with the marginal distribution of their symmetric joint distribution $\mathbb{W}$. Then for every $i \geq 1$,

$$\mu_{n,i} \to \mu_i(P_{\mathbb{W}}) \quad \text{as} \quad n \to \infty.$$

Proof. In case of a finite $\mathcal{X}$ (vertex set) we have a weighted graph, and we will show that the operator taking conditional expectation with respect to the joint distribution, determined by the edge-weights, corresponds to its normalized modularity matrix.

Indeed, let $\mathcal{X} = V$, $|V| = n$, and $G_n = (V, \mathbf{W})$ be an edge-weighted graph on the $n \times n$ weight matrix of the edges $\mathbf{W}$ with entries W_{ij}'s; now, they do not necessarily sum up to 1. (For the time being, n is kept fixed, so – for the sake of simplicity – we do not denote the dependence of $\mathbf{W}$ on n). Let the vertices be also weighted with special weights $\alpha_i(G_n) := \sum_{j=1}^{n} W_{ij}$, $i = 1, \dots, n$. Then the step-function graphon W_{G_n} is such that $W_{G_n}(x, y) = W_{ij}$ whenever $x \in I_i$ and $y \in I_j$, where the (not necessarily contiguous) intervals $I_1, \dots, I_n$ form a partition of [0,1] such that the length of I_i is $\alpha_i(G_n)/\alpha_{G_n}$ $(i = 1, \dots, n)$.

Let us transform $\mathbf{W}$ into a symmetric joint distribution $\mathbb{W}_n$ over $V \times V$. The entries $w_{ij} = W_{ij}/\alpha_{G_n}$ $(i, j = 1, \dots, n)$ embody this discrete joint distribution of random variables ξ and ξ' which are identically distributed with marginal distribution $d_1, \dots, d_n$, where $d_i = \alpha_i(G_n)/\alpha_{G_n}$ $(i = 1, \dots, n)$. With the previous notation $H = L^2(\xi)$, $H' = L^2(\xi')$, and the operator $P_{\mathbb{W}_n} : H' \to H$ taking conditional expectation is an integral operator with now discrete kernel $K_{ij} = \frac{w_{ij}}{d_i d_j}$. The fact that ψ, ψ' is an eigenfunction pair of $P_{\mathbb{W}_n}$ with eigenvalue λ means that

$$\frac{1}{d_i} \sum_{j=1}^{n} w_{ij} \psi'(j) = \sum_{j=1}^{n} \frac{w_{ij}}{d_i d_j} \psi'(j) d_j = \lambda \psi(i), \tag{4.15}$$

where $\psi(j) = \psi'(j)$ denotes the value of ψ or ψ' taken on with probability d_i (recall that ψ and ψ' are identically distributed). The above equation is equivalent to

$$\sum_{j=1}^{n} \frac{w_{ij}}{\sqrt{d_i}\sqrt{d_j}} \sqrt{d_j} \psi(j) = \lambda \sqrt{d_i} \psi(i).$$

Therefore, the vector of coordinates $\sqrt{d_i}\psi(i)$ $(i = 1, \dots, n)$ is a unit-norm eigenvector of the normalized modularity matrix with eigenvalue λ (note that the normalized modularity spectrum does not depend on the scale of the edge-weights, it is the same whether we use W_{ij}'s or w_{ij}'s as edge-weights). Consequently, the eigenvalues of the conditional expectation operator are the same as the eigenvalues of the normalized modularity matrix, and the possible values taken on by the eigenfunctions of the conditional expectation operator are the same as the coordinates of the transformed eigenvectors of the normalized modularity matrix forming the column vectors of the matrix $\mathbf{X}^*$ of the optimal $(k-1)$-dimensional representatives.

Let f be a stepwise constant function on [0,1], taking on value $\psi(i)$ on I_i. Then $\mathrm{Var}\psi = 1$ is equivalent to $\int_0^1 f^2(x)\,dx = 1$. Let K_{G_n} be the stepwise constant graphon defined as $K_{G_n}(x, y) = K_{ij}$ for $x \in I_i$ and $y \in I_j$. With this, the eigenvalue–eigenvector equation (4.15) looks like

$$\lambda f(x) = \int_0^1 K_{G_n}(x, y) f(y)\,dy.$$

The spectrum of K_{G_n} is the normalized modularity spectrum of G_n together with countably infinitely many 0's (it is of finite rank, and therefore, trivially compact), and because of the convergence of the weighted graph sequence G_n, in lack of dominant vertices, the sequence of graphons K_{G_n} also converges. Indeed, the $W_{G_n} \to W$ convergence in the cut metric means the convergence of the induced discrete distributions $\mathbb{W}_n$'s to the continuous $\mathbb{W}$. Since K_{G_n} and K are so-called copula transformations (see Nelsen (2006)) of those distributions; in lack of dominant vertices (this causes the convergence of the margins) they also converge, which in turn implies the $K_{G_n} \to K$ convergence in the cut metric.

Let K denote the limit graphon of K_{G_n} $(n \to \infty)$. This will be the kernel of the integral operator taking conditional expectation with respect to the joint distribution $\mathbb{W}$. It is easy to see that this operator is also a Hilbert–Schmidt operator, and therefore, compact. With these considerations the remainder of the proof is analogous to the proof of Theorem 6.7 of Borgs *et al.* (2012), where the authors prove that if the sequence (W_{G_n}) of graphons converges to the limit graphon W, then both ends of the spectra of the integral operators, induced by W_{G_n}'s as kernels, converge to the ends of the spectrum of the integral operator induced by W as kernel. We apply this argument for the spectra of the kernels induced by K_{G_n}'s and K.

Note that in Lovász and Szegedy (2011), kernel operators are also discussed, but not with our normalization.

Remark 4.6.2 *By using Theorem 4.2.2 (c), provided there are no dominant vertices, Theorem 4.6.1 implies that for any fixed positive integer k, the $(k-1)$-tuple of the largest absolute value eigenvalues of the normalized modularity matrix is testable.*

Theorem 4.6.3 *Suppose, there are constants $0 < \varepsilon < \delta \le 1$ such that the normalized modularity spectrum (with decreasing absolute values) of any G_n satisfies*

$$1 \ge |\mu_{n,1}| \ge \cdots \ge |\mu_{n,k-1}| \ge \delta > \varepsilon \ge |\mu_{n,k}| \ge \cdots \ge |\mu_{n,n}| = 0.$$

With the notions of Theorem 4.6.1, and assuming that there are no dominant vertices of G_n's, the subspace spanned by the transformed eigenvectors $\mathbf{D}^{-1/2}\mathbf{u}_1, \dots, \mathbf{D}^{-1/2}\mathbf{u}_{k-1}$ belonging to the $k-1$ largest absolute value eigenvalues of the normalized modularity matrix of G_n also converges to the corresponding $(k-1)$-dimensional subspace of $P_{\mathbb{W}}$. More precisely, if $\boldsymbol{P}_{n,k-1}$ denotes the projection onto the subspace spanned by the transformed eigenvectors belonging to $k-1$ largest absolute value eigenvalues of the normalized modularity matrix of G_n, and $\boldsymbol{P}_{k-1}$ denotes the projection onto the corresponding eigen-subspace of $P_{\mathbb{W}}$, then $\|\boldsymbol{P}_{n,k-1} - \boldsymbol{P}_{k-1}\| \to 0$ as $n \to \infty$ (in spectral norm).

Proof. If we apply the convergence fact $\mu_{n,i} \to \mu_i(P_{\mathbb{W}})$ for indices $i = k-1$ and k, we get that there will be a gap of order $\delta - \varepsilon - o(1)$ between $|\mu_{k-1}(P_{\mathbb{W}})|$ and $|\mu_k(P_{\mathbb{W}})|$ too.

Let $P_{\mathbb{W},n}$ denote the n-rank approximation of $P_{\mathbb{W}}$ (keeping its n largest absolute value eigenvalues, together with the corresponding eigenfunctions) in spectral norm.

The projection $\boldsymbol{P}_{k-1}$ ($k < n$) operates on the eigen-subspace spanned by the eigenfunctions belonging to the $k-1$ largest absolute value eigenvalues of $P_{\mathbb{W},n}$ in the same way as on the corresponding $(k-1)$-dimensional subspace determined by $P_{\mathbb{W}}$. With these considerations, we apply the perturbation theory of eigen-subspaces with the following unitary invariant norm: the Schatten 4-norm of the Hilbert–Schmidt operator A is $\|A\|_4 = (\sum_{i=1}^{\infty} \lambda_i^4(A))^{1/4}$ (see e.g., Borgs *et al.* (2008) and (A.15) for matrices). Our argument with the finite $(k-1)$ rank projections is the following. Denoting by $P_{\mathbb{W}_n}$ the integral operator belonging to the normalized modularity matrix of G_n (with kernel K_{G_n} introduced in the proof of Theorem 4.6.1),

$$\begin{aligned}\|\boldsymbol{P}_{n,k-1} - \boldsymbol{P}_{k-1}\| &= \|\boldsymbol{P}_{n,k-1}^{\perp}\boldsymbol{P}_{k-1}\| \le \|\boldsymbol{P}_{n,k-1}^{\perp}\boldsymbol{P}_{k-1}\|_4 \\ &\le \frac{c}{\delta - \varepsilon - o(1)}\|P_{\mathbb{W}_n} - P_{\mathbb{W},n}\|_4\end{aligned}$$

with constant c that is at most $\pi/2$ (see Theorem A.3.22). But

$$\|P_{\mathbb{W}_n} - P_{\mathbb{W},n}\|_4 \le \|P_{\mathbb{W}_n} - P_{\mathbb{W}}\|_4 + \|P_{\mathbb{W}} - P_{\mathbb{W},n}\|_4,$$

where the last term tends to 0 as $n \to \infty$, since the tail of the spectrum (taking the fourth power of the eigenvalues) of a Hilbert–Schmidt operator converges. For the convergence of the first term we use Lemma 7.1 of Borgs *et al.* (2008), which states that the Schatten 4-norm of an integral operator can be estimated from above by four times the cut-norm of the corresponding kernel. But the convergence in the cut distance of the corresponding kernels to zero follows from the considerations made in the proof of Theorem 4.6.1. This finishes the proof.

Remark 4.6.4 *As the k-variance depends continuously on the above subspaces (see the expansion (3.68) of s^2 in the proof of Theorem 3.3.3), Theorem 4.6.3 implies the testability of the k-variance as well.*

The above results suggest that in the absence of dominant vertices, even the normalized modularity matrix of a smaller part of the underlying weighted graph, selected at random with an appropriate procedure, is able to reveal its cluster structure. Hence, the gain regarding the computational time of this spectral clustering algorithm is twofold: we only use a smaller part of the graph and the spectral decomposition of its normalized modularity matrix runs in polynomial time in the reduced number of the vertices. Under the vertex- and cluster-balance conditions this method can give quite good approximations for the multiway cuts and helps us to find the number of clusters and identify the cluster structure.

Even if the spectrum of convergent graph sequences converges, the spectrum itself does not carry enough information for the cluster structure of the graph as stated in Lovász and T.-Sós (2008). However, together with the eigenvectors it must carry sufficient information; moreover, it suffices to consider the structural eigenvalues with their corresponding spectral subspaces.

4.7 Convergence of contingency tables

Now, we will extend the above theory to rectangular arrays with nonnegative, bounded entries. A statistic, defined on a contingency table, is testable if it can be consistently estimated based on a smaller, but still sufficiently large table which is selected randomly from the original one in an appropriate manner. By the above randomization, classical multivariate methods can be carried out on a smaller part of the array. This fact becomes important when our task is to discover the structure of large and evolving arrays, such as microarrays, social, and communication networks. Special block structures behind large tables are also discussed from the point of view of stability and spectra. In order to recover the structure of large rectangular arrays, classical methods of cluster and correspondence analysis may not be carried out on the whole table because of computational size limitations. In other situations, we want to compare contingency tables of different sizes. For the above reasons, convergence and distance of general normalized arrays is introduced.

Let $\boldsymbol{C} = \boldsymbol{C}_{m\times n}$ be a contingency table on row set $Row_C = \{1, \dots, m\}$ and column set $Col_C = \{1, \dots, n\}$. The nonnegative, real entries c_{ij}'s are thought of as associations between the rows and columns, and they are normalized such that $0 \le c_{ij} \le 1$. Sometimes we have *binary* tables of entries 0 or 1. We may assign positive weights $\alpha_1, \dots, \alpha_m$ to the rows and $\beta_1, \dots, \beta_n$ to the columns expressing the individual importance of the categories embodied by the rows and columns. (In correspondence analysis, these are the row- and column-sums.) A contingency table is called *simple* if all the row- and column-weights are equal to 1. Assume that C does not contain identically zero rows or columns, moreover C is dense in the sense that the number of nonzero entries is comparable with mn. Let $\mathcal{C}$ denote the set of such tables (with any natural numbers m and n).

Consider a simple binary table $\boldsymbol{F}_{a\times b}$ and maps $\Phi : Row_F \to Row_C$, $\Psi : Col_F \to Col_C$; further

$$\alpha_\Phi := \prod_{i=1}^{a} \alpha_{\Phi(i)}, \quad \beta_\Psi := \prod_{j=1}^{b} \beta_{\Psi(j)}, \quad \alpha_C := \sum_{i=1}^{m} \alpha_i, \quad \beta_C := \sum_{j=1}^{n} \beta_j.$$

Definition 4.7.1 *The* $\boldsymbol{F} \to \boldsymbol{C}$ *homomorphism density is*

$$t(\boldsymbol{F}, \boldsymbol{C}) = \frac{1}{(\alpha_C)^a (\beta_C)^b} \sum_{\Phi,\Psi} \alpha_\Phi \beta_\Psi \prod_{f_{ij}=1} c_{\Phi(i)\Psi(j)}.$$

If $\boldsymbol{C}$ is simple, then $t(\boldsymbol{F}, \boldsymbol{C}) = \frac{1}{m^a n^b} \sum_{\Phi,\Psi} \prod_{f_{ij}=1} c_{\Phi(i)\Psi(j)}$. If, in addition, C is binary too, then $t(\boldsymbol{F}, \boldsymbol{C})$ is the probability that a random map $\boldsymbol{F} \to \boldsymbol{C}$ is a homomorphism (preserves the 1's). The maps Φ and Ψ correspond to sampling a rows and b columns out of Row_C and Col_C with replacement, respectively. In case of simple $\boldsymbol{C}$ it means uniform sampling, otherwise the rows and columns are selected with probabilities proportional to their weights.

The following simple random binary table $\xi(a \times b, \boldsymbol{C})$ will play an important role in the definition of testable contingency table parameters. Select a rows and b columns of $\boldsymbol{C}$ with replacement, with probabilities α_i/α_C $(i = 1, \ldots, m)$ and β_j/β_C $(j = 1, \ldots, n)$, respectively. If the ith row and jth column of C are selected, they will be connected by 1 with probability c_{ij} and 0, otherwise, independently of the other selected row–column pairs, conditioned on the selection of the rows and columns. For large m and n, $\mathbb{P}(\xi(a \times b, \boldsymbol{C}) = F)$ is very close to $t(\boldsymbol{F}, \boldsymbol{C})$ that resembles a likelihood function. (The more precise formulation with induced and injective homomorphisms is to be found in Bolla (2010)).

Definition 4.7.2 *We say that the sequence $(\boldsymbol{C}_{m\times n})$ of contingency tables is convergent if the sequence $t(\boldsymbol{F}, \boldsymbol{C}_{m\times n})$ converges for any simple binary table $\boldsymbol{F}$ as $m, n \to \infty$.*

The convergence means that the tables $\boldsymbol{C}_{m\times n}$ become more and more similar in small details as they are probed by smaller 0-1 tables $(m, n \to \infty)$.

The limit object is a measurable function $U : [0, 1]^2 \to [0, 1]$ and we call it *contingon*, which is the non-symmetric generalization of a graphon (see Section 4.1) and was introduced in Bolla (2010). The step-function contingon U_C is assigned to $\boldsymbol{C}$ in the following way: the sides of the unit square are divided into intervals $I_1, \ldots, I_m$ and $J_1, \ldots, J_n$ of lengths $\alpha_1/\alpha_C, \ldots, \alpha_m/\alpha_C$ and $\beta_1/\beta_C, \ldots, \beta_n/\beta_C$, respectively; then over the rectangle $I_i \times J_j$ the step-function takes on the value c_{ij}.

In fact, the above convergence of contingency tables can be formulated in terms of the cut distance. First we define it for contingons.

Definition 4.7.3 *The cut distance between the contingons U and V is*

$$\delta_\square(U, V) = \inf_{\mu,\nu} \|U - V^{\mu,\nu}\|_\square \tag{4.16}$$

where the cut-norm of the contingon U is defined by

$$\|U\|_\square = \sup_{S,T\subset[0,1]} \left| \iint_{S\times T} U(x, y)\, dx\, dy \right|,$$

and the infimum in (4.16) is taken over all measure-preserving bijections $\mu, \nu : [0, 1] \to [0, 1]$, while $V^{\mu,\nu}$ denotes the transformed V after performing the measure preserving bijections μ and ν on the sides of the unit square, respectively.

An equivalence relation is defined over the set of contingons: two contingons belong to the same class if they can be transformed into each other by measure preserving map, that is, their cut distance is zero. In the sequel, we consider contingons modulo measure preserving maps, and by contingon we understand the whole equivalence class.

Definition 4.7.4 *The cut distance between the contingency tables $\boldsymbol{C}, \boldsymbol{C}' \in \mathcal{C}$ is*

$$\delta_\square(\boldsymbol{C}, \boldsymbol{C}') = \delta_\square(U_C, U_{C'}).$$

By the above remarks, this distance of $\boldsymbol{C}$ and $\boldsymbol{C}'$ is indifferent to permutations of the rows or columns of $\boldsymbol{C}$ and $\boldsymbol{C}'$. In the special case when $\boldsymbol{C}$ and $\boldsymbol{C}'$ are of the same size, $\delta_{\square}(\boldsymbol{C}, \boldsymbol{C}')$ is $\frac{1}{mn}$ times the usual cut distance of matrices, based on the cut-norm (see Definition 3.3.4).

The following reversible relation between convergent contingency table sequences and contingons also holds, as a rectangular analogue of Fact 4.1.2.

Fact 4.7.5 *For any convergent sequence $(\boldsymbol{C}_{m\times n}) \subset \mathcal{C}$ there exists a contingon such that $\delta_{\square}(U_{\boldsymbol{C}_{m\times n}}, U) \to 0$ as $m, n \to \infty$. Conversely, any contingon can be obtained as the limit of a sequence of contingency tables in $\mathcal{C}$. The limit of a convergent contingency table sequence is essentially unique: if $\boldsymbol{C}_{m\times n} \to U$, then also $\boldsymbol{C}_{m\times n} \to U'$ for precisely those contingons U' for which $\delta_{\square}(U, U') = 0$.*

It also follows that a sequence of contingency tables in $\mathcal{C}$ is convergent if and only if it is a Cauchy sequence in the metric $\delta_{\square}$.

A simple binary random $a \times b$ table $\xi(a \times b, U)$ can also be randomized based on the contingon U in the following way. Let $X_1, \dots, X_a$ and $Y_1, \dots, Y_b$ be i.i.d., uniformly distributed random numbers on [0,1]. The entries of $\xi(a \times b, U)$ are independent Bernoulli random variables, namely the entry in the ith row and jth column is 1 with probability $U(X_i, Y_j)$ and 0, otherwise. It is easy to see that the distribution of the previously defined $\xi(a \times b, \boldsymbol{C})$ and that of $\xi(a \times b, U_{\boldsymbol{C}})$ is the same. Further, $\delta_{\square}(\boldsymbol{C}_{m\times n}, \xi(a \times b, \boldsymbol{C}_{m\times n}))$ tends to 0 in probability, for fixed a and b as $m, n \to \infty$.

Note, that in the above way, we can theoretically randomize an infinite simple binary table $\xi(\infty \times \infty, U)$ out of the contingon U by generating countably infinitely many i.i.d. uniform random numbers on [0,1]. The distribution of the infinite binary array $\xi(\infty \times \infty, U)$ is denoted by $\mathbb{P}_U$. Because of the symmetry of the construction, this is an *exchangeable* array in the sense that the joint distribution of its entries is invariant under permutations of the rows and columns. Furthermore, any exchangeable binary array is a mixture of such $\mathbb{P}_U$'s. More precisely, the Aldous–Hoover representation theorem (see Diaconis and Janson (2008)) states that for every infinite exchangeable binary array ξ there exists a probability distribution μ (over the contingons) such that $\mathbb{P}(\xi \in A) = \int \mathbb{P}_U(A)\, \mu(dU)$.

A function $f : \boldsymbol{C} \to \mathbb{R}$ is called a *contingency table parameter* if it is invariant under isomorphism and scaling of the rows/columns. In fact, it is a statistic evaluated on the table, and hence, we are interested in contingency table parameters that are not sensitive to minor changes in the entries of the table.

Definition 4.7.6 *A contingency table parameter f is testable if for every $\varepsilon > 0$ there are positive integers a and b such that if the row- and column-weights of $\boldsymbol{C}$ satisfy*

$$\max_i \frac{\alpha_i}{\alpha_C} \le \frac{1}{a}, \qquad \max_j \frac{\beta_j}{\beta_C} \le \frac{1}{b}, \tag{4.17}$$

then

$$\mathbb{P}(|f(\boldsymbol{C}) - f(\xi(a \times b, \boldsymbol{C}))| > \varepsilon) \le \varepsilon.$$

Consequently, such a contingency table parameter can be consistently estimated based on a fairly large sample. Now, we introduce some equivalent statements of the testability, indicating that a testable parameter depends continuously on the whole table. This is the generalization of Theorem 4.2.2.

Theorem 4.7.7 *For a testable contingency table parameter f the following are equivalent:*

- *For every $\varepsilon > 0$ there are positive integers a and b such that for every contingency table $\boldsymbol{C} \in \mathcal{C}$ satisfying the condition (4.17),*

$$|f(\boldsymbol{C}) - \mathbb{E}(f(\xi(a \times b, \boldsymbol{C})))| \leq \varepsilon.$$

- *For every convergent sequence $(\boldsymbol{C}_{m\times n})$ of contingency tables with no dominant row- or column-weights, $f(\boldsymbol{C}_{m\times n})$ is also convergent $(m, n \to \infty)$.*
- *f is continuous in the cut distance.*

For example, in the case of a simple binary table, the singular spectrum is testable, since $\boldsymbol{C}_{m\times n}$ can be considered as part of the adjacency matrix of a bipartite graph on $m + n$ vertices, where Row_C and Col_C are the two independent vertex sets; further, the ith vertex of Row_C and the jth vertex of Col_C are connected by an edge if and only if $c_{ij} = 1$. The non-zero real eigenvalues of the symmetric $(m + n) \times (m + n)$ adjacency matrix of this bipartite graph are the numbers $\pm s_1, \ldots, \pm s_r$, where $s_1, \ldots, s_r$ are the non-zero singular values of $\boldsymbol{C}$, and $r \leq \min\{m, m\}$ is the rank of $\boldsymbol{C}$ (see Proposition A.3.29). Consequently, the convergence of the adjacency spectra implies the convergence of the singular spectra. Therefore, by Theorem 4.7.7, any property of a large contingency table based on its SVD (e.g., correspondence decomposition) can be concluded from a smaller part of it.

Using the notation of Section 3.2, analogously to the symmetric case, it can be proved that special blown-up tables (see Definition 3.2.2) burdened with a general kind of noise (see Definition 3.2.1) are convergent.

Proposition 4.7.8 *For any sequence $\boldsymbol{W}_{m\times n}$ of rectangular Wigner-noises*

$$\lim_{m,n\to\infty} \|U_{\boldsymbol{W}_{m\times n}}\|_\square = 0$$

almost surely, where $(U_{\boldsymbol{W}_{m\times n}})$ is the step-function contingon assigned to $\boldsymbol{W}_{m\times n}$.

Now, let us fix the pattern-matrix matrix $\boldsymbol{P}_{a\times b}$ and blow it up to obtain matrix $\boldsymbol{B}_{m\times n}$.

Proposition 4.7.9 *Let the block sizes of the blown-up matrix $\boldsymbol{B}_{m\times n}$ be $m_1, \ldots, m_a$ horizontally, and $n_1, \ldots, n_b$ vertically $\left(\sum_{i=1}^{a} m_i = m \text{ and } \sum_{j=1}^{b} n_j = n\right)$. Let $\boldsymbol{A}_{m\times n} = \boldsymbol{B}_{m\times n} + \boldsymbol{W}_{m\times n}$ and $m, n \to \infty$ is such a way that $m_i/m \to r_i$ $(i = 1, \ldots, a)$, $n_j/n \to q_j$ $(j = 1, \ldots, b)$, where r_i's and q_j's are fixed ratios. Under* *these conditions, the 'noisy' contingency table sequence $(\boldsymbol{A}_{m\times n})$ converges almost surely.*

In many applications we are looking for clusters of the rows and columns of a rectangular array such that the densities within the cross-products of the clusters be as homogeneous as possible. For example, in microarray analysis we are looking for clusters of genes and conditions such that genes of the same cluster equally influence conditions of the same cluster. The following theorem ensures the existence of such a structure with possibly many clusters. However, the number of clusters does not depend on the size of the array, it merely depends on the accuracy of the approximation. The following statement is a straightforward generalization of the weak regularity lemma 4.3.1.

Proposition 4.7.10 *For every $\varepsilon > 0$ and $\boldsymbol{C}_{m\times n} \in \mathcal{C}$ there exists a blown-up matrix $\boldsymbol{B}_{m\times n}$ of an $a \times b$ pattern matrix with $a + b \leq 4^{1/\varepsilon^2}$ (independently of m and n) such that $\delta_{\square}(\boldsymbol{C}, \boldsymbol{B}) \leq \varepsilon$.*

The statement can be proved by embedding $\boldsymbol{C}$ into the adjacency matrix of an edge-weighted bipartite graph. The statement itself is closely related to the testability of the following contingency table parameter. For fixed integers $1 \leq a \ll m$ and $1 \leq b \ll n$,

$$S_{a,b}^2(\boldsymbol{C}) = \min_{\substack{R_1,\dots,R_a \\ C_1,\dots,C_b}} \sum_{i=1}^{a} \sum_{j=1}^{b} \sum_{k\in R_i} \sum_{\ell\in C_j} (c_{k\ell} - \bar{c}_{i,j})^2 \tag{4.18}$$

where the minimum is taken over balanced a- and b-partitions $R_1, \dots, R_a$ and $C_1, \dots, C_b$ of Row_C and Col_C, respectively; further,

$$\bar{c}_{i,j} = \frac{1}{|R_i| \cdot |C_j|} \sum_{k\in R_i} \sum_{\ell\in C_j} c_{kl}$$

is the center of the bicluster $R_i \times C_j$ $(i = 1, \dots, a;\ j = 1, \dots, b)$. Note that, instead of $c_{k\ell}$, we may take $\alpha_k \beta_\ell c_{k\ell}$ in the row- and column-weighted case, provided there are no dominant rows/columns.

An objective function reminiscent of (4.18) is minimized in Hartigan (1972); Madeira and Oliveira (2004) in a more general block clustering problem, where not only Cartesian product type biclusters ($R_i \times C_j$, also called chess-board patterns in Kluger *et al.* (2003)) are allowed, but the contingency table is divided into disjoint rectangles forming as homogeneous blocks of the heterogeneous data as possible. (4.18) can as well be regarded as within-custer variance in a two-way ANOVA setup, see Section C.4 of the Appendix. In Madeira and Oliveira (2004), applications to microarrays is presented, where the rows correspond to genes, the columns to different conditions, whereas the entries are expression levels of genes under the distinct conditions. In this framework, biclusters identify subsets of genes sharing similar expression patterns across subsets of conditions. Recall that in Subsection 3.3.2 we looked for regular cluster pairs by means of spectral methods, but there we used

Cartesian product type biclusters, moreover, the number of row and column clusters was the same.

The *Gardner–Ashby's connectance* c_n of a not necessarily symmetric, quadratic array $\boldsymbol{A}_{n\times n}$ is the percentage of nonzero entries in the matrix, that is the ratio of actual row-column interactions to all possible ones in the network. In social and ecological models, a random array $\boldsymbol{A}_{n\times n}$ of independent entries is considered. Assume that the entries have symmetric distribution (consequently, zero expectation) and common variance σ_n^2, where σ_n is called *average interaction strength*. The *stability* of the system is characterized by the stability of the equilibrium solution $\mathbf{0}$ of the differential equation $d\mathbf{x}/dt = \boldsymbol{A}_{n\times n}\mathbf{x}$ (sometimes this is achieved by linearization techniques in the neighborhood of the equilibrium solution). Based on Wigner's famous semicircle law (see Theorem B.2.2), May (1972) proved that the equilibrium solution is stable in the $\sigma_n^2 nc_n < 1$, and unstable in the $\sigma_n^2 nc_n > 1$ case; further, the transition region between stability and instability becomes narrow as $n \to \infty$. Hence, it seems that high connectance and high interaction strength destroy stability, but only in this simple model. If $\boldsymbol{A}_{n\times n}$ is a block matrix, like a noisy matrix before, it has some structural, possibly complex eigenvalues (see Juhász (1996)). If all their real parts are negative, the system is stable, see Érdi and Tóth (1990). In fact, in many natural ecosystems and other networks the interactions are arranged in blocks.

References

Arora S, Karger D and Karpinski M 1995 Polynomial time approximation schemes for dense instances of NP-hard problems, in *Proc. 27th Annual ACM Symposium on the Theory of Computing (STOC 1995)*, pp. 284–293.

Ballester C, Calvó-Armengol A and Zenou Y 2006 Who's who in networks. Wanted: The key player. *Econometrica* **74** (5), 1403–1417.

Bazaraa MS and Shetty CM 1979 *Nonlinear Programming, Theory and Algorithms*. John Wiley & sons Ltd.

Bolla M 2010 Statistical inference on large contingency tables: convergence, testability, stability, in *Proc. 19th International Conference on Computational Statistics (COMPSTAT 2010), Paris* (eds Lechevallier Y and Saporta G), Physica-Verlag, Springer, pp. 817–824.

Bolla M, Kói T and Krámli A 2012 Testability of minimum balanced multiway cut densities. *Discret. Appl. Math.* **160**, 1019–1027.

Borgs C, Chayes JT, Lovász L, T.-Sós V and Szegedy B 2006 Graph limits and parameter testing. In *Proc. 38th Annual ACM Symposium on the Theory of Computing (STOC 2006)*, pp. 261–270.

Borgs C, Chayes JT, Lovász L *et al.* 2008 Convergent graph sequences I: Subgraph Frequencies, metric properties, and testing. *Advances in Math.* **219**, 1801–1851.

Borgs C, Chayes JT, Lovász L *et al.* 2011 Limits of randomly grown graph sequences. *European J. Comb.* **32**, 985–999.

Borgs C, Chayes JT, Lovász L, T.-Sós V and Vesztergombi K 2012 Convergent sequences of dense graphs II: Multiway cuts and statistical physics. *Ann. Math.* **176**, 151–219.

Chung F and Lu L. 2005 Concentration inequalities and martingale inequalities: A survey. *Internet Mathematics* **3**, 79–127.

Diaconis P and Janson S 2008 Graph limits and exchangeable random graphs. *Rend. Mat. Appl.* (VII. Ser.) **28**, 33–61.

Drineas P, Frieze A, Kannan R, Vempala S and Vinay V 2004 Clustering large graphs via the singular value decomposition. *Mach. Learn.* **56**, 9–33.

Elek G 2008 L^2-spectral invariants and convergent sequences of finite graphs. *J. Funct. Anal.* **254**, 2667–2689.

Érdi P and Tóth J 1990 What is and what is not stated by the May-Wigner theorem. *J. Theor. Biol.* **145**, 137–140.

Frieze A and Kannan R 1999 Quick approximation to matrices and applications. *Combinatorica* **19**, 175–220.

Goldreich O, Goldwasser S and Ron D 1998 Property testing and its connection to learning and approximation. *J. ACM* **45**, 653–750.

Hartigan JA 1972 Direct clustering of a data matrix. *J. Am. Stat. Assoc.* **67**, 123–129.

Juhász F 1996 On the structural eigenvalues of block random matrices. *Linear Algebra Appl.* **246**, 225–231.

Kluger Y, Basri R, Chang JT and Gerstein M 2003 Spectral biclustering of microarray data: coclustering genes and conditions. *Genome Res.* **13**, 703–716.

Lovász L 2008 Very large graphs, in *Current Developments in Mathematics* (eds Jerison D, Mazur B, Mrowka T, Schmid W, Stanley R and Yan ST), International Press, Somerville, MA, pp. 67–128.

Lovász L and Szegedy B 2006 Limits of dense graph sequences. *J. Comb. Theory B* **96**, 933–957.

Lovász L and Szegedy B 2007 Szemerédi's Lemma for the analyst. *Geom. Func. Anal.* **17**, 252–270.

Lovász L and Szegedy B 2011 Finitely forcible graphons. *J. Comb. Theory B.* **101**, 269–301.

Lovász L and T.-Sós V 2008 Generalized quasirandom graphs. *J. Comb. Theory B* **98**, 146–163.

Madeira SC and Oliveira AL 2004 Biclustering algorithms for biological data analysis: A survey. *IEEE-ACM Trans. Comput. Biol. Bioinform.* **1** (1), 24–45.

May RM 1972 Will a large complex system be stable? *Nature* **238**, 413–414.

Nelsen RB 2006 *An Introduction to Copulas*, Springer.

Reichardt J and Bornholdt S 2007 Partitioning and modularity of graphs with arbitrary degree distribution. *Phys. Rev. E* **76**, 015102(R).

5

Statistical learning of networks

In this chapter, classical and modern statistical methods are applied to find the underlying clusters of a given network or to classify graphs or contingency tables given some distinct prototypes. So far, we have used so-called nonparametric models, in that minimum multiway cuts, bicuts, and modularities were nonparametric statistics calculated on the graph. Although, in Chapter 4 we were speaking about parameter testing, the so-called testable parameters were, in fact, nonparametric statistics depending on the graph. We will also discuss parametric models in the classical sense, when the background distribution of the random graph depends on some parameters, and we use traditional methods to estimate them.

5.1 Parameter estimation in random graph models

We will discuss two basic types of parametric random graph models, and give algorithms for the maximum likelihood estimation of the parameters. Both models are capable of finding hidden partitions of the graph's vertices for a given number of clusters. Since these special clustering algorithms do not need any preliminary information on the clusters, they correspond to the unsupervised learning of the data at hand.

5.1.1 EM algorithm for estimating the parameters of the block-model

The so-called stochastic block-model, first introduced in Holland *et al.* (1981), later investigated by Bickel and Chen (2009); Karrer and Newman (2011); Rohe *et al.* (2011) among others, has already been discussed in Chapter 3. In fact, this is a generalized random graph model, formulated in terms of mixtures. The assumptions

Spectral Clustering and Biclustering: Learning Large Graphs and Contingency Tables, First Edition.
Marianna Bolla.

of the model are the following. Given a simple graph $G = (V, \boldsymbol{A})$ ($|V| = n$, with adjacency matrix $\boldsymbol{A}$) and k ($1 < k < n$), we are looking for the hidden k-partition $(V_1, \dots, V_k)$ of the vertices such that

- vertices are independently assigned to cluster V_a with probability π_a, $a = 1, \dots, k$; $\sum_{a=1}^{k} \pi_a = 1$
- given the cluster memberships, vertices of V_a and V_b are connected independently, with probability

$$\mathbb{P}(i \sim j \,|\, i \in V_a, j \in V_b) = p_{ab}, \quad 1 \leq a, b \leq k.$$

The parameters are collected in the vector $\boldsymbol{\pi} = (\pi_1, \dots, \pi_k)$ and the $k \times k$ symmetric matrix $\boldsymbol{P}$ of p_{ab}'s.

Our statistical sample is the $n \times n$ symmetric, 0-1 adjacency matrix $\boldsymbol{A} = (a_{ij})$ of G. There are no loops, so the diagonal entries are zeros. Based on $\boldsymbol{A}$, we want to estimate the parameters of the above block-model.

Using the theorem of mutually exclusive and exhaustive events, the likelihood function is the mixture of joint distributions of i.i.d. Bernoulli distributed entries:

$$\begin{aligned} &1/2 \sum_{1 \leq a,b \leq k} \pi_a \pi_b \prod_{i \in V_a, j \in V_b, i \neq j} p_{ab}^{a_{ij}} (1 - p_{ab})^{(1-a_{ij})} \\ &= 1/2 \sum_{1 \leq a,b \leq k} \pi_a \pi_b \cdot p_{ab}^{e_{ab}} (1 - p_{ab})^{(n_{ab} - e_{ab})}. \end{aligned}$$

This is the mixture of binomial distributions, where e_{ab} is the number of edges connecting vertices of V_a and V_b ($a \neq b$), while e_{aa} is twice the number of edges with both endpoints in V_a; further,

$$n_{ab} = |V_a| \cdot |V_b| \quad (a \neq b) \quad \text{and} \quad n_{aa} = |V_a| \cdot (|V_a| - 1) \quad (a = 1, \dots, k) \tag{5.1}$$

are the numbers of possible edges between V_a, V_b and within V_a, respectively.

Here $\boldsymbol{A}$ is the incomplete data specification as the cluster memberships are missing. Therefore, it is straightforward to use the *Expectation-Maximization*, briefly the EM algorithm, proposed by Dempster *et al.* (1977), also discussed for example, in Hastie, Tibshirani and Friedman (2001); McLachlan (1997) for parameter estimation from incomplete data. This special application for mixtures is sometimes called *collaborative filtering*, see Hofmann and Puzicha (1999); Ungar and Foster (1998).

First we complete our data matrix $\boldsymbol{A}$ with latent membership vectors $\boldsymbol{\Delta}_1, \dots, \boldsymbol{\Delta}_n$ of the vertices that are k-dimensional i.i.d. *Poly*$(1, \boldsymbol{\pi})$ (polynomially distributed) random vectors. More precisely, $\boldsymbol{\Delta}_i = (\Delta_{1i}, \dots, \Delta_{ki})$, where $\Delta_{ai} = 1$ if $i \in V_a$ and zero otherwise. Thus, the sum of the coordinates of any $\boldsymbol{\Delta}_i$ is 1, and $\mathbb{P}(\Delta_{ai} = 1) = \pi_a$.

Based on these, the likelihood function above is

$$1/2 \sum_{1 \leq a,b \leq k} \pi_a \pi_b \cdot p_{ab}^{\sum_{i \neq j} \Delta_{ai} \Delta_{bj} a_{ij}} \cdot (1 - p_{ab})^{\sum_{i \neq j} \Delta_{ai} \Delta_{bj} (1 - a_{ij})}$$

that is maximized in the alternating $\mathbf{E}$ and $\mathbf{M}$ steps of the EM algorithm.

Note that that the complete likelihood would be the squareroot of

$$\prod_{1\le a,b\le k} p_{ab}^{e_{ab}} \cdot (1-p_{ab})^{(n_{ab}-e_{ab})}$$
$$= \prod_{a=1}^{k}\prod_{i=1}^{n}\prod_{b=1}^{k}[p_{ab}^{\sum_{j:\, j\neq i}\Delta_{bj}a_{ij}} \cdot (1-p_{ab})^{\sum_{j:\, j\neq i}\Delta_{bj}(1-a_{ij})}]^{\Delta_{ai}} \tag{5.2}$$

that is valid only in the case of known cluster memberships.

Starting with initial parameter values $\boldsymbol{\pi}^{(0)}$, $\boldsymbol{P}^{(0)}$ and membership vectors $\boldsymbol{\Delta}_1^{(0)}, \dots, \boldsymbol{\Delta}_n^{(0)}$, the t-th step of the iteration is the following ($t = 1, 2, \dots$).

- **E**-step: we calculate the conditional expectation of each Δ_i conditioned on the model parameters and on the other cluster assignments obtained in step $t-1$ and collectively denoted by $M^{(t-1)}$. By the Bayes theorem, the responsibility of vertex i for cluster a is

$$\pi_{ai}^{(t)} = \mathbb{E}(\Delta_{ai} \mid M^{(t-1)}) = \frac{\mathbb{P}(M^{(t-1)}|\Delta_{ai}=1)\cdot \pi_a^{(t-1)}}{\sum_{b=1}^{k}\mathbb{P}(M^{(t-1)}|\Delta_{bi}=1)\cdot \pi_b^{(t-1)}}$$

  ($a = 1, \dots, k$; $i = 1, \dots, n$). For each i, $\pi_{ai}^{(t)}$ is proportional to the numerator, where

$$\mathbb{P}(M^{(t-1)}|\Delta_{ai}=1) = \prod_{b=1}^{k}(p_{ab}^{(t-1)})^{\sum_{j\neq i}\Delta_{bj}^{(t-1)}a_{ij}} \cdot (1-p_{ab}^{(t-1)})^{\sum_{j\neq i}\Delta_{bj}^{(t-1)}(1-a_{ij})}$$

  is the part of the likelihood (5.2) affecting vertex i under the condition $\Delta_{ai} = 1$.

- **M**-step: we maximize the truncated binomial likelihood

$$p_{ab}^{\sum_{i\neq j}\pi_{ai}^{(t)}\pi_{bj}^{(t)}a_{ij}} \cdot (1-p_{ab})^{\sum_{i\neq j}\pi_{ai}^{(t)}\pi_{bj}^{(t)}(1-a_{ij})}$$

  with respect to the parameter p_{ab}, for all a, b pairs separately. Obviously, the maximum is attained by the following estimators of p_{ab}'s comprising the symmetric matrix $\boldsymbol{P}^{(t)}$: $p_{ab}^{(t)} = \frac{\sum_{i,j:\, i\neq j}\pi_{ai}^{(t)}\pi_{bj}^{(t)}a_{ij}}{\sum_{i,j:\, i\neq j}\pi_{ai}^{(t)}\pi_{bj}^{(t)}}$ $(1 \le a \le b \le k)$, where edges connecting vertices of clusters a and b are counted fractionally, multiplied by the membership probabilities of their endpoints.

The maximum likelihood estimator of $\boldsymbol{\pi}$ in the t-th step is $\boldsymbol{\pi}^{(t)}$ of coordinates $\pi_a^{(t)} = \frac{1}{n}\sum_{i=1}^{n}\pi_{ai}^{(t)}$ ($a = 1, \dots, k$), while that of the membership vector $\boldsymbol{\Delta}_i$ is obtained by discrete maximization: $\Delta_{ai}^{(t)} = 1$ if $\pi_{ai}^{(t)} = \max_{b\in\{1,\dots,k\}} \pi_{bi}^{(t)}$, and 0 otherwise. (In case of ambiguity, the cluster with the smallest index is selected.) This choice of $\boldsymbol{\pi}$ will increase (or rather, not decrease) the likelihood function. Note that it is not necessary to assign vertices uniquely to the clusters, the responsibility π_{ai} of a vertex i can as well be regarded as the intensity of vertex i belonging to cluster a, giving rise to a fuzzy clustering.

According to the general theory of the EM algorithm, Csiszár and Shields (2004); Dempster *et al.* (1977), in exponential families (as in the present case), convergence to a local maximum can be guaranteed (depending on the starting values), but it runs in polynomial time in the number of vertices n. However, the speed and limit of the convergence depends on the starting clustering, which can be chosen by means of preliminary application of some nonparametric multiway cut algorithm of Section 5.2.

5.1.2 Parameter estimation in the α and β models

In the stochastic block-model of Subsection 5.1.1, within and between any pair of the vertex-clusters, edges came into existence with probability depending only on their endpoints' cluster memberships. However, there are more sophisticated models where the edge probabilities are not constant within the blocks.

Let us start with the one-cluster case. With different parametrization, Chatterjee *et al.* (2010) and Csiszár *et al.* (2011) introduced the following random graph model where the degree sequence is a sufficient statistic. We have a simple random graph on n vertices, there are no loops, and between distinct vertices edges come into existence independently, but not with the same probability. This random graph can uniquely be characterized by its $n \times n$ symmetric adjacency matrix $\boldsymbol{A} = (A_{ij})$ which has zero diagonal and the entries above the main diagonal are independent Bernoulli random variables whose parameters $p_{ij} = \mathbb{P}(A_{ij} = 1)$ obey the following rule. Actually, we formulate this rule for the $\frac{p_{ij}}{1-p_{ij}}$ ratios, the so-called *odds*:

$$\frac{p_{ij}}{1-p_{ij}} = \alpha_i \alpha_j \qquad i < j \tag{5.3}$$

where $\alpha_1, \dots, \alpha_n$ are positive parameters. This model is called *α-model* in Csiszár *et al.* (2011). With the parameter transformation $\beta_i = \ln \alpha_i$ $(i = 1, \dots n)$, it is equivalent to the *β-model* of Chatterjee *et al.* (2010) which applies to the *log-odds (or logits)*

$$\ln \frac{p_{ij}}{1-p_{ij}} = \beta_i + \beta_j \qquad i < j \tag{5.4}$$

with real parameters $\beta_1, \dots, \beta_n$. Of course, the probability p_{ij} or $1 - p_{ij}$ can be expressed in terms of the parameters, such as

$$p_{ij} = \frac{\alpha_i \alpha_j}{1+\alpha_i \alpha_j} = \frac{e^{\beta_i+\beta_j}}{1+e^{\beta_i+\beta_j}} \quad \text{and} \quad 1-p_{ij} = \frac{1}{1+\alpha_i \alpha_j} = \frac{1}{1+e^{\beta_i+\beta_j}} \tag{5.5}$$

which formulas will be intensively used in the subsequent calculations.

We are looking for the maximum likelihood estimate of the parameter $\boldsymbol{\alpha} = (\alpha_1, \ldots, \alpha_n)$ or $\boldsymbol{\beta} = (\beta_1, \ldots, \beta_n)$ based on the observed simple graph as a statistical sample. (It may seem that we have a one-element sample here; however, there are $\binom{n}{2}$ independent random variables in the background.) First we will prove that in the above random graph model the degree sequence is a sufficient statistic. Let $\mathbf{D} = (D_1, \ldots, D_n)$ denote the degree-vector of the above random graph, where $D_i = \sum_{j=1}^{n} A_{ij}$ $(i = 1, \ldots n)$. The random vector $\mathbf{D}$, as a function of the sample entries A_{ij}'s, is a statistic. Moreover, it is a sufficient statistic for the parameter $\boldsymbol{\alpha}$, or equivalently, $\boldsymbol{\beta}$. Roughly speaking, a sufficient statistic itself contains all the information – that can be retrieved from the sample – for the parameter. More precisely, a statistic is sufficient when the conditional distribution of the sample, given the statistic, does not depend on the parameter. By the Neyman–Fisher factorization theorem (see e.g., Rao (1973)), a statistic is sufficient if and only if the likelihood function of the sample can be factorized into two factors: one does not include the parameter, and the other (which includes the parameter) only contains the sample entries compressed into this sufficient statistic. In this way, we can recognize sufficient statistics that are far from being unique.

Consider the Neyman-Fisher factorization of the likelihood function (joint probability of A_{ij}'s) in our case. Because of the symmetry of $\boldsymbol{A}$, this is

$$
\begin{aligned}
L_{\boldsymbol{\alpha}}(\boldsymbol{A}) &= \prod_{i=1}^{n-1} \prod_{j=i+1}^{n} p_{ij}^{A_{ij}} (1 - p_{ij})^{1-A_{ij}} \\
&= \left(\prod_{i=1}^{n} \prod_{j=1}^{n} p_{ij}^{A_{ij}} (1 - p_{ij})^{1-A_{ij}} \right)^{1/2} \\
&= \left(\prod_{i=1}^{n} \prod_{j=1}^{n} \left(\frac{p_{ij}}{1 - p_{ij}} \right)^{A_{ij}} \prod_{i=1}^{n} \prod_{j=1}^{n} (1 - p_{ij}) \right)^{1/2} \\
&= \left(\prod_{i=1}^{n} \alpha_i^{\sum_{j=1}^{n} A_{ij}} \prod_{j=1}^{n} \alpha_j^{\sum_{i=1}^{n} A_{ij}} \right)^{1/2} \left(\prod_{i \neq j} (1 - p_{ij}) \right)^{1/2} \\
&= \left(\prod_{i \neq j} \frac{1}{1 + \alpha_i \alpha_j} \right)^{1/2} \cdot \left(\prod_{i=1}^{n} \alpha_i^{D_i} \prod_{j=1}^{n} \alpha_j^{D_j} \right)^{1/2} \\
&= \left(\prod_{i<j} \frac{1}{1 + \alpha_i \alpha_j} \right) \left(\prod_{i=1}^{n} \alpha_i^{D_i} \right) = C_{\boldsymbol{\alpha}} \prod_{i=1}^{n} \alpha_i^{D_i}
\end{aligned}
$$

where we used (5.5) and the facts that $A_{ij} = A_{ji}$, $p_{ij} = p_{ji}$ $(i > j)$ and $A_{ii} = 0$, $p_{ii} = 0$ $(i = 1, \ldots, n)$. Here the constant $C_{\boldsymbol{\alpha}} = \prod_{i<j} \frac{1}{1+\alpha_i \alpha_j}$ only dependson $\boldsymbol{\alpha}$, and the whole likelihood function depends on the A_{ij}'s merely through D_i's.

Therefore, $\mathbf{D}$ is a sufficient statistic. What is the other factor? It is constantly 1, indicating that the conditional joint distribution of the entries – given $\mathbf{D}$ – is uniform, but we will not make use of this fact. Just note that Newman and Park (2004) call the uniform distribution on graphs with fixed degree sequence microcanonical. The whole model comes from the so-called log-linear way of model building, see Lauritzen (1988).

Let (a_{ij}) be the matrix of the sample realizations (the adjacency entries of the observed graph), $d_i = \sum_{j=1}^{n} a_{ij}$ be the actual degree of vertex i ($i = 1, \dots, n$) and $\mathbf{d} = (d_1, \dots, d_n)$ be the observed degree-vector. The above factorization also indicates that the joint distribution of the entries belongs to the exponential family (see e.g., Rao (1973)), and hence, the maximum likelihood estimate $\hat{\boldsymbol{\alpha}}$ (or equivalently, $\hat{\boldsymbol{\beta}}$) is derived from the fact that with it, the observed degree d_i is equal to the expected one, which is $\mathbb{E}(D_i) = \sum_{i=1}^{n} p_{ij}$. Therefore, $\hat{\boldsymbol{\alpha}}$ is the solution of the following so-called *likelihood equation*:

$$d_i = \sum_{j \neq i}^{n} \frac{\alpha_i \alpha_j}{1 + \alpha_i \alpha_j} \quad (i = 1, \dots, n). \tag{5.6}$$

Note that the same system of equation is obtained by differentiating the likelihood function with respect to the parameters and equating the partial derivatives to zero. Of course, the maximum likelihood estimate $\hat{\boldsymbol{\beta}}$ is easily obtained from $\hat{\boldsymbol{\alpha}}$ via taking the logarithms of its coordinates.

Before discussing the solution of the system of equations (5.6), let us see what conditions a sequence of nonnegative integers should satisfy so that it could be realized as the degree sequence of a graph.

Definition 5.1.1 *The sequence $d_1, \dots, d_n$ of nonnegative integers is called graphic if there is a simple graph on n vertices such that its vertex-degrees are the numbers $d_1, \dots, d_n$ in some order.*

Without loss of generality, d_i's can be enumerated in non-increasing order. The following theorem, proved by Erdős and Gallai (1960), gives a necessary and sufficient condition for a sequence to be graphic.

Theorem 5.1.2 (Erdős–Gallai theorem) *The sequence $d_1 \geq d_2 \geq \cdots \geq d_n \geq 0$ of integers is graphic if and only if it satisfies the following two conditions:*

$$\sum_{i=1}^{n} d_i \quad \text{is even}$$

and

$$\sum_{i=1}^{k} d_i \leq k(k-1) + \sum_{i=k+1}^{n} \min\{k, d_i\}, \quad k = 1, \dots, n-1. \tag{5.7}$$

Note that for nonnegative (not necessarily integer) real sequences a continuous analogue of (5.7) is derived in Chatterjee *et al.* (2010). For given n, the convex hull of all possible graphic degree sequences is a polytope, to be denoted by $\mathcal{D}_n$. Its extreme points are the so-called threshold graphs, see Mahadev and Peled (1995) for details.

For given $n \geq 2$, Csiszár *et al.* (2011) prove that if $\mathbf{d} \in \mathcal{D}_n$ is an interior point, then the maximum likelihood equation (5.6) has a unique solution. They recommend the following algorithm and prove that the iteration of it converges to this unique solution. To motivate the iteration, we rewrite (5.6) as

$$d_i = \alpha_i \sum_{j \neq i} \frac{1}{\frac{1}{\alpha_j} + \alpha_i} \qquad (i = 1, \ldots, n).$$

Then the iteration is as follows.

Algorithm: for the *maximum likelihood estimation* in the α-model

---

Input: adjacency matrix of a simple graph.

1. Calculate the graphic degree sequence $d_1, \ldots, d_n$.

2. Initialize: $\alpha_1^{(0)}, \ldots, \alpha_n^{(0)}$.

3. Iterate: **for** t=1,2,...
   $\alpha_i^{(t)} = \frac{d_i}{\sum_{j \neq i} \frac{1}{\frac{1}{\alpha_j^{(t-1)}} + \alpha_i^{(t-1)}}} \quad (i = 1, 2, \ldots, n)$
   until convergence.

Output: maximum likelihood estimates $\alpha_1, \ldots, \alpha_n$ or $\beta_1 = \ln \alpha_1, \ldots, \beta_n = \ln \alpha_n$.

Under mild assumptions on the parameter $\boldsymbol{\beta}$, Chatterjee *et al.* (2010) prove the consistency of the maximum likelihood estimate and also give the uniform accuracy for the $|\hat{\beta}_i - \beta_i|$ differences (depending only on n). The authors discuss the topic in the more general framework of the convergent graph sequences and graphons of Chapter 4. Namely, they prove that if a random graph is uniformly chosen with a given degree sequence, then sequences of such graphs have limits in the sense of Definition 4.1.1.

This kind of exponential model traces back to Rasch (1960) and was used for psychological and educational measurements, and later market research. The frequently cited Rasch model involves categorical data, mainly binary variables, therefore the underlying random object can be thought of as a contingency table. According to the Rasch model, the entries of an $m \times n$ contingency table are independent Bernoulli random variables, where for the parameter p_{ij} of the entry in the (i, j) position the following holds:

$$\ln \frac{p_{ij}}{1 - p_{ij}} = \beta_i - \delta_j \qquad i = 1, \ldots m; \; j = 1, \ldots, n \tag{5.8}$$

with real parameters $\beta_1, \dots, \beta_m$ and $\delta_1, \dots, \delta_n$. Based on a binary table, the author also derived maximum likelihood estimation for the parameters. For those who are interested in psychology, we tell that Rasch (1960) investigated contingency tables where the rows corresponded to persons and the columns to items of some psychological test, whereas the jth entry of the ith row was 1 if person i answered test item j correctly, and 0 otherwise. He also gave a description of the parameters: β_i was the ability of person i, while δ_j the difficulty of test item j. Therefore, in view of the model equation (5.8), the more intelligent the person and the less difficult the test, the larger the success to failure ratio was on a logarithmic scale.

Without this psychological meaning, the model works for bipartite graphs with equivalent parameter sets $\beta_1, \dots, \beta_m$ and $\gamma_1, \dots, \gamma_n$ in the following way:

$$\ln \frac{p_{ij}}{1 - p_{ij}} = \beta_i + \gamma_j \quad (i \in A,\ j \in B) \tag{5.9}$$

where A and B are disjoint independent sets of the underlying bipartite graph's vertices.

The Rasch model can as well be adapted to directed graphs in the following way. Using the notation of Subsection 3.3.3, a simple directed graph on n vertices is given by its $n \times n$ (usually not symmetric) adjacency matrix $\mathbf{A}$ with zero diagonal; further, $A_{ij} = 1$ if there is an $i \to j$ edge, and 0 otherwise ($i \neq j$). In the model, the non-diagonal entries are independent Bernoulli random variables with $\mathbb{P}(A_{ij} = 1) = p_{ij}$ such that

$$\ln \frac{p_{ij}}{1 - p_{ij}} = \beta_i + \gamma_j, \quad i \neq j$$

where now the in- and out-degrees are the sufficient statistics for the real parameters $\beta_1, \dots, \beta_n$ and $\gamma_1, \dots, \gamma_n$; furthermore, the maximum likelihood estimation works likewise as in the undirected case.

Going further, the model can be generalized for the k-cluster case. Given $1 \le k \le n$, we are looking for k-partition $(V_1, \dots, V_k)$ of the vertices such that

- vertices are independently assigned to cluster V_a with probability π_a, $a = 1, \dots, k$; $\sum_{a=1}^{k} \pi_a = 1$
- given the cluster memberships, vertices $i \in V_a$ and $j \in V_b$ are connected independently, with probability p_{ij} such that

  $$\ln \frac{p_{ij}}{1 - p_{ij}} = \beta_{ib} + \beta_{ja}$$

  for any $1 \le a, b \le k$ pair.

The parameters are collected in the vector $\boldsymbol{\pi} = (\pi_1, \dots, \pi_k)$ and the $n \times k$ matrix of β_{ia}'s. To estimate the parameters, the *EM algorithm* can again be used. In the $\mathbf{E}$-step we calculate the responsibilities of the vertices for the clusters (see Subsection 5.1.1). In the **M**-step, for the within-cluster edges we use the parameter estimation of model (5.4), obtaining estimates of β_{ia}'s ($i \in V_a$) in each cluster separately ($a = 1, \dots, k$); whereas for the inter-cluster edges we use the bipartite graph model (5.9) in the following way. For $a \neq b$, edges connecting vertices of V_a and V_b form a bipartite graph, based on which the parameters β_{ib} ($i \in V_a$) and β_{ja} ($j \in V_b$) are estimated either with the method of Rasch or that of the simple graph algorithm extended to the bipartite case. With the estimated parameters we go back to the **M**-step, etc. At the end, we relocate a vertex into the cluster for which it has the largest responsibility. By the general theory of the EM algorithm, since we are in the exponential family, the iteration will converge.

Note that here the parameter β_{ib} for $i \in V_a$ embodies the affinity of vertex i of cluster V_a towards vertices of cluster V_b; and likewise, β_{ja} for $j \in V_b$ embodies the affinity of vertex j of cluster V_b towards vertices of cluster V_a. By the model, this affinities are added together on the level of the log-odds. This so-called k-β model, introduced in Csiszár *et al.* (2012), is applicable to social networks, where attitudes of persons in the same social group (say, a) are the same toward members of another social group (say, b), though, this attitude also depends on the person in group a, and vice versa. The model may also be applied to biological networks, where the clusters consist, for example, of different functioning cells or units of the brain.

There are a lot of other parametric and semiparametric random graph models in the literature. Among those, we discussed the model of Chaudhuri *et al.* (2012) in Subsection 3.1.5. This is an extended planted partition model, and the authors also propose an algorithm to find the clusters and estimate the parameters involved.

5.2 Nonparametric methods for clustering networks

Now we are not interested in the parameter estimation of the block-model described in Subsection 5.1.1, we are rather interested to find k clusters of the vertices, for given k. To this end, nonparametric statistics, like multiway cuts or the Newman–Girvan modularity is minimized or maximized over the k-partitions of the vertices. Under certain balancing conditions, we proved the testability of some of these statistics, called testable graph parameters in Chapter 4. Bickel and Chen (2009) investigated the consistency of the clusterings obtained by the following modularities.

The *likelihood modularity* is, in fact, the logarithm of the squareroot of the likelihood function in (5.2), where – with the notations (5.1) – the relative frequency $\frac{e_{ab}}{n_{ab}}$ is substituted for p_{ab}, yielding

$$LM(P_k, \mathbf{A}) = 1/2 \sum_{1 \le a,b \le k} \left[e_{ab} \ln \frac{e_{ab}}{n_{ab}} + (n_{ab} - e_{ab}) \ln \left(1 - \frac{e_{ab}}{n_{ab}} \right) \right]$$

($\mathbf{A}$ is the $n \times n$ adjacency matrix of a simple graph). This modularity is maximized with respect to the k-partitions $P_k = (V_1, \dots, V_k)$ of the vertices, or equivalently with respect to the hidden membership vector $\boldsymbol{m} \in \mathbb{R}^n$ such that its ith coordinate is the membership of vertex i: $m_i = a$ if and only if $i \in V_a$. Under mild conditions, Bickel and Chen (2009) prove that by maximizing the likelihood modularity with respect to $\boldsymbol{m}$, theoretically one gets a consistent estimate of the true membership vector (that of the block-model). This also means that we can appropriately decrease the misclassification probabilities as $n \to \infty$. To maximize $LM(P_k, \boldsymbol{A})$ in practice, the following iteration could be used. Starting with an initial k-partition, in the first step we estimate the parameters p_{ab} by recalculating the $\frac{e_{ab}}{n_{ab}}$ ratios between and within the clusters. In the second step, we find new memberships of the vertices by relocating them to the cluster where their neighborhood (edges emanating from them) have the largest likelihood. Then we repeat these steps until convergence, which can be expected, since the likelihood modularity is increased (not decreased) in each step, and the likelihood is bounded from above. We can think of this algorithm as the greedy version of the EM algorithm described in Subsection 5.1.1, which is a bit tedious with estimating a lot of parameters through the iteration process.

The Newman–Girvan modularity (see Definition 2.4.1 for edge-weighted graphs) has the following form for the simple graph with adjacency matrix $\boldsymbol{A}$:

$$M(P_k, \boldsymbol{A}) = 2e \sum_{a=1}^{k} \left[\frac{e_{aa}}{2e} - \left(\frac{d_a}{2e} \right)^2 \right]$$

where $d_a = \sum_{b=1}^{k} e_{ab}$, whereas $2e = \sum_{a=1}^{k} d_a$ is twice the number of edges among all pairs of vertices.

Bickel and Chen (2009) show that maximizing the Newman–Girvan modularity not always results in a consistent estimate, but it may be consistent in the submodel

$$\sqrt{|V_a|} \cdot e_{aa} > \sum_{b \neq a} \sqrt{|V_b|} \cdot e_{ab}, \quad a = 1, \dots, k.$$

Note that the Newman–Girvan modularity focuses on the diagonal blocks and capable to find clusters of intra-cluster connections larger than expected under independent attachment of the vertices (see Section 2.4 for details).

In the sequel we will summarize the spectral clustering algorithms applicable to find a clustering close to those obtainable by minimizing some normalized cuts of edge-weighted graphs or bicuts of contingency tables, or just minimizing the pairwise discrepancy of the clusters. Minimization over k-partitions of vertices (or rows and columns) is NP-complete, therefore SD or SVD of matrices (depending on the objective function) is used. Of course, these spectral methods – as continuous relaxations of the discrete problem – give only an approximate solution to the original problem, but in polynomial time (in n). Concerning the accuracy of these approximations a lot of statements were proved in Chapters 2 and 3. These statements – via inspecting the signs of the structural eigenvalues – also give us suggestions about the optimum choice of k and the nature of the clusters themselves.

A drawback of the spectral methods is that the structural eigenvalues with corresponding eigenvectors are only capable of revealing fundamental clusters (this is why they are related to balanced clusters in Chapter 4), while small clusters are hidden behind the near-zero eigenvalues. It seems straightforward to enter eigenvectors belonging to 'small' eigenvalues into the representation based classification, but it would cause complications: partly because in case of 'large' graphs there are too many small eigenvalues (the normalized modularity spectrum has tendency accumulate around zero) and partly because small eigenvalues can be indications of small clusters and low degree vertices at the same time. These considerations are valid for large and dense enough graphs. For sparse ones, a so-called core of the graph can be separated which is used to find the underlying clusters.

5.2.1 Spectral clustering of graphs and biclustering of contingency tables

Algorithms, using the results of Section 2.2 and 3.3 are introduced to find minimum normalized or regular cuts of edge-weighted graphs. We will give a pseudocode which can be used for different purposes after preliminary inspection of the spectrum. Given a connected edge-weighted graph $G = (V, \boldsymbol{W})$ on n vertices, first we calculate the SD of its normalized modularity matrix $\boldsymbol{M}_D$. Denote by

$$1 \geq |\mu_1| \geq |\mu_2| \geq \cdots \geq |\mu_n| = 0$$

the eigenvalues in decreasing absolute values, as in Theorem 3.3.3. If n is 'very large', it suffices to find some leading eigenvalues. For this purpose fast numerical algorithms are available, for example the Lánczos method (see Golub and van Loan (1989)). Then select a k such that there is a gap between $|\mu_{k-1}|$ and $|\mu_k|$. We distinguish between the three following cases.

- If $\mu_1, \ldots, \mu_{k-1}$ are all positive, the output of the algorithm will be an approximation for the minimum normalized k-way cut of G with intra-cluster edge densities significantly higher than the inter-cluster ones (community structure).
- If $\mu_1, \ldots, \mu_{k-1}$ are all negative, the output of the algorithm will be an approximation for the maximum normalized k-way cut of G with intra-cluster edge densities significantly lower than the inter-cluster ones (anticommunity structure).
- If there are both positive and negative ones among $\mu_1, \ldots, \mu_{k-1}$, the output of the algorithm will be a clustering of the vertices with relatively homogeneous edge-densities within the clusters and between any pairs of them (clustering with 'small' discrepancy). Some of the clusters or pairs of the clusters may have low, some of them may have high intra- or inter-cluster edge-density, as special cases.

If there is no noticeable gap in the spectrum, we may select a relatively 'small' k such that $|\mu_k|$ and $|\mu_{k+1}|$ do not differ too much. We can also select k, starting from 2 up to a reasonable value.

Algorithm: finding *homogeneous clusters* of an edge-weighted graph

---

Input: $n \times n$ edge-weight matrix $\boldsymbol{W}$ and the number of clusters k.

1. Calculate the degree-vector $\mathbf{d}$, degree-matrix $\boldsymbol{D}$, and the normalized modularity matrix $\boldsymbol{M}_D = \boldsymbol{D}^{-1/2}\boldsymbol{W}\boldsymbol{D}^{-1/2} - \sqrt{\mathbf{d}}\sqrt{\mathbf{d}}^T$.

2. Compute the $k-1$ largest absolute value eigenvalues and the corresponding eigenvectors $\mathbf{u}_1, \dots, \mathbf{u}_{k-1}$ of $\boldsymbol{M}_D$.

3. Find representatives $\mathbf{r}_1, \dots, \mathbf{r}_n$ of the vertices as row vectors of the matrix $\boldsymbol{D}^{-1/2}\mathbf{u}_1, \dots, \boldsymbol{D}^{-1/2}\mathbf{u}_{k-1}$.

4. Cluster the points $\mathbf{r}_1, \dots, \mathbf{r}_n$ by the weighted k-means algorithm into k clusters.

Output: clusters $V_1, \dots, V_k$ of the vertices.

The weighted k-means algorithm of Section C.5 is as follows.

Algorithm: *weighted k-means clustering*

Input: finite dimensional points $\mathbf{r}_1, \dots, \mathbf{r}_n$ with weights $d_1, \dots, d_n$ and the number of clusters k.

1. Initialize: $V_1^{(0)}, \dots, V_k^{(0)}$, the clusters of $\{1, \dots, n\}$.

2. Iterate: **for** t=1,2,...
   a. calculate the cluster centers $\mathbf{c}_a^{(t)} = \frac{1}{\sum_{j \in V_a^{(t-1)}} d_j} \sum_{j \in V_a^{(t-1)}} d_j \mathbf{r}_j \quad (a = 1, \dots, k)$;

   b. relocate the points: $\mathbf{r}_j$ is assigned to cluster $V_a^{(t)}$ for which $\|\mathbf{r}_j - \mathbf{c}_a^{(t)}\|$ is minimum
   until convergence.

Output: clusters $V_1, \dots, V_k$ of $\{1, \dots, n\}$.

Now we will formulate a similar algorithm to find homogeneous biclustering of a non-decomposable contingency table $\boldsymbol{C}$ on row-set $Rov = \{1, \dots, m\}$ and column-set $Col = \{1, \dots, n\}$. First we inspect the singular spectrum $1 = s_0 > s_1 \geq s_2 \geq \dots \geq s_r > 0$ of the normalized table $\boldsymbol{C}_{corr}$ of rank r. Then we disregard the singular value 1, and select a k such that there is a gap between s_{k-1} and s_k.

Algorithm: finding *homogeneous biclustering* of a contingency table

---

Input: $m \times n$ non-decomposable contingency table $\boldsymbol{C}$ and the number of clusters k.

1. Compute the row- and column-sums, $d_{row,1}, \dots, d_{row,m}$ and $d_{col,1}, \dots, d_{col,n}$; form the diagonal matrices $\boldsymbol{D}_{row} = \text{diag}(d_{row,1}, \dots, d_{row,m})$ and $\boldsymbol{D}_{col} = \text{diag}(d_{col,1}, \dots, d_{col,n})$; form the normalized contingency table $\boldsymbol{C}_{corr} = \boldsymbol{D}_{row}^{-1/2} \boldsymbol{C} \boldsymbol{D}_{col}^{-1/2}$.

2. Compute the $k-1$ largest singular values (disregarding the 1) and the corresponding left and right singular vectors of $\boldsymbol{C}_{corr}$: $\mathbf{v}_1, \dots, \mathbf{v}_{k-1}$ and $\mathbf{u}_1, \dots, \mathbf{u}_{k-1}$;

3. find representatives $\mathbf{r}_1, \dots, \mathbf{r}_m$ of the rows as row vectors of the matrix $(\boldsymbol{D}_{row}^{-1/2}\mathbf{v}_1, \dots, \boldsymbol{D}_{row}^{-1/2}\mathbf{v}_{k-1})$; find representatives $\mathbf{q}_1, \dots, \mathbf{q}_n$ of the columns as row vectors of the matrix $(\boldsymbol{D}_{col}^{-1/2}\mathbf{u}_1, \dots, \boldsymbol{D}_{col}^{-1/2}\mathbf{u}_{k-1})$;

4. cluster the points $\mathbf{r}_1, \dots, \mathbf{r}_m$ by the weighted k-means algorithm with weights $d_{row,1}, \dots, d_{row,m}$ into k clusters; cluster the points $\mathbf{q}_1, \dots, \mathbf{q}_n$ by the weighted k-means algorithm with weights $d_{col,1}, \dots, d_{col,n}$ into k clusters.

Output: clusters $R_1, \dots, R_k$ of the row-set $\{1, \dots, m\}$ and clusters $C_1, \dots, C_k$ of the column-set $\{1, \dots, n\}$.

Note that the same algorithm can be used to find simultaneously regular in- and out-vertex cluster pairs of a directed graph. With the notation of Subsection 3.3.3, instead of $\boldsymbol{C}$, we use the $n \times n$ (usually not symmetric) edge-weight matrix of zero diagonal and the entry w_{ij} is the weight of the $i \to j$ edge. The normalized matrix will be $\boldsymbol{W}_{corr} = \boldsymbol{D}_{out}^{-1/2} \boldsymbol{W} \boldsymbol{D}_{in}^{-1/2}$, where $\boldsymbol{D}_{out}$ and $\boldsymbol{D}_{in}$ are diagonal matrices containing of the out- and in-degrees (row and column sums of $\boldsymbol{W}$) in their main diagonals.

5.2.2 Clustering of hypergraphs

With the notation of Subsection 1.1.2, let $H = (V, E)$ be a hypergraph with vertex-set $V = \{v_1, v_2, \dots, v_n\}$, edge-set $E = \{e_1, e_2, \dots, e_m\}$, and $m \times n$ incidence matrix $\boldsymbol{B}$ of entries $b_{ij} = \mathcal{I}(v_j \in e_i)$. For example, $v_1, v_2, \dots, v_n$ are binary random variables representing some properties (say, symptoms) taking the values 0-1 (no-yes) and $e_1, e_2, \dots, e_m$ is a sample for them consisting of objects having or not these properties (say, patients having or not the symptoms). In the microarray setup (see Madeira and Oliveira (2004)), the vertices are conditions, the edges genes, and the incidence relation indicates whether the given gene is expressed under the given condition or not (binary table). Usually, $n \ll m$.

Let $E' \subset E$ be a *sub-sample*. The sub-hypergraph $H' = (V, E')$ is called the *hypergraph of the edge-cluster* E'. Let $0 = \lambda_0(H') \le \lambda_1(H') \le \cdots \le \lambda_{n-1}(H')$ be the Laplacian spectrum of H' (see Definition 1.1.4), whereas the $n \times n$ matrix $\mathbf{X}^*(H')$ contains the corresponding pairwise orthonormal eigenvectors in its columns. According to the Representation theorem for hypergraphs (see Theorem 1.1.5), for any integer d $(1 \le d \le n)$ the $d \times n$ matrix $\mathbf{X}_d^*(H')$ – obtained from $\mathbf{X}^*(H')$ by retaining the eigenvectors corresponding to $\lambda_0(H'), \lambda_1(H') \ldots, \lambda_{d-1}(H')$ – determines an optimal d-dimensional representation of H'. Furthermore, this is the representation with which the sum of the cost of edges of E' (see (1.7)) is minimal, and it is equal to

$$Q(\mathbf{X}_d^*(H')) = \sum_{e \in E'} Q(e, \mathbf{X}_d^*(H')) = \sum_{j=1}^{d-1} \lambda_j(H').$$

The cost of the embedding of H' is defined by

$$K(H') = \min_{d \in \{1,\ldots,n\}} [c2^{n-d} + Q(\mathbf{X}_d^*(H'))]$$

where $c > 0$ is a constant (chosen previously depending on the size of problem). Since the first term is a decreasing, while the second one is an increasing function of d, this discrete minimization problem always has a solution. The dimension d^* giving the minimum is called the *dimension of the edge-cluster* E'.

Let $\mathcal{S}$ denote the set of all partitions of E into non-empty disjoint sub-samples. Our purpose is to find a partition $S \in \mathcal{S}$ consisting of sub-samples E_i-s for which the objective function $K = \sum_i K(H_i)$ is minimal, where $H_i = (V, E_i)$ is the hypergraph of the edge-cluster E_i.

Now let k be a fixed integer, $(1 \le k \le n)$. We will relax the problem to minimize over k-partitions $(E_1, \ldots, E_k) \in \mathcal{S}_k$ of E. With the previous notation, the following cost function is constructed: $Q = \sum_{i=1}^{k} Q_{d_i}(H_i)$, where

$$Q_{d_i}(H_i) := c2^{n-d_i} + Q(\mathbf{X}_{d_i}^*(H_i)), \quad (i = 1, \ldots, k).$$

To minimize the cost function Q – with respect to the k-partitions of the edges and the dimensions of the edge-clusters – the following iteration is introduced. First let us choose k disjoint clusters $E_1, \ldots, E_k$ of the edges (e.g., by the k-means algorithm). Then iterate the following two steps:

1. Fixing the clusters $E_1, \ldots, E_k$, the spectra and optimal Euclidean representations of the sub-hypergraps $H_i = (V, E_i)$ $(i = 1, \ldots, k)$ are calculated. The function $Q_{d_i}(H_i)$ is minimized with respect to the dimension d_i $(1 \le d_i \le n)$

for each $i = 1, \ldots, k$ separately. A unique d_i^* giving the i-th minimum always exists, and for it

$$Q_{d_i^*}(H_i) = c2^{n-d_i^*} + \sum_{j=1}^{d_i^*} \lambda_j(H_i) \quad (i = 1, \ldots, k)$$

holds. In this step the cost function Q is decreased. Until this moment the minimization took place within the clusters.

2. In the next step the edges are relocated between the clusters as follows. Fixing the d_i^*-dimensional optimal Euclidean representations $\mathbf{X}^*_{d_i^*}(H_i)$-s, an edge e is replaced into the cluster E_i, for which $Q(e, \mathbf{X}^*_{d_i^*}(H_i))$ is minimal (see (1.7)). If the minimum is taken for more than one i, let us replace e into the cluster E_i with the smallest index i. In this step Q is also decreased.

In this way a new disjoint classification $E_1^*, \ldots, E_k^*$ of the edges is obtained. From now on, we go back to Step 1 with starting classification $E_1^*, \ldots, E_k^*$, etc.

Since the cost function Q is in each step decreased and in Steps 1 and 2 discrete minimizations are performed, the algorithm must converge to a local minimum of Q in finite steps. As a new step of the iteration, a minimization with respect to k could be introduced, but it would be very time-demanding. We rather inspect the whole Laplacian spectrum of H at the beginning, and try to find a k at a gap in the spectrum and adjust the constant c to it.

During the iteration some edge-clusters may become empty. Also the hypergraph $H_i = (V, E_i)$ may contain isolated vertices (this results in additional zero eigenvalues). Let us denote by V_i the set of the non-isolated vertices of H_i. Provided H has no isolated vertices, $\cup_{i=1}^k V_i = V$ and $V_1, \ldots, V_k$ are not necessarily disjoint subsets of the vertices. V_i is called the *characteristic property-association* of the sub-sample E_i. In this way, the algorithm finds disjoint clusters of the hyperedges, and corresponding (usually overlapping) clusters of the vertices. For other biclustering and block clustering algorithms of similar flavor see Hartigan (1972); Kluger *et al.* (2003); Madeira and Oliveira (2004), however, they do not use spectral methods.

5.3 Supervised learning

In the supervised learning setup, the classes are specified by an expert, and we want to reproduce them based on the data. This problem is the task of discriminant analysis, see Duda *et al.* (2000) or Section C.7 for a summary. In the context of graphs and contingency tables, two types of classification problems emerge.

In the first type, the objects to be classified are vertices of a graph, or rows and columns of a contingency table. Given a classification by an expert, we want to imitate the same classification based on the pairwise relations (edge-weights) of the vertices or entries of the contingency table. When the expert-made classification has something to do with the similarity structure of the graph, there is a hope that the discriminant analysis based algorithm will give back the classes with relatively 'small' proportion of missclassifications. For example, if the vertices are financial units (say, banks) and the edge-weights are proportional to the amount of transactions between them, then the algorithm is capable of giving back classes which have intensive intra-class transactions. When the expert gives classes such that a class is comprised of financial units of the same economic group, then there is a hope that the algorithm will reproduce the expert-made classification with good performance. However, if the classes are countries or some other historical formations, one might not expect to reproduce them based on pairwise relations between units of them.

To perform the discriminant analysis, we must have finite dimensional data points assigned to the vertices of the graph or rows and columns of a contingency table. For this purpose, we can use the representation techniques introduced in Chapter 1. The dimension is chosen after we have inspected the normalized Laplacian or modularity spectrum of the edge-weighted graph (or singular spectrum of the normalized contingency table). We may also have some practical considerations, for example, the dimension cannot be too 'large' if the number of vertices (or size of the contingency table) is already 'large'. In fact, the classical discriminant analysis of Fisher is based on maximizing (C.8), via the between- and within-class covariances of the Gaussian distributed data. The assumption of multivariate normality is usually not met for the representatives. However, with the notation of Section 2.1 we can overcome this difficulty as follows. Let us decompose the edge-weight matrix as $\boldsymbol{W} = \boldsymbol{W}_b + \boldsymbol{W}_w$, where $\boldsymbol{W}_b$ and $\boldsymbol{W}_w$ denote the edge-weights retaining the between- and within-cluster ones, in the classification made by the expert, respectively. Let as denote by $\boldsymbol{D}_b$ and $\boldsymbol{D}_w$ the degree-matrices, accordingly. Then the matrices $\boldsymbol{D}_b + \boldsymbol{W}_b$ and $\boldsymbol{D}_w + \boldsymbol{W}_w$ are like between- and within-cluster covariance matrices (the signless Laplacians of Section 1.1), with which we can perform the classical discrimination. For further approaches and applications, see for example Yan *et al.* (2007).

In the second type problems, the objects to be classified are edge-weighted graphs or contingency tables themselves. For example, we want to classify pictures, fingerprints, or microarrays. A preliminary classification is made by an expert, for example the classes are collections of pictures of separate individuals. Our purpose is to recognize the person based on a new picture. In this context, it is important that we could embed the graphs (or contingency tables) into finite dimensional spaces or to define distances of them (even if they are not of the same size).

The book Riesen and Bunke (2010) is entirely devoted to graph-based pattern recognition and shows how classical methods, originally developed for statistical pattern recognition, can be used on this area. To this end, the authors define a new class of graph embedding procedures based on dissimilarity representation of graphs. The key idea is to use the dissimilarities of an input graph to a number of training graphs, termed prototypes, as its coordinates in a vectorial description. A number of methods using the edit distance of graphs is proposed. The edit distance is, roughly speaking, the minimum number of operations – vertex and edge deletions and insertions – transforming a graph into another one. Because of high computational complexity, the application of these methods is limited. There are many real-life examples in the book. Among others, the authors use letter-, digit-, fingerprint-, protein-, and webpage graphs which can be directed or undirected, weighted or unweighted. To give a simple example: one wants to recognize letters. For this purpose the sections of letters are transformed into graphs, and each letter of the alphabet has a prototype. After this, an ugly handwritten letter is classifed to a letter of the alphabet based on its dissimilarities to these prototypes via usual techniques of supervised learning, when training and test data are separated.

The key idea of the classification of graphs or contingency tables themselves is that we would be able to calculate distances between them. The edit distance is mainly used for simple graphs of the same size (see Riesen and Bunke (2010)). The advantage of the cut distance, introduced in Chapter 4, is that – through the graphon or contingon extension – it can be calculated between edge-, even vertex-weighted graphs of different sizes (or between general contingency tables of different sizes with special row- and column-weights). Based on these distances, the test graphs or contingency tables can be assigned to one or more given prototypes (like the graph of pictures where the vertices are the pixels, see Figure 2.3; or microarrays corresponding to typical states). Going further, based on these distances, pairwise similarities between the graphs or contingency tables of different sizes can be calculated, and in this way we can build a graph whose vertices are graphs and the edge-weights are pairwise similarities between them. Based on low dimensional Euclidean representation of this new graph, techniques of discriminant analysis can be used, akin to the first type problems.

References

Bickel PJ and Chen A 2009 A nonparametric view of network models and Newman-Girvan and other modularities. *Proc. Natl. Acad. Sci. USA* **106** (50), 21068–21073.

Chatterjee S, Diaconis P and Sly A 2010 Random graphs with a given degree sequence. *Ann. Stat.* **21**, 1400–1435.

Chaudhuri K, Chung F and Tsiatas A 2012 Spectral clustering of graphs with general degrees in the extended planted partition model, in *JMLR Workshop and Conference Proceedings. Proc. 25th Annual Conference on Learning Theory (COLT 2012)* (eds Mannor S, Srebro N and Williamson RC), Vol. 23, Edinbourgh, Scotland, pp. 35.1–35.23.

Csiszár I and Shields P 2004 Information Theory and Statistics: A Tutorial. *Foundations and Trends in Communications and Information Theory*, Volume 1, Issue 4, Now Publishers, USA.

Csiszár V, Hussami P, Komlós J, Móri TF, Rejtő L and Tusnády G 2011 When the degree sequence is a sufficient statistic. *Acta Math. Hung.* **134**, 45–53.

Csiszár V, Hussami P, Komlós J, Móri TF, Rejtő L and Tusnády G 2012 Testing goodness of fit of random graph models. *Algorithms* **5**, 629–635.

Dempster AP, Laird NM and Rubin DB 1977 Maximum likelihood from incomplete data via the EM algorithm. *J. R. Statist. Soc. B* **39**, 1–38.

Duda RO, Hart, PE and Stork DG 2000 *Pattern Classification*, John Wiley & Sons, Ltd.

Erdős P and Gallai T 1960 Graphs with given degree of vertices (in Hungarian). *Matematikai Lapok* **11**, 264–274.

Golub GH and Van Loan CF 1996 *Matrix Computations*, 3rd edn Johns Hopkins University Press, Baltimore, MD.

Hartigan JA 1972 Direct clustering of a data matrix. *J. Am. Stat. Assoc.* **67**, 123–129.

Hastie T, Tibshirani R and Friedman J 2001 *The Elements of Statistical Learning. Data Mining, Inference, and Prediction*, Springer.

Hofmann T and Puzicha J 1999 Latent class models for collaborative filtering. In *Proc. 16th International Joint Congress on Artificial Intelligence (IJCAI 99)* (ed. Dean T), Morgan Kaufmann Publications, San Francisco CA, Vol. 2, pp. 688–693.

Holland P, Laskey KB and Leinhardt S 1981 Stochastic blockmodels: some first steps. *J. Am. Stat. Assoc.* **76**, 33–50.

Karrer B and Newman MEJ 2011 Stochastic blockmodels and community structure in networks. *Phys. Rev. E* **83**, 016107.

Kluger Y, Basri R, Chang JT and Gerstein M 2003 Spectral biclustering of microarray data: coclustering genes and conditions. *Genome Res.* **13**, 703–716.

Lauritzen SL 1988 *Extremal families and systems of sufficient statistics*, Lecture Notes in Statistics 49, Springer.

Madeira SC and Oliveira AL 2004 Biclustering algorithms for biological data analysis: A survey. *IEEE-ACM Trans. Comput. Biol. Bioinform.* **1** (1), 24–45.

Mahadev NVR and Peled UN 1995 Threshold graphs and related topics. *Ann. Discrete. Math.* **56**. North-Holland, Amsterdam.

McLachlan GJ 1997 *The EM Algorithm and Extensions*, John Wiley & Sons, Ltd.

Newman MEJ and Park J 2004 Statistical mechanics of networks. *Phys. Rev. E* **70**, 066117.

Rao CR 1973 *Linear Statistical Inference and Its Applications*, John Wiley & Sons, Ltd.

Rasch G 1960 *Studies in Mathematical Psychology: I. Probabilistic Models for Some Intelligence and Attainment Tests*, Nielsen and Lydiche, Oxford, UK.

Riesen K and Bunke H 2010 Graph classification and clustering based on vector space embedding, in *Ser. Machine Perception and Artificial Intelligence* (eds Bunke H and Wang PSP), Vol 77. World Scientific.

Rohe K, Chatterjee S and Yu B 2011 Spectral clustering and the high-dimensional stochastic blockmodel. *Ann. Stat.* **39** (4), 1878–1915.

Ungar LH and Foster DP 1998 A formal statistical approach to collaborative filtering, in *Proc. Conference on Automatical Learning and Discovery (CONALD 98)*.

Yan S, Xu D, Zhang B, Zhang H-J, Yang Q and Lin S 2007 Graph embedding and extensions: a general framework for dimensionality reduction. *IEEE Trans. Pattern Anal. Mach. Intell.* **29** (1), 40–51.

Appendix A

Linear algebra and some functional analysis

A.1 Metric, normed vector, and Euclidean spaces

Theory of real matrices will be discussed in the more general framework of linear operators between Hilbert spaces. Linear operators, sometimes between infinite dimensional spaces, will be intensively used throughout the book. For this purpose, we first introduce the notion of some abstract spaces. We suppose that the reader is familiar with the notion of a *vector space* (in other words, linear space). Here we only consider real vector spaces, for the elements of which the operations of addition and the multiplication with a real number are defined together with the usual commutativity, associativity, and distributivity properties; further, there is a distinguished zero element which is, in fact, equal to the real zero times any element of the vector space.

Definition A.1.1 *$(\mathcal{X}, d)$ is a metric space if a metric d is defined on the set $\mathcal{X}$ by the following properties: $\forall x, y, z \in \mathcal{X}$*

- $d(x, y) = d(y, x) \geq 0$,
- $d(x, y) = 0$ *if and only if* $x = y$,
- $d(x, y) \leq d(x, z) + d(z, y)$ *(triangle inequality).*

By means of the metric d, a natural topology can be established via the notion of open and closed sets; further, convergence facts and continuity of real valued functions on $\mathcal{X}$ can be stated. We say that the set $K \subset \mathcal{X}$ is *complete* if every Cauchy sequence

Spectral Clustering and Biclustering: Learning Large Graphs and Contingency Tables, First Edition.
Marianna Bolla.

in it converges to a limit belonging to K. The metric space $(\mathcal{X}, d)$ is complete if $\mathcal{X}$ itself is complete. The set $K \subset \mathcal{X}$ is *compact* if every (infinite) sequence in K has (at least one) point of accumulation in K, and it is *precompact* if for any $r > 0$ it can be covered with finitely many spheres of radius r. Any compact set in a metric space is closed; further, the compactness of the subset K of a metric space is equivalent to any of the following facts:

- K is precompact and complete;
- Any open cover of K has a finite cover. (If we cover K with open sets, then a finite number of them can be selected which also cover K).

In a complete metric space a set is compact if and only if it is precompact and closed. The metric space $(\mathcal{X}, d)$ itself is compact if $\mathcal{X}$ is compact. The Weierstrass theorem states the following.

Theorem A.1.2 *Any continuous function $\mathcal{X} \to \mathbb{R}$ defined on a compact, non-empty metric space is bounded, and it takes on a maximal and minimal value.*

A metric space is called *separable* if it contains a countable dense set.

A *normed vector space* is a vector space endowed with a metric compatible with its linear structure.

Definition A.1.3 *$(\mathcal{X}, \|.\|)$ is a normed vector space if $\mathcal{X}$ is a vector space and the norm $\| \cdot \|$ defined on it satisfies the following conditions: $\forall x, y \in \mathcal{X}$ and $c \in \mathbb{R}$*

- $\|x\| \geq 0$,
- $\|x\| = 0$ *if and only if* $x = 0$,
- $\|cx\| = |c| \cdot \|x\|$,
- $\|x + y\| \leq \|x\| + \|y\|$ *(triangle inequality).*

A normed vector space is also a metric space with the metric $d(x, y) = \|x - y\|$. A complete normed vector space is called *Banach space*, see von Neumann (1935). It can be shown that every finite dimensional normed vector space is complete, and hence, it is a Banach space. A subspace of a normed vector space consists of a subset of $\mathcal{X}$ which is a subspace of the original vector space together with the same norm restricted to the subset.

Definition A.1.4 *A real Euclidean space (in other words, inner product space) is a vector space $\mathcal{X}$ endowed with a bilinear function $\langle ., . \rangle$ (called inner product) which satisfies the following requirements: $\forall x, y, z \in \mathcal{X}$ and $a, b \in \mathbb{R}$*

- $\langle x, x \rangle \geq 0$,
- $\langle x, x \rangle = 0$ *if and only if* $x = 0$,

- $\langle x, y\rangle = \langle y, x\rangle$,
- $\langle ax + by, z\rangle = a\langle x, z\rangle + b\langle y, z\rangle$.

We will only consider real Euclidean spaces, and simply call them Euclidean spaces. Any Euclidean space is a normed vector space with the natural norm

$$\|x\| = \sqrt{\langle x, x\rangle}.$$

This norm satisfies the *Cauchy–Schwarz inequality* as follows.

Proposition A.1.5 *For any two elements x, y in a Euclidean space*

$$|\langle x, y\rangle| \leq \|x\| \cdot \|y\|$$

holds, where equality is attained if and only if x and y are linearly dependent.

If we think of x and y as vectors, and neither of them is the zero element of the underlying Euclidean space, then their linear dependence means that $y = cx$ with some real constant $c \neq 0$. Visually, they are parallel vectors (possibly of opposite direction). The meaning of two- or three-dimensional perpendicular vectors can also be generalized. We say that x and y are *orthogonal* if $\langle x, y\rangle = 0$ (the zero element is at the same time orthogonal and parallel to any other element of the Euclidean space). The notion of angle extends to the following: the *angle* between the non-zero elements x and y is

$$\cos^{-1} \frac{\langle x, y\rangle}{\|x\| \cdot \|y\|}.$$

Further, for given non-zero x, any element y can uniquely be decomposed as $y = y_1 + y_2$, where y_1 is parallel and y_2 is orthogonal to x. Namely,

$$y_1 = \frac{\langle x, y\rangle}{\|x\|^2} x$$

and $y_2 = y - y_1$.

A.2 Hilbert spaces

A *Hilbert space* is a complete Euclidean space. In fact, every Euclidean space can be completed (in other words, imbedded) into a Hilbert space of which it is a dense subspace. Therefore, Euclidean spaces are also called pre-Hilbert spaces. This distinction between a Hilbert and pre-Hilbert space makes sense only in the infinite dimensional case; every finite dimensional Euclidean space is a Hilbert space, and of course a normed vector space, Banach space, and metric space, at the same time. Whenever we write $\mathbb{R}^n$ we think of the n-dimensional real vector space in all of these senses. Every subspace of a Euclidean space is again a Euclidean space with the induced inner product, while every closed subspace of a Hilbert space is a Hilbert space.

It can be shown that all separable Hilbert spaces are isomorphic as they are, in fact, L^2 spaces. We can think of them as countably infinite dimensional vectors. Let us see some examples.

(a) Countably infinite dimensional vectors $x = (x_1, x_2, \ldots)$ form a so-called l^2 space if $\sum_{i=1}^{\infty} x_i^2 < \infty$. In this case $\langle x, y \rangle = \sum_{i=1}^{\infty} x_i y_i$ (which is finite by the Cauchy–Schwarz inequality) and $\|x\| = (\sum_{i=1}^{\infty} x_i^2)^{1/2}$.

(b) The set $C([a, b])$ of continuous, real valued functions on the closed interval $[a, b]$ form a Euclidean space with the inner product

$$\langle f, g \rangle = \int_a^b f(x)g(x)\,dx, \quad f, g \in C([a, b]).$$

However, this space is not complete, therefore $C([a, b])$ is not a Hilbert space; see the example on page 6 of Komornik (2003).

(c) The $L^2(\mathcal{X})$ space of real-valued, square-integrable functions with respect to some finite measure μ on a compact set $\mathcal{X}$ is a separable Hilbert space with the inner product

$$\langle f, g \rangle = \int_{\mathcal{X}} f(x)g(x)\,d\mu(x), \quad f, g \in L^2(\mathcal{X}).$$

The induced norm is $\|f\| = \sqrt{\int_{\mathcal{X}} f^2(x)\,d\mu(x)}$.

It can be shown that if μ is the Lebesgue measure on $\mathbb{R}$, then $L^2([a, b])$ is the completion of the pre-Hilbert space $C([a, b])$ into a Hilbert space.

The separability of a Hilbert space $\mathcal{H}$ means that – as a vector space – it is spanned by countably many elements of it, the linear independence of which can be assumed (such a set is called a basis); or else – by virtue of the Schmidt orthogonalization procedure – there exists an *orthonormal basis* $b_1, b_2, \dots \in \mathcal{H}$ such that $\langle b_i, b_j \rangle = \delta_{ij}$ ($\forall i, j$), where δ_{ij} is the Kronecker-delta. For example, in (a), the infinite dimensional vectors $e_1, e_2, \ldots$ form an orthonormal basis if the ith component of e_i is 1, and the others are all zeros (this is the canonical basis); in (c), there are many orthonormal bases, and they can be constructed by means of orthogonal systems of polynomials or trigonometric functions.

The linearity of a Hilbert space (which is defined over $\mathbb{R}$) is somehow compatible with the set of real-valued linear functions – called functionals – on it. A *linear functional* F on the Hilbert space $\mathcal{H}$ assigns the real number $F(x)$ to any $x \in \mathcal{H}$ and this assignment is linear. The continuity (in the usual sense) of such an F is equivalent to its boundedness, that is, there is a constant C such that

$$|F(x)| \leq C\|x\|, \quad \forall x \in \mathcal{H}.$$

The *dual* of a Hilbert space consists of the continuous linear functionals on it, which also form a Hilbert space. The famous Riesz–Fréchet representation theorem, published by the two authors in the same issue of the *C R Acad Sci, Paris* (see Fréchet (1907); Riesz (1907)), states the following.

Theorem A.2.1 (Riesz–Fréchet representation theorem) *A Hilbert space and its dual are isomorphic. Namely, the effect of any $F : \mathcal{H} \to \mathbb{R}$ continuous linear functional can be written as*

$$F(x) = \langle x, y \rangle, \quad \forall x \in \mathcal{H}$$

with an appropriate $y \in \mathcal{H}$. Moreover, y is uniquely determined by F.

In the sequel, we will investigate linear maps between two Hilbert spaces $\mathcal{H}$ and $\mathcal{H}'$ which are bounded, that is, there is a constant $C \geq 0$ such that

$$\|Ax\| \leq C\|x\|, \quad \forall x \in \mathcal{H}. \tag{A.1}$$

The $A : \mathcal{H} \to \mathcal{H}'$ bounded linear maps are called *operators*, and the boundedness of them also implies their continuity in the usual sense. The *operator norm* of A is the smallest bound C in (A.1), and by the linearity, it has the equivalent form

$$\|A\| = \sup_{\|x\|_{\mathcal{H}}=1} \|Ax\|_{\mathcal{H}'}.$$

We will be mainly interested in *completely continuous* (with other words, *compact*) operators. We will see that they can be well approximated by means of finite rank ones, hence the important facts about them are reminiscent of those of linear algebra. Here the *rank* of an operator is the dimension of its range (image space) in $\mathcal{H}'$.

Definition A.2.2 *The operator $A : \mathcal{H} \to \mathcal{H}'$ is said to be compact or completely continuous if it maps bounded sets of $\mathcal{H}$ into precompact sets of $\mathcal{H}'$; or equivalently, for any bounded sequence $(x_n) \subset \mathcal{H}$, the image sequence $(Ax_n) \subset \mathcal{H}'$ has a convergent subsequence.*

Frequently, $\mathcal{H} = \mathcal{H}'$, and in this situation, we can define the *identity* operator I such that $Ix = x$, $\forall x \in \mathcal{H}$. The identity operator in an infinite dimensional Hilbert space is not compact, since the sequence (e_n) of an orthonormal basis is bounded, still, it does not have a convergent subsequence.

Any operator between finite dimensional Hilbert spaces is compact, and it can be identified with a matrix, as we will see in the linear algebra part. Analogously, an operator between the separable Hilbert spaces $\mathcal{H}$ and $\mathcal{H}'$ is described with a special infinite matrix in the following way. Let $(e_i) \subset \mathcal{H}$ and $(f_j) \subset \mathcal{H}'$ be canonical bases and for the real numbers a_{ij} the condition

$$\sum_{i=1}^{\infty} \sum_{j=1}^{\infty} a_{ij}^2 < \infty \tag{A.2}$$

holds. Then the formula

$$A\left(\sum_{i=1}^{\infty} c_i e_i\right) = \sum_{j=1}^{\infty}\left(\sum_{i=1}^{\infty} a_{ji} c_i\right) f_j$$

defines a compact operator $A : \mathcal{H} \to \mathcal{H}'$. Such an operator is called a *Hilbert–Schmidt operator*. Soon, we will see that there are compact operators which are not of Hilbert–Schmidt type.

Definition A.2.3 *The adjoint A^* of the operator $A : \mathcal{H} \to \mathcal{H}'$ is defined by*

$$\langle Ax, y\rangle_{\mathcal{H}'} = \langle x, A^* y\rangle_{\mathcal{H}}, \quad \forall x \in \mathcal{H},\ y \in \mathcal{H}'.$$

Therefore, A^* is a $\mathcal{H}' \to \mathcal{H}$ operator, and if necessary, we will denote in the lower index the Hilbert space in which the inner product is understood. The inner product $\langle Ax, y\rangle_{\mathcal{H}}$ is called *bilinear form* as it is linear (and also continuous) in both variables x and y.

Definition A.2.4 *The operator $A : \mathcal{H} \to \mathcal{H}$ is said to be self-adjoint if $A = A^*$.*

Recall that we only deal with Hilbert spaces over $\mathbb{R}$, therefore we sometimes call a self-adjoint operator symmetric, especially in the finite rank case.

Compact operators can be decomposed as weighted sums of special rank one operators, where the weights tend to zero, and therefore keeping the most important ones, we obtain low rank approximation of them. More precisely, the following statements are true.

The *spectral theorem of Hilbert and Schmidt* states a very important property of compact, self-adjoint operators in a separable Hilbert space.

Theorem A.2.5 (Hilbert-Schmidt theorem) *Let $\mathcal{H}$ be a separable Hilbert space and $A : \mathcal{H} \to \mathcal{H}$ be a compact, self-adjoint operator. Then there exist an orthonormal basis $(\psi_i) \subset \mathcal{H}$ and a sequence (λ_i) of real numbers such that*

$$A\psi_i = \lambda_i \psi_i, \quad i = 1, 2, \ldots. \tag{A.3}$$

Further, if $\mathcal{H}$ has infinite dimension, then $\lim_{i\to\infty} \lambda_i = 0$.

The element $\psi_i \in \mathcal{H}$ in (A.3) is called an *eigenfunction* corresponding to the *eigenvalue* λ_i $(i = 1, 2, \ldots)$. If there are no multiple eigenvalues then the unit-norm eigenfunctions are unique up to orientation (multiplication with ± 1). To multiple eigenvalues there corresponds a unique eigenspace of dimension equal to the multiplicity of this eigenvalue. Within the eigenspace any orthonormal system can embody the eigenfunctions corresponding to the equal eigenvalues.

The Hilbert–Schmidt theorem gives rise to the following *spectral decomposition* (SD) of the self-adjoint, compact operator A:

$$A = \sum_{i=1}^{\infty} \lambda_i \langle ., \psi_i \rangle \psi_i. \tag{A.4}$$

Consequently, the effect of A on $x \in \mathcal{H}$ can be written as

$$Ax = \sum_{i=1}^{\infty} \lambda_i \langle x, \psi_i \rangle \psi_i$$

meaning the pointwise convergence of the sequence $\sum_{i=1}^{n} \lambda_i \langle x, \psi_i \rangle \psi_i$. This implies the convergence of (A.4) in operator norm as well. That is, with the notation $A_n = \sum_{i=1}^{n} \lambda_i \langle ., \psi_i \rangle \psi_i$, the relation $\|A - A_n\| \to 0$ also holds as $n \to \infty$. Thus, in the self-adjoint case, we have shown that compact operators can be approached by finite rank ones.

Note that the spectral theorem naturally extends to non-compact, self-adjoint $\mathcal{H} \to \mathcal{H}$ operators which can be decomposed as $A + cI$, where A is compact, self-adjoint and c is a non-zero constant. Then the eigenvalues of $A + cI$ are the numbers $\lambda_i + c$ with the eigenfunctions ψ_i $(i = 1, 2, \dots)$, and the spectrum of $A + cI$ converges to the constant c.

The whole machinery can be extended to general compact operators as follows.

Theorem A.2.6 *Let $\mathcal{H}$ and $\mathcal{H}'$ be separable Hilbert spaces and $A : \mathcal{H} \to \mathcal{H}'$ be a compact operator. Then there exist orthonormal bases $(\psi_i) \subset \mathcal{H}$ and $(\phi_i) \subset \mathcal{H}'$ together with a sequence (s_i) of nonnegative real numbers such that*

$$A\psi_i = s_i \phi_i, \quad A^* \phi_i = s_i \psi_i, \quad i = 1, 2, \dots. \tag{A.5}$$

Further, if $\mathcal{H}$ and $\mathcal{H}'$ have infinite dimension, then $\lim_{i \to \infty} s_i = 0$.

The elements $\psi_i \in \mathcal{H}$ and $\phi_i \in \mathcal{H}'$ in (A.5) are called *singular function pairs* (or left and right singular functions) corresponding to the *singular value* s_i $(i = 1, 2, \dots)$. Concerning the uniqueness the same can be said as in case of the SD. We remark that the singular values of a self-adjoint operator are the absolute values of its eigenvalues. In case of a positive eigenvalue, the left and right singular functions are the same (they coincide with the corresponding eigenfunction with any, but the same, orientation). In case of a negative eigenvalue, the left and right singular functions are opposite (any of them can be the corresponding eigenfunction of bivalent orientation). In case of a zero singular value the orientation is immaterial as it does not contribute to the decomposition of the underlying operator.

The above theorem gives rise to the following *singular value decomposition* (SVD) of the compact operators A and A^*:

$$A = \sum_{i=1}^{\infty} s_i \langle ., \psi_i \rangle_{\mathcal{H}} \phi_i, \quad A^* = \sum_{i=1}^{\infty} s_i \langle ., \phi_i \rangle_{\mathcal{H}'} \psi_i. \tag{A.6}$$

Consequently, the effect of A on $x \in \mathcal{H}$ can be written as

$$Ax = \sum_{i=1}^{\infty} \lambda_i \langle x, \psi_i \rangle \phi_i,$$

while the effect of A^* on $y \in \mathcal{H}'$ as

$$A^* y = \sum_{i=1}^{\infty} \lambda_i \langle y, \phi_i \rangle_{\mathcal{H}'} \phi_i.$$

The convergence in (A.6) is also meant in spectral norm. That is, with the notation $A_n = \sum_{i=1}^{n} s_i \langle ., \psi_i \rangle_{\mathcal{H}} \phi_i$, the convergence fact $\|A - A_n\| \to 0$ gives rise to a finite rank approximation of a general compact operator.

For a Hilbert–Schmidt operator, $\sum_{i=1}^{\infty} s_i^2 < \infty$ (also $\sum_{i=1}^{\infty} \lambda_i^2 < \infty$ if it is self-adjoint) in accord with the requirement (A.2). These convergences imply that $\lim_{i\to\infty} s_i = 0$ (also $\lim_{i\to\infty} \lambda_i = 0$ in the self-adjoint case). Therefore a Hilbert-Schmidt operator is always compact, but the converse is not true: if only $\lim_{i\to\infty} s_i = 0$ holds and $\sum_{i=1}^{\infty} s_i^2 = \infty$, then our operator is compact but not Hilbert–Schmidt.

It can be shown that for a general compact operator with SVD in (A.6)

$$\|A\| = \max_i s_i$$

and for a self-adjoint one with SD in (A.4)

$$\|A\| = \max_i |\lambda_i|.$$

Therefore, the operator norm of compact operators is also called *spectral norm.*

We will frequently use the following propositions for compact operators.

Proposition A.2.7 *Let $A : \mathcal{H} \to \mathcal{H}'$ be a compact operator with SVD in (A.6). Assume that its singular values are enumerated in non-increasing order ($s_1 \geq s_2 \geq \dots$). Then*

$$\max_{\|x\|_{\mathcal{H}}=1,\ \|y\|_{\mathcal{H}'}=1} \langle Ax, y \rangle = s_1$$

and it is attained with the choice $x = \psi_1$ and $y = \phi_1$ (uniquely if $s_1 > s_2$). This was the $k = 1$ case. Further, for $k = 2, 3, \dots$

$$\max_{\substack{\|x\|_{\mathcal{H}}=1,\ \|y\|_{\mathcal{H}'}=1 \\ \langle x, \psi_i \rangle_{\mathcal{H}}=0\,(i=1,\dots,k-1) \\ \langle y, \phi_i \rangle_{\mathcal{H}'}=0\,(i=1,\dots,k-1)}} \langle Ax, y \rangle = s_k$$

and it is attained with the choice $x = \psi_k$ and $y = \phi_k$ (uniquely if $s_k > s_{k+1}$).

The following are also true.

Proposition A.2.8 *Let $A : \mathcal{H} \to \mathcal{H}'$ be a compact operator with SVD in (A.6). Assume that its singular values are enumerated in non-increasing order and $k > 0$ is an integer such that $s_k > s_{k+1}$. Then*

$$\max_{\substack{\langle x_i, x_j \rangle_{\mathcal{H}} = \delta_{ij} \\ \langle y_i, y_j \rangle_{\mathcal{H}'} = \delta_{ij}}} \sum_{i=1}^{k} \langle Ax_i, y_i \rangle = \sum_{i=1}^{k} s_i$$

and it is attained with the choice $x_i = \psi_i$ and $y_i = \phi_i$ $(i = 1, \ldots, k)$.

Proposition A.2.9 *Let $A : \mathcal{H} \to \mathcal{H}$ be a self-adjoint compact operator with SD in (A.4). Assume that its eigenvalues are enumerated in non-increasing order ($\lambda_1 \geq \lambda_2 \geq \ldots$). Then*

$$\max_{\|x\|=1} \langle Ax, x \rangle = \lambda_1$$

and it is attained with the choice $x = \psi_1$ (uniquely if $\lambda_1 > \lambda_2$). This was the $k = 1$ case. Further, for $k = 2, 3, \ldots$

$$\max_{\substack{\|x\|=1 \\ \langle x, \psi_i \rangle = 0 \, (i=1,\ldots,k-1)}} \langle Ax, x \rangle = \lambda_k$$

and it is attained with the choice $x = \psi_k$ (uniquely if $\lambda_k > \lambda_{k+1}$).

Proposition A.2.10 *Let $A : \mathcal{H} \to \mathcal{H}$ be a self-adjoint compact operator with SD in (A.4). Assume that its eigenvalues are enumerated in non-increasing order and $k > 0$ is an integer such that $\lambda_k > \lambda_{k+1}$. Then*

$$\max_{\langle x_i, x_j \rangle = \delta_{ij}} \sum_{i=1}^{k} \langle Ax_i, x_i \rangle = \sum_{i=1}^{k} \lambda_i$$

and it is attained with the choice $x_i = \psi_i$ $(i = 1, \ldots, k)$.

Integral operators between L^2 spaces (see Riesz (1909); Riesz and Sz.-Nagy (1952)) are important examples of Hilbert–Schmidt operators. In Example (c) we introduced the $\mathcal{H} = L^2(\mathcal{X})$ space of real-valued, square-integrable functions with respect to some finite measure μ on the compact set $\mathcal{X}$. Likewise, consider $\mathcal{H}' = L^2(\mathcal{Y})$ with the finite measure ν, and let $K : \mathcal{X} \times \mathcal{Y} \to \mathbb{R}$ be a *kernel* such that for it

$$\int_{\mathcal{X}} \int_{\mathcal{Y}} K^2(x, y) \, \mu(dx) \, \nu(dy) < \infty \tag{A.7}$$

holds. With the kernel K, the *integral operator* $A_K : H' \to H$ is defined in the following way: to the function $g \in \mathcal{H}'$ the operator A_K assigns the function $f \in \mathcal{H}$ such that

$$f(x) = (A_K g)(x) = \int_{\mathcal{Y}} K(x, y) g(y)\, \nu(dy), \quad x \in \mathcal{X}.$$

It is easy to see that A_K is linear, further it is a Hilbert–Schmidt operator, therefore compact with

$$\|A_K\| \leq \|K\|_2 \tag{A.8}$$

where $\|K\|_2$ is the squareroot of the finite expression in (A.7), that is the L^2-norm of K in the product space.

Say, A_K has the SVD

$$A_K = \sum_{i=1}^{\infty} s_i \langle ., \phi_i \rangle_{H'} \psi_i \tag{A.9}$$

where for the singular values

$$\sum_{i=1}^{\infty} s_i^2 = \|K\|_2^2 < \infty$$

also holds, implying that $\lim_{i\to\infty} s_i = 0$ (if s_i's are really countably infinitely many).

It is easy to see that the adjoint of A_K is the integral operator $A_K^* : \mathcal{H} \to \mathcal{H}'$ with SVD

$$A_K^* = \sum_{i=1}^{\infty} s_i \langle ., \psi_i \rangle_H \phi_i,$$

and

$$\|A_K\| = \|A_K^*\| = s_1.$$

Remark that $\|K\|_2 = (\sum_{i=1}^{\infty} s_i^2)^{1/2}$ is called the *Hilbert–Schmidt norm* of A_K (in finite dimension it will be called Frobenius norm) denoted by $\|A_K\|_2$, and for it, $\|A_K\|_2 \geq \|A_K\|$ due to Inequality (A.8).

With a *symmetric kernel* $K : \mathcal{X} \times \mathcal{X} \to \mathbb{R}$ ($K(x, y) = K(y, x)\ \forall x, y$), the induced integral operator A_K becomes self-adjoint, therefore it also admits an SD

$$A_K = \sum_{i=1}^{\infty} \lambda_i \langle ., \psi_i \rangle \psi_i$$

with real eigenvalues such that $\sum_{i=1}^{\infty} \lambda_i^2 < \infty$ and corresponding orthonormal eigenvectors $\psi_1, \psi_2, \ldots$. Under some additional condition, the kernel itself can be expanded; see the upcoming Mercer theorem.

Definition A.2.11 *The integral operator A_K corresponding to the symmetric kernel K is said to be positive semidefinite if it has all nonnegative eigenvalues.*

Note that in Section 1.5 we will introduce an equivalent definition for a positive definite kernel.

The SD of the self-adjoint A_K gives rise to the following, so-called Karhunen–Loève expansion of K: $K(x, y) = \sum_{i=1}^{\infty} \lambda_i \psi_i(x)\psi_i(y)$, where the convergence is understood in the L^2-norm of the product space. The following theorem states more, namely, uniform pointwise convergence.

Theorem A.2.12 (Mercer theorem) *If K is a symmetric, continuous kernel of a positive semidefinite integral operator on $L^2(\mathcal{X})$, where $\mathcal{X}$ is some compact space, then it can be expanded into the following uniformly convergent series:*

$$K(x, y) = \sum_{i=1}^{\infty} \lambda_i \psi_i(x)\psi_i(y), \quad \forall x, y \in \mathcal{X}$$

by the eigenfunctions ψ_i and the eigenvalues $\lambda_i \geq 0$ of the integral operator induced by the kernel K.

Finally, we will use the following *uniform boundedness principle* concerning a collection of linear operators between Hilbert spaces. In fact, the forthcoming Banach–Steinhaus theorem is stated more generally, for normed vector spaces.

Theorem A.2.13 (Banach–Steinhaus theorem) *Let $\mathcal{X}$ be a Banach space and $\mathcal{Y}$ a normed vector space; further, let $\mathcal{A}$ be a collection of $\mathcal{X} \to \mathcal{Y}$ continuous linear operators. If for all $x \in \mathcal{X}$ we have*

$$\sup_{A \in \mathcal{A}} \|Ax\| < \infty,$$

then

$$\sup_{A \in \mathcal{A}} \|A\| < \infty$$

too.

A.3 Matrices

From now on, we will confine ourselves to finite dimensional real Euclidean spaces which are also Hilbert spaces. Linear operations between these spaces can be described by matrices of real entries. To stress that the elements of the Euclidean space are finite dimensional vectors, we will use bold-face lower-case letters, further vectors are treated as column-vectors. The inner product of the vectors $\mathbf{x}, \mathbf{y} \in \mathbb{R}^n$ is therefore written with matrix multiplication, like $\mathbf{x}^T\mathbf{y}$, where ^{T} stands for the transposition, hence $\mathbf{x}^T$ is a row-vector. Matrices will be denoted by bold-face upper-case letters.

An $m \times n$ matrix $\boldsymbol{A} = (a_{ij})$ of real entries a_{ij}'s corresponds to an $\mathbb{R}^n \to \mathbb{R}^m$ linear transformation (operator). Its transpose, $\boldsymbol{A}^T$, is an $n \times m$ matrix. An $n \times n$ matrix is called quadratic and it maps $\mathbb{R}^n$ into itself. The identity matrix is denoted by $\boldsymbol{I}$ or $\boldsymbol{I}_n$ if we want to refer to its size.

The quadratic matrix $\boldsymbol{A}$ is *symmetric* if $\boldsymbol{A} = \boldsymbol{A}^T$ and *orthogonal* if $\boldsymbol{A}\boldsymbol{A}^T = \boldsymbol{I}$. The orthogonal matrix $\boldsymbol{P}$ is a *permutation matrix* if, in each row and column, exactly one of its entries differs from 0, and this non-zero entry is 1.

The $n \times n$ matrix $\boldsymbol{A}$ has an inverse if and only if its determinant, $|\boldsymbol{A}| \neq 0$, and its inverse is denoted by $\boldsymbol{A}^{-1}$. In this case, the linear transformation corresponding to $\boldsymbol{A}^{-1}$ undoes the effect of the $\mathbb{R}^n \to \mathbb{R}^n$ transformation corresponding to $\boldsymbol{A}$, that is, $\boldsymbol{A}^{-1}\mathbf{y} = \mathbf{x}$ if and only if $\boldsymbol{A}\mathbf{x} = \mathbf{y}$ for any $\mathbf{y} \in \mathbb{R}^n$. It is important that in case of an invertible (*regular*) matrix $\boldsymbol{A}$, the *range* (or image space) of $\boldsymbol{A}$ – denoted by $\mathcal{R}(\boldsymbol{A})$ – is the whole $\mathbb{R}^n$, and in exchange, the kernel of $\boldsymbol{A}$ (the subspace of vectors that are mapped into the zero vector by $\boldsymbol{A}$) consists of the only $\mathbf{0}$.

Note that for an $m \times n$ matrix $\boldsymbol{A}$, its range is

$$\mathcal{R}(\boldsymbol{A}) = \mathrm{Span}\{\mathbf{a}_1, \ldots, \mathbf{a}_n\}$$

where $\mathbf{a}_1, \ldots, \mathbf{a}_n$ are the column vectors of $\boldsymbol{A}$ for which fact the notation $\boldsymbol{A} = (\mathbf{a}_1, \ldots, \mathbf{a}_n)$ will be used; further, $\mathrm{Span}\{\ldots\}$ is the subspace spanned by the vectors in its argument. The *rank* of $\boldsymbol{A}$ is the dimension of its range:

$$\mathrm{rank}(\boldsymbol{A}) = \dim \mathcal{R}(\boldsymbol{A}),$$

and it is also equal to the maximum number of linearly independent rows of $\boldsymbol{A}$; trivially, $\mathrm{rank}(\boldsymbol{A}) \leq \min\{m, n\}$. In case of $m = n$, $\boldsymbol{A}$ is regular if and only if $\mathrm{rank}(\boldsymbol{A}) = n$, and *singular*, otherwise.

An orthogonal matrix $\boldsymbol{A}$ is always regular and $\boldsymbol{A}^{-1} = \boldsymbol{A}^T$; further its rows (or columns) constitute a complete orthonormal set in $\mathbb{R}^n$. Let k $(1 \leq k < n)$ be an integer; an $n \times k$ matrix $\boldsymbol{A}$ is called *suborthogonal* if its columns form a (not complete) orthonormal set in $\mathbb{R}^n$. For such an $\boldsymbol{A}$, the relation $\boldsymbol{A}^T\boldsymbol{A} = \boldsymbol{I}_k$ holds, but $\boldsymbol{A}\boldsymbol{A}^T \neq \boldsymbol{I}_n$. In fact, the $n \times n$ matrix $\boldsymbol{P} = \boldsymbol{A}\boldsymbol{A}^T$ is symmetric and idempotent ($\boldsymbol{P}^2 = \boldsymbol{P}$), hence, it corresponds to the orthogonal projection onto $\mathcal{R}(\boldsymbol{A})$. The *trace* of the $n \times n$ matrix $\boldsymbol{A}$ is

$$\mathrm{tr}(\boldsymbol{A}) = \sum_{i=1}^{n} a_{ii}.$$

How will the above matrix–matrix and matrix–scalar functions look if the underlying matrix is a product? If $\boldsymbol{A}$ and $\boldsymbol{B}$ can be multiplied together ($\boldsymbol{A}$ is $m \times n$ and $\boldsymbol{B}$ is $n \times k$ type), then their product corresponds to the succession of linear operations $\boldsymbol{B}$ and $\boldsymbol{A}$ in this order, therefore

$$(\boldsymbol{A}\boldsymbol{B})^T = \boldsymbol{B}^T\boldsymbol{A}^T$$

and if $\boldsymbol{A}$ and $\boldsymbol{B}$ are regular $n \times n$ matrices, then so is $\boldsymbol{A}\boldsymbol{B}$, and

$$(\boldsymbol{A}\boldsymbol{B})^{-1} = \boldsymbol{B}^{-1}\boldsymbol{A}^{-1}.$$

Further, $(A^{-1})^T = (A^T)^{-1}$, and vice versa. If A and B are $n \times n$ matrices, then

$$|AB| = |A| \cdot |B|.$$

Therefore, the determinant of the product of several matrices of the same size does not depend on the succession of the matrices; however, the matrix multiplication is usually not commutative. The trace is commutative in the following sense: if A is an $n \times k$ and B is a $k \times n$ matrix, then

$$\text{tr}(AB) = \text{tr}(BA).$$

For several factors, the trace is accordingly cyclically commutative:

$$\text{tr}(A_1 A_2 \dots A_n) = \text{tr}(A_2 \dots A_n A_1) = \dots = \text{tr}(A_n A_1 \dots A_{n-1})$$

when, of course, the sizes of the factors are such that the successive multiplications in $A_1 \dots A_n$ can be performed and the number of rows in A_1 is equal to the number of columns in A_n. Further,

$$\text{rank}(AB) \leq \min\{\text{rank}(A), \text{rank}(B)\},$$

consequently, the rank cannot be increased in course of matrix multiplications.

Given an $n \times n$ symmetric real matrix A, the quadratic form in the variables $x_1, \dots, x_n$ is the homogeneous quadratic function of these variables:

$$\sum_{i=1}^{n} \sum_{j=1}^{n} a_{ij} x_i x_j = \mathbf{x}^T A\mathbf{x},$$

where $\mathbf{x} = (x_1, \dots, x_n)^T$, hence the matrix multiplication results in a scalar. The possible signs of a quadratic form (with different $\mathbf{x}$'s) characterize the underlying matrix. Accordingly, they fall into exactly one of the following categories.

Definition A.3.1 *Let A be $n \times n$ symmetric real matrix.*

- *A is positive (negative) definite if $\mathbf{x}^T A\mathbf{x} > 0$ ($\mathbf{x}^T A\mathbf{x} < 0$), $\forall \mathbf{x} \neq \mathbf{0}$.*
- *A is positive (negative) semidefinite if $\mathbf{x}^T A\mathbf{x} \geq 0$ ($\mathbf{x}^T A\mathbf{x} \leq 0$), $\forall \mathbf{x} \in \mathbb{R}^n$, and $\mathbf{x}^T A\mathbf{x} = \mathbf{0}$ for at least one $\mathbf{x} \neq \mathbf{0}$.*
- *A is indefinite if $\mathbf{x}^T A\mathbf{x}$ takes on both positive and negative values (with different, non-zero $\mathbf{x}$'s).*

The positive and negative definite matrices are all regular, whereas the positive and negative semidefinite ones are singular. The indefinite matrices can be either regular or singular. To more easily characterize the definiteness of symmetric matrices, we will use their eigenvalues.

In analogy with the spectral theory of compact operators, the notion of an eigenvalue and eigenvector is introduced: λ is an eigenvalue of the $n \times n$ real matrix $\boldsymbol{A}$ with corresponding eigenvector $\mathbf{u} \neq \mathbf{0}$ if $\boldsymbol{A}\mathbf{u} = \lambda\mathbf{u}$. If $\mathbf{u}$ is an eigenvector of $\boldsymbol{A}$, it is easy to see that for $c \neq 0$, $c\mathbf{u}$ is also an eigenvector with the same eigenvalue. Therefore, it is better to speak about *eigen-directions* instead of eigenvectors; or else, we will consider specially normalized, for example unit-norm eigenvectors, when only the orientation is bivalent. It is well known that an $n \times n$ matrix $\boldsymbol{A}$ has exactly n eigenvalues (with multiplicities) which are (possibly complex) roots of the characteristic polynomial $|\boldsymbol{A} - \lambda \boldsymbol{I}|$. Knowing the eigenvalues, the corresponding eigenvectors are obtained by solving the system of linear equations $(\boldsymbol{A} - \lambda \boldsymbol{I})\mathbf{u} = \mathbf{0}$ which must have a non-trivial solution due to the choice of λ. In fact, there are infinitely many solutions (in case of single eigenvalues they are constant multiples of each other). An eigenvector corresponding to a complex eigenvalue must also have complex coordinates, but in case of our main interest (the symmetric matrices) this cannot occur.

The notion of an eigenvalue and eigenvector extends to matrices of complex entries in the same way. As for the allocation of the eigenvalues of a quadratic matrix (even of complex entries), the following result is known.

Theorem A.3.2 (Gersgorin disc theorem) *Let $\boldsymbol{A}$ be an $n \times n$ matrix of entries $a_{ij} \in \mathbb{C}$. The Gersgorin disks of $\boldsymbol{A}$ are the following regions of the complex plane:*

$$D_i = \{z \in \mathbb{C} : |z - a_{ii}| \leq \sum_{j \neq i} |a_{ij}|\}, \quad i = 1, \dots n.$$

Let $\lambda_1, \dots, \lambda_n$ denote the (possibly complex) eigenvalues of $\boldsymbol{A}$. Then

$$\{\lambda_1, \dots, \lambda_n\} \subset \cup_{i=1}^n D_i.$$

Furthermore, any connected component of the set $\cup_{i=1}^n D_i$ contains as many eigenvalues of $\boldsymbol{A}$ as the number of discs that form this component.

We will introduce the notion of *normal matrices* which admit a spectral decomposition (SD) similar to that of compact operators. The real matrix $\boldsymbol{A}$ is called normal if $\boldsymbol{A}\boldsymbol{A}^T = \boldsymbol{A}^T\boldsymbol{A}$. Among real matrices, only the symmetric, anti-symmetric ($\boldsymbol{A}^T = -\boldsymbol{A}$), and orthogonal matrices are normal. Normal matrices have the following important spectral property: to their eigenvalues there corresponds an orthonormal set of eigenvectors; choosing this as a new basis, the matrix becomes *diagonal* (all the off-diagonal entries are zeros). Here we only state the analogous version of the Hilbert–Schmidt theorem A.2.5 for symmetric matrices which, in addition, have all real eigenvalues, and consequently, eigenvectors of real coordinates.

Theorem A.3.3 *The $n \times n$ symmetric, real matrix $\boldsymbol{A}$ has real eigenvalues $\lambda_1 \geq \dots \geq \lambda_n$ (with multiplicities), and the corresponding eigenvectors $\mathbf{u}_1, \dots, \mathbf{u}_n$ can be chosen such that they constitute a complete orthonormal set in $\mathbb{R}^n$.*

This so-called spectral decomposition theorem implies the following SD of the $n \times n$ symmetric matrix $\boldsymbol{A}$:

$$\boldsymbol{A} = \sum_{i=1}^{n} \lambda_i \mathbf{u}_i \mathbf{u}_i^T = \boldsymbol{U} \boldsymbol{\Lambda} \boldsymbol{U}^T, \tag{A.10}$$

where $\boldsymbol{\Lambda} = \text{diag}(\lambda_1, \ldots, \lambda_n)$ is the diagonal matrix containing the eigenvalues – called *spectrum* – in its main diagonal, while $\boldsymbol{U} = (\mathbf{u}_1, \ldots, \mathbf{u}_n)$ is the orthogonal matrix containing the corresponding eigenvectors of $\boldsymbol{A}$ in its columns in the order of the eigenvalues. Of course, permuting the eigenvalues in the main diagonal of $\boldsymbol{\Lambda}$, and the columns of $\boldsymbol{U}$ accordingly, will lead to the same SD; however – if not otherwise stated – we will enumerate the real eigenvalues in non-increasing order. Concerning the uniqueness of the above SD we can state the following: the unit-norm eigenvector corresponding to a single eigenvalue is unique (up to orientation), whereas to an eigenvalue with multiplicity m there corresponds a unique m-dimensional so-called *eigen-subspace* within which any orthonormal set can be chosen for the corresponding eigenvectors.

It is easy to verify that for the eigenvalues of the symmetric matrix $\boldsymbol{A}$

$$\sum_{i=1}^{n} \lambda_i = \text{tr}(\boldsymbol{A}) \quad \text{and} \quad \prod_{i=1}^{n} \lambda_i = |\boldsymbol{A}|$$

hold. Therefore $\boldsymbol{A}$ is singular if and only if it has a 0 eigenvalue, and

$$r = \text{rank}(\boldsymbol{A}) = \text{rank}(\boldsymbol{\Lambda}) = |\{i : \lambda_i \neq 0\}|;$$

moreover, $\mathcal{R}(\boldsymbol{A}) = \text{Span}\{\mathbf{u}_i : \lambda_i \neq 0\}$. Therefore, the SD of $\boldsymbol{A}$ simplifies to

$$\sum_{i=1}^{r} \lambda_i \mathbf{u}_i \mathbf{u}_i^T.$$

Its spectrum also determines the definiteness of $\boldsymbol{A}$ in the following manner.

Proposition A.3.4 *Let $\boldsymbol{A}$ be $n \times n$ symmetric real matrix.*

- *$\boldsymbol{A}$ is positive (negative) definite if and only if all of its eigenvalues are positive (negative).*
- *$\boldsymbol{A}$ is positive (negative) semidefinite if and only if all of its eigenvalues are nonnegative (nonpositive), and its spectrum includes the zero.*
- *$\boldsymbol{A}$ is indefinite if its spectrum contains at least one positive and one negative eigenvalue.*

The matrix of an orthogonal projection $\boldsymbol{P}_F$ onto the r-dimensional subspace $F \subset \mathbb{R}^n$ has the following SD (only the $r < n$ case is of importance, since in the $r = n$ case $\boldsymbol{P}_F = \boldsymbol{I}_n$):

$$\boldsymbol{P}_F = \sum_{i=1}^{r} \mathbf{u}_i \mathbf{u}_i^T = \boldsymbol{A}\boldsymbol{A}^T,$$

where $\mathbf{u}_1, \ldots, \mathbf{u}_r$ is any orthonormal set in F which is the eigen-subspace corresponding to the eigenvalue 1 of multiplicity r. Note that the eigenspace corresponding to the other eigenvalue 0 of multiplicity $n - r$ is the orthogonal complementary subspace $F^{\perp}$ of F in $\mathbb{R}^n$, but it has no importance, as only the eigenvectors in the first r columns of $\boldsymbol{U}$ enter into the above SD of $\boldsymbol{P}_F$. With the notation $\boldsymbol{A} = (\mathbf{u}_1, \ldots, \mathbf{u}_r)$, the SD of $\boldsymbol{P}_F$ simplifies to $\boldsymbol{A}\boldsymbol{A}^T$, indicating that $\boldsymbol{A}$ is a suborthogonal matrix.

Now the analogue of Theorem A.2.6 for rectangular matrices is formulated.

Theorem A.3.5 *Let* $\boldsymbol{A}$ *be an* $m \times n$ *rectangular matrix of real entries,* $\text{rank}(\boldsymbol{A}) = r \leq \min\{m, n\}$. *Then there exists an orthonormal set* $\{\mathbf{v}_1, \ldots, \mathbf{v}_r\} \subset \mathbb{R}^m$ *and* $\{\mathbf{u}_1, \ldots, \mathbf{u}_r\} \subset \mathbb{R}^n$ *together with the positive real numbers* $s_1 \geq s_2 \geq \cdots \geq s_r > 0$ *such that*

$$\boldsymbol{A}\mathbf{u}_i = s_i \mathbf{v}_i, \quad \boldsymbol{A}^* \mathbf{v}_i = s_i \mathbf{u}_i, \quad i = 1, 2, \ldots, r. \tag{A.11}$$

The elements $\mathbf{v}_i \in \mathbb{R}^m$ and $\mathbf{u}_i \in \mathbb{R}^n$ $(i = 1, \ldots, r)$ in (A.11) are called *relevant singular vector pairs* (or left and right singular vectors) corresponding to the *singular value* s_i $(i = 1, 2, \ldots, r)$. The transformations in (A.11) give a one-to one mapping between $\mathcal{R}(\boldsymbol{A})$ and $\mathcal{R}(\boldsymbol{A}^T)$, all the other vectors of $\mathbb{R}^n$ and $\mathbb{R}^m$ are mapped into the zero vector of $\mathbb{R}^m$ and $\mathbb{R}^n$, respectively. However, the left and right singular vectors can appropriately be completed into a complete orthonormal set $\{\mathbf{v}_1, \ldots, \mathbf{v}_m\} \subset \mathbb{R}^m$ and $\{\mathbf{u}_1, \ldots \mathbf{u}_n\} \subset \mathbb{R}^n$, respectively, such that, the extra vectors so introduced in the kernel subspaces in $\mathbb{R}^m$ and $\mathbb{R}^n$ are mapped into the zero vector of $\mathbb{R}^n$ and $\mathbb{R}^m$, respectively. With the orthogonal matrices $\boldsymbol{V} = (\mathbf{v}_1, \ldots, \mathbf{v}_m)$ and $\boldsymbol{U} = (\mathbf{u}_1, \ldots \mathbf{u}_n)$, the following SVD of $\boldsymbol{A}$ and $\boldsymbol{A}^T$ holds:

$$\boldsymbol{A} = \boldsymbol{V}\boldsymbol{S}\boldsymbol{U}^T = \sum_{i=1}^{r} s_i \mathbf{v}_i \mathbf{u}_i^T \quad \text{and} \quad \boldsymbol{A}^T = \boldsymbol{U}\boldsymbol{S}^T\boldsymbol{V}^T = \sum_{i=1}^{r} s_i \mathbf{u}_i \mathbf{v}_i^T, \tag{A.12}$$

where $\boldsymbol{S}$ is an $m \times n$ so-called *generalized diagonal matrix* which contains the singular values $s_1, \ldots, s_r$ in the first r positions of its main diagonal (starting from the upper left corner) and zeros otherwise. We remark that there are other equivalent forms of the above SVD depending on whether $m < n$ or $m \geq n$. For example, in the $m < n$ case, $\boldsymbol{V}$ can be an $m \times m$ orthogonal, $\boldsymbol{S}$ an $m \times m$ diagonal, and $\boldsymbol{U}$ an $n \times m$ suborthogonal matrix with the same relevant entries. Concerning the uniqueness of the SVD the following can be stated: to a single positive singular value there corresponds a unique singular vector pair (of course, the orientation of the left and right singular vectors can be changed at the same time). To a positive singular value of multiplicity say $k > 1$ a k-dimensional left and right so-called *isotropic subspace* corresponds, within which, any k-element orthonormal sets can embody the left and right singular vectors with orientation such that the requirements in (A.11) are met.

We also remark that the singular values of a symmetric matrix are the absolute values of its eigenvalues. In the case of a positive eigenvalue, the left and right singular vectors are the same (they coincide with the corresponding eigenvector with any, but the same, orientation). In the case of a negative eigenvalue, the left and right singular vectors are opposite (any of them is the corresponding eigenvector which have a bivalent orientation). In the case of a zero singular value the orientation is immaterial, as it does not contribute to the SVD of the underlying matrix. Numerical algorithms for SD and SVD of real matrices are presented in Golub and van Loan (1996); Wilkinson (1965).

Assume that the $m \times n$ matrix $\boldsymbol{A}$ of rank r has SVD (A.12). It is easy to see that the matrices $\boldsymbol{A}\boldsymbol{A}^T$ and $\boldsymbol{A}^T\boldsymbol{A}$ are positive semidefinite (possibly, positive definite) matrices of rank r, and their SD is

$$\boldsymbol{A}\boldsymbol{A}^T = \boldsymbol{V}(\boldsymbol{S}\boldsymbol{S}^T)\boldsymbol{V}^T = \sum_{i=1}^{r} s_i^2 \mathbf{v}_i \mathbf{v}_i^T \quad \text{and} \quad \boldsymbol{A}^T\boldsymbol{A} = \boldsymbol{U}(\boldsymbol{S}^T\boldsymbol{S})\boldsymbol{U}^T = \sum_{i=1}^{r} s_i^2 \mathbf{u}_i \mathbf{u}_i^T$$

where the diagonal matrices $\boldsymbol{S}\boldsymbol{S}^T$ and $\boldsymbol{S}^T\boldsymbol{S}$ both contain the numbers $s_1^2, \ldots, s_r^2$ in the leading positions of their main diagonals as non-zero eigenvalues.

These facts together also imply that the only positive singular value of a sub-orthogonal matrix is the 1 with multiplicity of its rank.

Definition A.3.6 *We say that the $n \times n$ symmetric matrix $\boldsymbol{G} = (g_{ij})$ is a Gram-matrix if its entries are inner products: there is a dimension $d > 0$ and vectors $\mathbf{x}_1, \ldots, \mathbf{x}_n \in \mathbb{R}^d$ such that*

$$g_{ij} = \mathbf{x}_i^T \mathbf{x}_j, \quad i, j = 1, \ldots n.$$

Proposition A.3.7 *The symmetric matrix $\boldsymbol{G}$ is a Gram-matrix if and only if it is positive semidefinite or positive definite.*

Proof. If $\boldsymbol{G}$ is a Gram-matrix, then it can be decomposed as $\boldsymbol{G} = \boldsymbol{A}\boldsymbol{A}^T$, where $\boldsymbol{A}^T = (\mathbf{x}_1, \ldots, \mathbf{x}_n)$. With this,

$$\mathbf{x}^T\boldsymbol{G}\mathbf{x} = \mathbf{x}^T\boldsymbol{A}\boldsymbol{A}^T\mathbf{x} = (\boldsymbol{A}^T\mathbf{x})^T(\boldsymbol{A}^T\mathbf{x}) = \|\boldsymbol{A}^T\mathbf{x}\|^2 \geq 0, \quad \forall \mathbf{x} \in \mathbb{R}^n.$$

Conversely, if $\boldsymbol{G}$ is positive semidefinite (or positive definite) with rank $r \leq n$, then its SD – using (A.10) – can be written as

$$\boldsymbol{G} = \sum_{i=1}^{r} \lambda_i \mathbf{u}_i \mathbf{u}_i^T$$

with non-negative real eigenvalues. Let the $n \times r$ matrix $\boldsymbol{A}$ be defined as

$$\boldsymbol{A} = (\sqrt{\lambda_1}\mathbf{u}_1, \ldots, \sqrt{\lambda_r}\mathbf{u}_r). \tag{A.13}$$

Then the row vectors of the matrix $\boldsymbol{A}$ will be r-dimensional vectors reproducing $\boldsymbol{G}$. Of course, such a decomposition is not unique: first of all, instead of $\boldsymbol{A}$ the matrix $\boldsymbol{A}\boldsymbol{Q}$ will also do, where $\boldsymbol{Q}$ is an arbitrary $r \times r$ orthogonal matrix (obviously, $\mathbf{x}_i$'s can be rotated); and $\mathbf{x}_i$'s can also be put in a higher ($d > r$) dimension with attaching any (but the same) number of zero coordinates to them.

The spectral norm (operator norm) of an $m \times n$ real matrix $\boldsymbol{A}$ of rank r, with positive singular values $s_1 \geq \cdots \geq s_r > 0$, is

$$\|\boldsymbol{A}\| = \max_{\|\mathbf{x}\|=1} \|\boldsymbol{A}\mathbf{x}\| = s_1,$$

and its *Frobenius norm*, denoted by $\|.\|_2$, is

$$\|\boldsymbol{A}\|_2 = \left(\sum_{i=1}^{m}\sum_{j=1}^{n} a_{ij}^2\right)^{1/2} = \sqrt{\mathrm{tr}(\boldsymbol{A}\boldsymbol{A}^T)} = \sqrt{\mathrm{tr}(\boldsymbol{A}^T\boldsymbol{A})} = \left(\sum_{i=1}^{r} s_i^2\right)^{1/2}.$$

The Frobenius norm is sometimes called Euclidean norm and corresponds to the Hilbert–Schmidt norm of operators between separable Hilbert spaces. Accordingly, the operator $\mathbb{R}^n \to \mathbb{R}^m$ corresponding to $\boldsymbol{A}$, can be considered as an integral operator with kernel (a_{ij}), hence, $\|\boldsymbol{A}\|_2$ is the L^2-norm of this kernel, which is expressed in the above formula. For a symmetric real matrix $\boldsymbol{A}$,

$$\|\boldsymbol{A}\| = \max_{\|\mathbf{x}\|=1} \|\boldsymbol{A}\mathbf{x}\| = \max_i |\lambda_i| \quad \text{and} \quad \|\boldsymbol{A}\|_2 = \left(\sum_{i=1}^{r} \lambda_i^2\right)^{1/2}.$$

Obviously, for a real matrix $\boldsymbol{A}$ of rank r,

$$\|\boldsymbol{A}\| \leq \|\boldsymbol{A}\|_2 \leq \sqrt{r}\|\boldsymbol{A}\|. \tag{A.14}$$

More generally, a matrix norm is called *unitary invariant* if

$$\|\boldsymbol{A}\|_{\mathrm{un}} = \|\boldsymbol{Q}\boldsymbol{A}\boldsymbol{R}\|_{\mathrm{un}}$$

with any $m \times m$ and $n \times n$ orthogonal matrices $\boldsymbol{Q}$ and $\boldsymbol{R}$, respectively. It is easy to see that a unitary invariant norm of a real matrix merely depends on its singular values (or eigenvalues if it is symmetric). For example, the spectral and Frobenius norms are such, and the *Schatten 4-norm* defined by

$$\|\boldsymbol{A}\|_4 = \left(\sum_{i=1}^{r} s_i^4\right)^{\frac{1}{4}} \tag{A.15}$$

is also unitary invariant (we will use it in Chapter 4).

By means of SD or SVD we are able to define so-called *generalized inverses* of singular square or rectangular matrices: in fact, any matrix that undoes the effect of the underlying linear transformation between the ranges of $\boldsymbol{A}^T$ and $\boldsymbol{A}$ will do. A generalized inverse is far from unique as any transformation operating on the kernels can be added. However, the following *Moore–Penrose inverse* is uniquely defined and it coincides with the usual inverse if exists.

Definition A.3.8 *The Moore–Penrose inverse of the $n \times n$ symmetric matrix with SD (A.10) is*

$$A^+ = \sum_{i=1}^{r} \frac{1}{\lambda_i} \mathbf{u}_i \mathbf{u}_i^T = U\Lambda^+ U^T,$$

where $\Lambda^+ = \mathrm{diag}(\frac{1}{\lambda_1}, \dots, \frac{1}{\lambda_r}, 0, \dots, 0)$ *is the diagonal matrix containing the reciprocals of the non-zero eigenvalues, otherwise zeros, in its main diagonal.*

The Moore–Penrose inverse of the $m \times n$ real matrix is the $n \times m$ matrix A^+ with SVD (A.12)

$$A^+ = \sum_{i=1}^{r} \frac{1}{s_i} \mathbf{u}_i \mathbf{v}_i^T = US^+ V^T,$$

where S^+ is $n \times m$ generalized diagonal matrix containing the reciprocals of the non-zero singular values of A in the leading positions, otherwise zeros, in its main diagonal.

Note that, analogously, any analytic function f of the symmetric real matrix A can be defined by its SD, $A = U\Lambda U^T$, in the following way:

$$f(A) := U f(\Lambda) U^T \tag{A.16}$$

where $f(\Lambda) = \mathrm{diag}(f(\lambda_1), \dots, f(\lambda_n))$, of course, only if every eigenvalue is in the domain of f. In this way, for a positive semidefinite (or positive definite) A, its squareroot is

$$A^{1/2} = U\Lambda^{1/2} U^T, \tag{A.17}$$

and for a regular A its inverse is obtained by applying the $f(x) = x^{-1}$ function to it:

$$A^{-1} = U\Lambda^{-1} U^T.$$

For a singular A, the Moore–Penrose inverse is obtained by using Λ^+ instead of Λ^{-1}. Accordingly, for a positive semidefinite matrix, its $-1/2$ power is defined as the squareroot of A^+.

We will frequently use the following propositions, called *separation theorems* for singular values and eigenvalues. These are special cases of Propositions A.2.7, A.2.8, A.2.9, and A.2.10. For the applications of these theorems in multivariate statistical analysis see Rao (1979).

Proposition A.3.9 *Let A be an $m \times n$ real matrix with SVD in (A.12). Assume that its non-zero singular values are enumerated in non-increasing order ($s_1 \geq s_2 \geq \dots s_r > 0$). Then*

$$\max_{\substack{\mathbf{x} \in \mathbb{R}^n, \mathbf{y} \in \mathbb{R}^m \\ \|\mathbf{x}\|=1, \|\mathbf{y}\|=1}} \mathbf{y}^T A \mathbf{x} = s_1$$

and it is attained with the choice $x = \mathbf{u}_1$ *and* $y = \mathbf{v}_1$ *(uniquely if* $s_1 > s_2$*). This was the* $k = 1$ *case. Further, for* $k = 2, 3, \ldots, r$

$$\max_{\substack{\mathbf{x}\in\mathbb{R}^n,\ \mathbf{y}\in\mathbb{R}^m \\ \|x\|=1,\ \|y\|=1 \\ \mathbf{x}^T\mathbf{u}_i=0\,(i=1,\ldots,k-1) \\ \mathbf{y}^T\mathbf{v}_i=0\,(i=1,\ldots,k-1)}} \mathbf{y}^T \boldsymbol{A}\mathbf{x} = s_k$$

and it is attained with the choice $\mathbf{x} = \mathbf{u}_k$ *and* $y = \mathbf{v}_k$ *(uniquely if* $s_k > s_{k+1}$*).*

The following are also true.

Proposition A.3.10 *Let* $\boldsymbol{A}$ *be an* $m \times n$ *real matrix with SVD in (A.12). Assume that its singular values are enumerated in non-increasing order and* $k > 0$ *is an integer such that* $s_k > s_{k+1}$*. Then*

$$\max_{\substack{\boldsymbol{X}\text{ is } n\times k,\ \boldsymbol{Y}\text{ is } m\times k \\ \boldsymbol{X}^T\boldsymbol{X}=\boldsymbol{I}_k,\ \boldsymbol{Y}^T\boldsymbol{Y}=\boldsymbol{I}_k}} \operatorname{tr}(\boldsymbol{Y}^T\boldsymbol{A}\boldsymbol{X}) = \max_{\substack{\mathbf{x}_i^T\mathbf{x}_j=\delta_{ij} \\ \mathbf{y}_i^T\mathbf{y}_j=\delta_{ij}}} \sum_{i=1}^{k} \mathbf{y}_i^T\boldsymbol{A}\mathbf{x}_i = \sum_{i=1}^{k} s_i$$

and it is attained with the suborthogonal matrices $\boldsymbol{X} = (\mathbf{x}_1, \ldots, \mathbf{x}_k)$ *and* $\boldsymbol{Y} = (\mathbf{y}_1, \ldots, \mathbf{y}_k)$ *such that* $\mathbf{x}_i = \mathbf{u}_i$ *and* $\mathbf{y}_i = \mathbf{v}_i$ $(i = 1, \ldots, k)$*.*

Proposition A.3.11 *Let* $\boldsymbol{A}$ *be* $n \times n$ *real symmetric matrix with SD in (A.10). Assume that its eigenvalues are enumerated in non-increasing order* $(\lambda_1 \geq \lambda_2 \geq \cdots \geq \lambda_n)$*. Then*

$$\max_{\|\mathbf{x}\|=1} \mathbf{x}^T\boldsymbol{A}\mathbf{x} = \lambda_1$$

and it is attained with the choice $\mathbf{x} = \mathbf{u}_1$ *(uniquely if* $\lambda_1 > \lambda_2$*). This was the* $k = 1$ *case. Further, for* $k = 2, 3, \ldots, n$

$$\max_{\substack{\mathbf{x}\in\mathbb{R}^n,\ \|\mathbf{x}\|=1 \\ \mathbf{x}^T\mathbf{u}_i=0\,(i=1,\ldots,k-1)}} \mathbf{x}^T\boldsymbol{A}\mathbf{x} = \lambda_k$$

and it is attained with the choice $\mathbf{x} = \mathbf{u}_k$ *(uniquely if* $\lambda_k > \lambda_{k+1}$*).*

Proposition A.3.12 *Let* $\boldsymbol{A}$ *be* $n \times n$ *real symmetric matrix with SD in (A.10). Assume that its eigenvalues are enumerated in non-increasing order and* $k > 0$ *is an integer such that* $\lambda_k > \lambda_{k+1}$*. Then*

$$\max_{\substack{\boldsymbol{X}\text{ is } n\times k \\ \boldsymbol{X}^T\boldsymbol{X}=\boldsymbol{I}_k}} \operatorname{tr}(\boldsymbol{X}^T\boldsymbol{A}\boldsymbol{X}) = \max_{\substack{\mathbf{x}_i\in\mathbb{R}^n\,(i=1,\ldots,k) \\ \mathbf{x}_i^T\mathbf{x}_j=\delta_{ij}}} \sum_{i=1}^{k} \mathbf{x}_i^T\boldsymbol{A}\mathbf{x}_i = \sum_{i=1}^{k} \lambda_i$$

and it is attained with the suborthogonal matrix $\boldsymbol{X} = (\mathbf{x}_1, \ldots, \mathbf{x}_k)$ *such that* $\mathbf{x}_i = \mathbf{u}_i$ $(i = 1, \ldots, k)$*.*

Applying the above statements for the symmetric real matrix $-\boldsymbol{A}$, similar statements can be proved. Note that these minima and maxima are attained as we optimize over convex sets.

Proposition A.3.13 *Let $\boldsymbol{A}$ be an $n \times n$ real symmetric matrix with SD in (A.10). Assume that its eigenvalues are enumerated in non-increasing order ($\lambda_1 \geq \cdots \geq \lambda_{n-1} \geq \lambda_n$). Then*

$$\min_{\|x\|=1} \mathbf{x}^T \boldsymbol{A}\mathbf{x} = \lambda_n$$

and it is attained with the choice $\mathbf{x} = \mathbf{u}_n$ (uniquely if $\lambda_n < \lambda_{n-1}$). This was the $k = 1$ case. Further, for $k = 2, 3, \ldots, n$

$$\min_{\substack{\mathbf{x}\in\mathbb{R}^n,\ \|\mathbf{x}\|=1 \\ \mathbf{x}^T\mathbf{u}_i=0\,(i=n-k+2,\ldots,n)}} \mathbf{x}^T \boldsymbol{A}\mathbf{x} = \lambda_{n-k+1}$$

and it is attained with the choice $\mathbf{x} = \mathbf{u}_{n-k+1}$ (uniquely if $\lambda_{n-k+2} < \lambda_{n-k+1}$).

Proposition A.3.14 *Let $\boldsymbol{A}$ be $n \times n$ real symmetric matrix with SD in (A.10). Assume that its eigenvalues are enumerated in non-increasing order and $k > 0$ is an integer such that $\lambda_{n-k+1} < \lambda_{n-k}$. Then*

$$\min_{\substack{\mathbf{X}\text{ is } n\times k \\ \mathbf{X}^T\mathbf{X}=\boldsymbol{I}_k}} \mathrm{tr}(\mathbf{X}^T\boldsymbol{A}\mathbf{X}) = \min_{\substack{\mathbf{x}_i\in\mathbb{R}^n\,(i=1,\ldots,k) \\ \mathbf{x}_i^T\mathbf{x}_j=\delta_{ij}}} \sum_{i=1}^{k} \mathbf{x}_i^T\boldsymbol{A}\mathbf{x}_i = \sum_{i=n-k+1}^{n} \lambda_i$$

and it is attained with the suborthogonal matrix $\mathbf{X} = (\mathbf{x}_1, \ldots, \mathbf{x}_k)$ such that $\mathbf{x}_i = \mathbf{u}_{n-k+i}$ $(i = 1, \ldots, k)$.

Many of the above propositions follow from the forthcoming so-called *minimax principle*.

Theorem A.3.15 (Courant–Fischer–Weyl theorem) *Let $\boldsymbol{A}$ be an $n \times n$ symmetric real matrix with eigenvalues $\lambda_1 \geq \cdots \geq \lambda_n$. Then*

$$\lambda_k = \max_{\substack{F\subset\mathbb{R}^n \\ \dim(F)=k}} \min_{\substack{\mathbf{x}\in F \\ \|\mathbf{x}\|=1}} \mathbf{x}^T\boldsymbol{A}\mathbf{x} = \min_{\substack{F\subset\mathbb{R}^n \\ \dim(F)=n-k+1}} \max_{\substack{\mathbf{x}\in F \\ \|\mathbf{x}\|=1}} \mathbf{x}^T\boldsymbol{A}\mathbf{x} \quad (k = 1, \ldots, n).$$

The statement naturally extends to singular values of rectangular matrices.

Theorem A.3.16 *Let $\boldsymbol{A}$ be an $m \times n$ real matrix with positive singular values $s_1 \geq \cdots \geq s_r$, where $r = \mathrm{rank}(\boldsymbol{A})$. Then*

$$s_k = \max_{\substack{F\subset\mathbb{R}^n \\ \dim(F)=k}} \min_{\mathbf{x}\in F} \frac{\|\boldsymbol{A}\mathbf{x}\|}{\|\mathbf{x}\|} \quad (k = 1, \ldots, r).$$

Theorem A.3.15 is, in turn, implied by the upcoming separation theorem. In the sequel, we will denote by $\lambda_i(.)$ the ith largest eigenvalue of the symmetric matrix in the argument (they are enumerated in non-increasing order).

Theorem A.3.17 (Cauchy–Poincaré separation theorem) *Let $\boldsymbol{A}$ be an $n \times n$ symmetric real matrix and $\boldsymbol{B}$ be an $n \times k$ suborthogonal matrix ($k \leq n$). Then*

$$\lambda_i(\boldsymbol{A}) \geq \lambda_i(\boldsymbol{B}^T \boldsymbol{A} \boldsymbol{B}) \geq \lambda_{i+n-k}(\boldsymbol{A}), \quad i = 1, \ldots, k.$$

The first inequality is attained with equality if $\boldsymbol{B}$ contains the eigenvectors corresponding to the k largest eigenvalues of $\boldsymbol{A}$ in its columns; whereas, the second inequality is attained with equality if $\boldsymbol{B}$ contains the eigenvectors corresponding to the k smallest eigenvalues of $\boldsymbol{A}$ in its columns.

Note that the first inequality makes sense for a k such that $\lambda_k > \lambda_{k+1}$, whereas the second inequality makes sense for a k such that $\lambda_{n-k+1} < \lambda_{n-k}$.

The Cauchy–Poincaré theorem implies the following important inequalities due to H. Weyl.

Theorem A.3.18 (Weyl's perturbation theorem) *Let $\boldsymbol{A}$ and $\boldsymbol{C}$ be $n \times n$ symmetric matrices. Then*

$$\lambda_j(\boldsymbol{A} + \boldsymbol{C}) \leq \lambda_i(\boldsymbol{A}) + \lambda_{j-i+1}(\boldsymbol{C}) \quad \text{if} \quad i \leq j,$$
$$\lambda_j(\boldsymbol{A} + \boldsymbol{C}) \geq \lambda_i(\boldsymbol{A}) + \lambda_{j-i+n}(\boldsymbol{C}) \quad \text{if} \quad i \geq j.$$

The above inequalities give rise to the following perturbation result for symmetric matrices we will intensively use in Chapter 1. Here we consider symmetric matrices such that $\boldsymbol{A} = \boldsymbol{B} + \boldsymbol{C}$, where $\boldsymbol{C}$ is a 'small' perturbation on $\boldsymbol{B}$.

Theorem A.3.19 *Let $\boldsymbol{A}$ and $\boldsymbol{B}$ be $n \times n$ symmetric matrices. Then*

$$|\lambda_i(\boldsymbol{A}) - \lambda_i(\boldsymbol{B})| \leq \|\boldsymbol{A} - \boldsymbol{B}\|, \quad i = 1, \ldots, n.$$

A similar statement is valid for rectangular matrices.

Theorem A.3.20 *Let $\boldsymbol{A}$ and $\boldsymbol{B}$ be $m \times n$ real matrices with singular values $s_1(\boldsymbol{A}) \geq \cdots \geq s_{\min\{m,n\}}(\boldsymbol{A})$ and $s_1(\boldsymbol{B}) \geq \cdots \geq s_{\min\{m,n\}}(\boldsymbol{B})$. Then*

$$|s_i(\boldsymbol{A}) - s_i(\boldsymbol{B})| \leq \|\boldsymbol{A} - \boldsymbol{B}\|, \quad i = 1, \ldots, \min\{m, n\}.$$

Applying the above theorems for rank k matrices $\boldsymbol{B}$ we can solve the following optimization problems stated in a more general form, for rectangular matrices.

Theorem A.3.21 *Let $\boldsymbol{A}$ be an arbitrary $m \times n$ real matrix with SVD $\sum_{i=1}^{r} s_i \mathbf{u}_i \mathbf{v}_i^T$, where r is the rank of $\boldsymbol{A}$. Then for any positive integer $k \leq r$ such that $s_k > s_{k+1}$,*

$$\min_{\substack{\boldsymbol{B} \text{ is } m\times n \\ \text{rank}(\boldsymbol{B})=k}} \|\boldsymbol{A} - \boldsymbol{B}\| = s_{k+1} \quad \text{and} \quad \min_{\substack{\boldsymbol{B} \text{ is } m\times n \\ \text{rank}(\boldsymbol{B})=k}} \|\boldsymbol{A} - \boldsymbol{B}\|_2 = \left(\sum_{i=k+1}^{r} s_i^2 \right)^{1/2}$$

hold, and both minima are attained with the matrix $\boldsymbol{B}_k = \sum_{i=1}^{k} s_i \mathbf{u}_i \mathbf{v}_i^T$.

Note that $\boldsymbol{B}_k$ is called the *best rank k approximation* of $\boldsymbol{A}$, and the aforementioned theorem guarantees that it is the best approximation both in spectral and Frobenius norm. In fact, it is true for any unitary invariant norm:

$$\min_{\substack{\boldsymbol{B} \text{ is } m\times n \\ \text{rank}(\boldsymbol{B})=k}} \|\boldsymbol{A} - \boldsymbol{B}\|_{\text{un}} = \|\boldsymbol{A} - \boldsymbol{B}_k\|_{\text{un}}.$$

The next theorem is about perturbation of eigenvectors of symmetric matrices. In fact, the original theorem, due to Davis and Kahan (1970), applies to complex self-adjoint (Hermitian) matrices and to spectral subspaces spanned by eigenvectors corresponding to a set of eigenvectors separated from the others, see Stewart and Sun (1990). Here we follow the formalism of Bhatia (1997).

Theorem A.3.22 (Perturbation of spectral subspaces) *Let $\boldsymbol{A}$ and $\boldsymbol{B}$ be symmetric matrices; S_1 and S_2 are subsets of $\mathbb{R}$ or $\mathbb{C}$ such that* $\text{dist}(S_1, S_2) = \delta > 0$. *Let $\boldsymbol{P}_{\boldsymbol{A}}(S_1)$ and $\boldsymbol{P}_{\boldsymbol{B}}(S_2)$ be orthogonal projections onto the subspace spanned by the eigenvectors of the matrix in the lower index, corresponding to the eigenvalues within the subset in the argument. Then with any unitary invariant norm:*

$$\|\boldsymbol{P}_A(S_1)\boldsymbol{P}_B(S_2)\|_{\text{un}} \leq \frac{c_1}{\delta} \|\boldsymbol{P}_A(S_1)(\boldsymbol{A} - \boldsymbol{B})\boldsymbol{P}_B(S_2)\|_{\text{un}} \leq \frac{c_1}{\delta} \|\boldsymbol{A} - \boldsymbol{B}\|_{\text{un}}$$

where c_1 is a positive constant.

Remark A.3.23 *In another context, Sz.-Nagy (1953) proved that $c_1 = \pi/2$. When S_1 and S_2 are separated by an annulus, then the constant improves to $c_1 = 1$; further, with the Frobenius norm, $c_1 = 1$ will always do, see Bhatia (1997). If $\boldsymbol{P}_A(S_1)$ and $\boldsymbol{P}_B^{\perp}(S_2)$ project onto subspaces of the same dimension, then either the spectral- or the Frobenius norm of $\boldsymbol{P}_A(S_1)\boldsymbol{P}_B(S_2)$ can be expressed in terms of the sines of the so-called canonical (principal) angles between these subspaces and $\|\boldsymbol{P}_A(S_1)\boldsymbol{P}_B(S_2)\|_2$ is considered as the distance between them. This is why the special case of Theorem A.3.22, when Frobenius norm is used, is called the Davis–Kahan* $\sin(\boldsymbol{\theta})$ *theorem.*

We will also need the following statements concerning relations between singular values, eigenvalues, and traces.

Proposition A.3.24 *Let* $\boldsymbol{A}$ *be an* $n \times n$ *matrix and* $\boldsymbol{Q}$ *be an* $n \times n$ *orthogonal one. Then* $\operatorname{tr}(\boldsymbol{A}\boldsymbol{Q})$ *is maximal if* $\boldsymbol{A}\boldsymbol{Q}$ *is symmetric, and in this case the trace of this symmetric matrix is equal to the sum of the singular values of* $\boldsymbol{A}$.

Proposition A.3.25 *Let* $\boldsymbol{A}$ *and* $\boldsymbol{B}$ *be* $n \times n$ *symmetric, positive semidefinite matrices with eigenvalues* $\lambda_i(\boldsymbol{A})$*'s and* $\lambda_i(\boldsymbol{B})$*'s. Then*

$$\operatorname{tr}(\boldsymbol{A}\boldsymbol{B}) \leq \sum_{i=1}^{n} \lambda_i(\boldsymbol{A}) \cdot \lambda_i(\boldsymbol{B}),$$

with equality if and only if $\boldsymbol{A}$ *and* $\boldsymbol{B}$ *commute, that is,* $\boldsymbol{A}\boldsymbol{B} = \boldsymbol{B}\boldsymbol{A}$.

Note that a necessary and sufficient condition for the symmetric $\boldsymbol{A}$ and $\boldsymbol{B}$ commute is that they have the same system of eigenvectors (possibly, eigenspaces).

Proposition A.3.26 *Let* $\boldsymbol{A}$ *and* $\boldsymbol{B}$ *be* $n \times n$ *real matrices with singular values* $s_1(\boldsymbol{A}) \geq \cdots \geq s_n(\boldsymbol{A}) \geq 0$ *and* $s_1(\boldsymbol{B}) \geq \cdots \geq s_n(\boldsymbol{B}) \geq 0$. *Then*

$$\prod_{i=1}^{k} s_i(\boldsymbol{A}\boldsymbol{B}) \leq \prod_{i=1}^{k} [s_i(\boldsymbol{A}) \cdot s_i(\boldsymbol{B})], \quad k = 1, \ldots, n.$$

Especially, for $k = 1$, this implies that

$$s_{\max}(\boldsymbol{A}\boldsymbol{B}) \leq s_{\max}(\boldsymbol{A}) \cdot s_{\max}(\boldsymbol{B}),$$

which is not surprising, since the maximal singular value is the operator norm of the matrix.

The next part will be devoted to the Perron–Frobenius theory of matrices with nonnegative entries. First we define the notion of the irreducibility for a quadratic matrix, and a similar notion for rectangular matrices.

Definition A.3.27 *A quadratic matrix* $\boldsymbol{A}$ *is called reducible if there exists an appropriate permutation of its rows and columns, or equivalently, there exists a permutation matrix (see Section A.3)* $\boldsymbol{P}$ *such that, with it,* $\boldsymbol{A}$ *can be transformed into the following block-matrix form:*

$$\boldsymbol{P}\boldsymbol{A}\boldsymbol{P}^T = \begin{pmatrix} \boldsymbol{B} & \boldsymbol{O} \\ \boldsymbol{D} & \boldsymbol{C} \end{pmatrix} \quad \text{or} \quad \boldsymbol{P}\boldsymbol{A}\boldsymbol{P}^T = \begin{pmatrix} \boldsymbol{B} & \boldsymbol{D} \\ \boldsymbol{O} & \boldsymbol{C} \end{pmatrix},$$

where $\boldsymbol{A}$ *and* $\boldsymbol{B}$ *are quadratic matrices, whereas* $\boldsymbol{O}$ *is the zero matrix of appropriate size. A quadratic matrix is called irreducible if it is not reducible.*

Note that the eigenvalues of a quadratic matrix are unaffected under the same permutation of its rows and columns, while the coordinates of the corresponding eigenvectors are subject to the same permutation. Since in Definition A.3.27, the same permutation is applied to the rows and columns, and the spectrum of the involved block-matrix consists of the spectra of $\boldsymbol{B}$ and $\boldsymbol{C}$, the SD of a reducible matrix can be traced back to the SD of some smaller matrices. Now we want to create a similar notion for rectangular matrices in terms of their SVD.

Definition A.3.28 *An $m \times n$ real matrix $\boldsymbol{A}$ is called decomposable if there exist appropriate permutations of its rows and columns, or equivalently, there exist permutation matrices $\boldsymbol{P}$ and $\boldsymbol{Q}$ of sizes $m \times m$ and $n \times n$, respectively, such that, with them, $\boldsymbol{A}$ can be transformed into the following block-matrix form:*

$$\boldsymbol{P}\boldsymbol{A}\boldsymbol{Q}^T = \begin{pmatrix} \boldsymbol{B} & \boldsymbol{O} \\ \boldsymbol{O} & \boldsymbol{C} \end{pmatrix}.$$

The real matrix $\boldsymbol{A}$ is called non-decomposable if it does not decompose.

Note that the singular values of a real matrix are unaffected under appropriate permutation of its rows and columns, while the coordinates of the corresponding singular vectors are subject to the same permutations. Since in Definition A.3.28, the singular spectrum of the matrix $\boldsymbol{A}$ is composed of the singular spectra of $\boldsymbol{B}$ and $\boldsymbol{C}$, only SVD of non-decomposable matrices is of importance.

When we apply Definition A.3.28 to a symmetric matrix $\boldsymbol{A}$, the matrices $\boldsymbol{B}$ and $\boldsymbol{C}$ of Definition A.3.27 are also symmetric, whereas $\boldsymbol{D}$ becomes the zero matrix. As the singular values of a symmetric matrix are the absolute values of its eigenvalues, a reducible symmetric matrix is also decomposable. However, the converse is not true: there exist quadratic (even symmetric) real matrices which are decomposable, yet still irreducible. The easiest example is the

$$\begin{pmatrix} 0 & 1 \\ 1 & 0 \end{pmatrix}$$

matrix which has eigenvalues $1, -1$ and singular values $1, 1$. The former ones cannot be concluded from the spectra of smaller matrices, while the latter ones can, as the above matrix can be transformed into the identity matrix with interchanging its rows or columns (but not both of them). Nonetheless, for positive semidefinite matrices, the two notions are equivalent. We also remark that the rectangular matrix $\boldsymbol{A}$ is non-decomposable if and only if the symmetric matrix $\boldsymbol{A}\boldsymbol{A}^T$ (or, equivalently, $\boldsymbol{A}^T\boldsymbol{A}$) is irreducible.

The block-matrix of the following statement (see e.g., Bhatia (1997)) is also decomposable and irreducible at the same time.

Proposition A.3.29 *Let $\boldsymbol{A}$ be an $m \times n$ real matrix of rank r. Then the $(m+n) \times (m+n)$ symmetric matrix*

$$\widetilde{\boldsymbol{A}} = \begin{pmatrix} \boldsymbol{O} & \boldsymbol{A} \\ \boldsymbol{A}^T & \boldsymbol{O} \end{pmatrix}$$

has the following spectrum: its largest and smallest non-zero eigenvalues are

$$\lambda_i(\widetilde{\boldsymbol{A}}) = -\lambda_{n+m-i+1}(\widetilde{\boldsymbol{A}}) = s_i(\boldsymbol{A}), \quad i = 1, \ldots, r,$$

while the others are zeros, where $\lambda_i(.)$ and $s_i(.)$ denote the ith largest eigenvalue and singular value of the matrix in the argument, respectively.

Hence, the spectrum of $\widetilde{\boldsymbol{A}}$ is symmetric about the 0, and its rank is $2r$; moreover, the relevant eigenvectors can be composed of the singular vector pairs of $\boldsymbol{A}$.

The subsequent theorems apply to matrices of nonnegative entries.

Theorem A.3.30 (Frobenius theorem) *Any irreducible, quadratic real matrix of nonnegative entries has a single positive eigenvalue among its maximum absolute value ones with corresponding eigenvector of all positive coordinates.*

Remark A.3.31 *More precisely, there may be $k \geq 1$ complex eigenvalues of maximum absolute value r, allocated along the circle of radius r in the complex plane. In fact, those complex numbers are vertices of a regular k-gone, but the point is that exactly one of these vertices is allocated on the positive part of the real axis; see Rózsa (1974) for the proof.*

The Perron theorem is the specialized version of the Frobenius theorem, applicable to matrices of strictly positive entries.

Theorem A.3.32 (Perron theorem) *Any irreducible, quadratic real matrix of positive entries has only one maximum absolute value eigenvalue which is positive with multiplicity one, and the corresponding eigenvector has all positive coordinates.*

As a byproduct of the proof of the above theorems, the following useful bounds for the maximum absolute value positive eigenvalue – guaranteed by the Frobenius theorem – can be obtained.

Proposition A.3.33 *Let $\boldsymbol{A}$ be an irreducible $n \times n$ real matrix of nonnegative entries and introduce the following notation for the maxima and minima of the row-sums of $\boldsymbol{A}$:*

$$m := \min_{i \in \{1,\ldots,n\}} \sum_{j=1}^{n} a_{ij} \quad \text{and} \quad M := \max_{i \in \{1,\ldots,n\}} \sum_{j=1}^{n} a_{ij}.$$

Then the single positive eigenvalue λ with maximum absolute value admits the following lower and upper bound:

$$m \leq \lambda \leq M,$$

where either the lower or the upper bound is attained if and only if $m = M$, that is, the row-sums of $\boldsymbol{A}$ have a constant value.

Finally, we introduce the *Kronecker-sum* and *Kronecker-product* of matrices.

Definition A.3.34 *Let* $\boldsymbol{A}_i$ *be* $n_i \times n_i$ *matrix* $(i = 1, \dots, k)$, $n := \sum_{i=1}^{k} n_i$. *The Kronecker-sum of* $\boldsymbol{A}_1, \dots, \boldsymbol{A}_k$ *is the* $n \times n$ *block-diagonal matrix* $\boldsymbol{A}$ *the diagonal blocks of which are the matrices* $\boldsymbol{A}_1, \dots, \boldsymbol{A}_k$ *in this order. We use the notation* $\boldsymbol{A} = \boldsymbol{A}_1 \oplus \dots \oplus \boldsymbol{A}_k$ *for it.*

Definition A.3.35 *Let* $\boldsymbol{A}$ *be* $p \times n$ *and* $\boldsymbol{B}$ *be* $q \times m$ *real matrix. Their Kronecker-product, denoted by* $\boldsymbol{A} \otimes \boldsymbol{B}$, *is the following* $pq \times nm$ *block-matrix: it has pn blocks each of which is a* $q \times m$ *matrix such that the block indexed by* (i, j) *is the matrix* $a_{ij}\boldsymbol{B}$ $(i = 1, \dots, p;\ j = 1, \dots, n)$.

This product is associative, for the addition distributive, but usually not commutative. If $\boldsymbol{A}$ is $n \times n$ and $\boldsymbol{B}$ is $m \times m$ quadratic matrix, then

$$|\boldsymbol{A} \otimes \boldsymbol{B}| = |\boldsymbol{A}|^m \cdot |\boldsymbol{B}|^n;$$

further, if both are regular, then so is their Kronecker-product. Namely,

$$(\boldsymbol{A} \otimes \boldsymbol{B})^{-1} = \boldsymbol{A}^{-1} \otimes \boldsymbol{B}^{-1}.$$

It is also useful to know that – provided $\boldsymbol{A}$ and $\boldsymbol{B}$ are symmetric – the spectrum of $\boldsymbol{A} \otimes \boldsymbol{B}$ consists of the real numbers

$$\alpha_i \beta_j \quad (i = 1, \dots, n;\ j = 1, \dots, m),$$

where α_i's and β_j's are the eigenvalues of $\boldsymbol{A}$ and $\boldsymbol{B}$, respectively.

References

Bhatia R 1997 *Matrix Analysis*, Springer.

Davis C and Kahan WM 1970 The rotation of eigenvectors by a perturbation, III. *SIAM J. Numer. Anal.* **7**, 1–46.

Fréchet M 1907 Sur les ensembles de fonctions et les opérations linéaires. *C. R. Acad. Sci. Paris* **144**, 1414–1416.

Golub GH and van Loan CF 1996 *Matrix Computations*, 3rd edn Johns Hopkins University Press, Baltimore, MD.

Komornik V 2003 *Lecture Notes on Real Analysis II (in Hungarian)*, Typotex, Budapest.

Rao CR 1979 Separation theorems for singular values of matrices and their applications in multivariate analysis. *J. Multivariate Anal.* **9**, 362–377.

Riesz F 1907 Sur une espèce de géométrie analytique des systèmes de fonctions sommables. *C. R. Acad. Sci. Paris* **144**, 1409–1411.

Riesz F 1909 Sur les opérations fonctionnelles linéaires. *C. R. Acad. Sci. Paris* **149**, 974–977.

Riesz F and Sz.-Nagy B 1952 *Leçons d'analyse fonctionnelle*, Academic Publishing House, Budapest.

Rózsa P 1974 *Linear Algebra and Its Applications (in Hungarian)*, Műszaki Könyvkiadó, Budapest.

Stewart GW and Sun J-g. 1990 *Matrix Pertubation Theory*, Academic Press.

Sz.-Nagy B 1953 Über die Ungleichnung von H. Bohr. *Math. Nachr.* **9**, 255–259.

von Neumann J 1935 On complete topological spaces. *Trans. Amer. Math. Soc.* **37**, 1–20.

Wilkinson JH 1965 *The Algebraic Eigenvalue Problem*, Clarendon Press, Oxford.

Appendix B

Random vectors and matrices

B.1 Random vectors

Random vectors are vector-valued random variables with distribution characterized by the joint distribution of their coordinates. Scalar valued random variables will be denoted by upper-case letters, whereas random vectors by bold-face upper-case ones (usually with letters at the end of the alphabet).

Consider first a two-variate joint distribution for introducing the notion of conditional expectation. We will distinguish between the following two cases depending on the coordinates of the random vector (X, Y).

(a) Both X and Y are so-called categorical variables (they have finitely many values which cannot be compared on any scale, like hair-color and eye-color, medical diagnoses or possible answers to a questionnaire). Say, X takes on m possible values $x_1, \dots, x_m$, while Y takes on n possible ones $y_1, \dots, y_n$. The joint distribution of X and Y is defined by the probabilities $p_{ij} = \mathbb{P}(X = x_i, Y = y_j)$ such that $\sum_{i=1}^{m} \sum_{j=1}^{n} p_{ij} = 1$. These are usually estimated from the frequency counts collected in an $m \times n$ rectangular array, called a *contingency table*. The marginal distributions of X and Y are given by the probabilities

$$p_{i.} = \sum_{j=1}^{n} p_{ij} \quad \text{and} \quad p_{.j} = \sum_{i=1}^{m} p_{ij},$$

respectively. The conditional distribution of Y given $X = x_i$ is defined by the conditional probabilities $\frac{p_{ij}}{p_{i.}}$ for $j = 1, \dots n$, and the *conditional expectation*

Spectral Clustering and Biclustering: Learning Large Graphs and Contingency Tables, First Edition.
Marianna Bolla.

of Y under the same condition is

$$\mathbb{E}(Y|X = x_i) = \sum_{j=1}^{n} y_j \frac{p_{ij}}{p_{i.}} = \frac{1}{p_{i.}} \sum_{j=1}^{n} y_j p_{ij}, \quad i = 1, \ldots, m.$$

Note that neither the conditional distribution nor the conditional expectation of Y depends on the actual value of X. Making use of this property, we can define the $\mathbb{E}(Y|X)$ random variable which takes on the possible value $\mathbb{E}(Y|X = x_i)$ with probability p_i for $i = 1, \ldots, m$. (We may say that the random variable $\mathbb{E}(Y|X)$ is measurable with respect to the σ-algebra generated by X, but we do not want to go into measure theoretical considerations within the framework of this book.) The most important fact is that $\mathbb{E}(Y|X)$ is a measurable function of X.

(b) Both X and Y are absolutely continuous random variables (e.g., body-height and body-weight, or two clinical measurements), then we use the joint density $f(x, y)$ in the calculations. By means of the marginal density $f_X(x) = \int f(x, y)\, dy$ we define the conditional distribution of Y given that X takes on the value x by means of the conditional density $\frac{f(x,y)}{f_X(x)}$, and the *conditional expectation* of Y under the same condition is

$$\mathbb{E}(Y|X = x) = \int y \frac{f(x, y)}{f_X(x)}\, dy = \frac{1}{f_X(x)} \int y f(x, y)\, dy.$$

The conditional expectation of Y given X is the random variable $\mathbb{E}(Y|X)$ which is again a measurable function of X.

In both cases the conditional expectation $\mathbb{E}(Y|X)$ provides the best least-square approximation of Y in terms of measurable functions of X in the following sense:

$$\min_{t=t(X)} \mathbb{E}(Y - t(X))^2 = \mathbb{E}(Y - \mathbb{E}(Y|X))^2.$$

The conditional expectation $E(X|Y)$ can be defined likewise, akin to the conditional expectation of a subset of coordinates on another subset of a random vector with several coordinates.

In fact, we take conditional expectations in everyday life. For example, if we have recorded students' grades in two subjects which measure similar abilities, and we have lost the grade of a student in subject Y, then we can conclude for it, based on his or her grade in subject X in the following way. We take the average Y-grade of other students who have the same X-grade as the student in question. In this way, we take the conditional expectation of the (unknown) Y-grade given the (known) X-grade. (Of course, grades are coded with integers and the conditional expectation is rounded). Or, in other situations, we conclude for the (unknown) age of a person through the average (known) age of those who are similar to him/her in other respects.

In multivariate statistics, the most frequently used multivariate distribution is the *multivariate normal (Gaussian) distribution.*

Definition B.1.1 *We say that* $\mathbf{Z}$ *is a p-dimensional standard normal vector if its coordinates are independent standard normal variables. Let* $\boldsymbol{A}$ *be a* $p \times p$ *regular real matrix and* $\boldsymbol{\mu} \in \mathbb{R}^p$ *be a vector. Then the linear transformation* $\mathbf{X} = \boldsymbol{A}\mathbf{Z} + \boldsymbol{\mu}$ *defines a p-dimensional random normal vector. We use the notation* $\mathbf{Z} \sim \mathcal{N}_p(\mathbf{0}, \boldsymbol{I}_p)$ *and* $\mathbf{X} \sim \mathcal{N}_p(\boldsymbol{\mu}, \boldsymbol{C})$*, where* $\boldsymbol{\mu}$ *is called the expectation vector and* $\boldsymbol{C} = \boldsymbol{A}\boldsymbol{A}^T$ *the covariance matrix of* $\mathbf{X}$.

It is easy to verify that the entries of $\boldsymbol{C}$ are

$$c_{ij} = \mathrm{Cov}(X_i, X_j) = \mathrm{Cov}(X_j, X_i), \quad i, j = 1, \ldots, n.$$

We remark that a linear transformation with a singular $\boldsymbol{A}$ would result in a degenerated p-variate normal distribution which, in fact, is realized in a lower, namely, rank($\boldsymbol{A}$)-dimensional subspace. Hence, we will only deal with regular $\boldsymbol{A}$ and $\boldsymbol{C}$. Note that rank($\boldsymbol{C}$) = rank($\boldsymbol{A}$), and the p-variate p.d.f. of $\mathbf{X}$ is

$$f(\mathbf{x}) = \frac{1}{(2\pi)^{\frac{p}{2}} |\boldsymbol{C}|^{\frac{1}{2}}} e^{-\frac{1}{2}(\mathbf{x}-\boldsymbol{\mu})^T \boldsymbol{C}^{-1}(\mathbf{x}-\boldsymbol{\mu})}, \quad \mathbf{x} \in \mathbb{R}^p. \tag{B.1}$$

The level surfaces (contours of equal density) of f are ellipsoids, which are spheres if and only if the components of $\mathbf{X}$ are independent with equal variances.

Conversely, given the parameters $\boldsymbol{\mu}$ and $\boldsymbol{C}$, $\mathbf{X}$ can be transformed into a p-dimensional standard normal vector $\mathbf{Z}$ in the following way. $\boldsymbol{C}$, being a Gram-matrix, can be (not uniquely) decomposed as $\boldsymbol{A}\boldsymbol{A}^T$ with a regular $p \times p$ matrix $\boldsymbol{A}$. For example, $\boldsymbol{A} = \boldsymbol{C}^{1/2}$ will do, but $\boldsymbol{A}\boldsymbol{Q}$ is also convenient with any $p \times p$ orthogonal matrix $\boldsymbol{Q}$. Then the formula

$$\mathbf{Z} = \boldsymbol{A}^{-1}(\mathbf{X} - \boldsymbol{\mu})$$

defines a p-dimensional standard normal vector. Note that using $\boldsymbol{A}\boldsymbol{Q}$ instead of $\boldsymbol{A}$ will result in an orthogonal rotation of $\mathbf{Z}$ which has the same distribution as $\mathbf{Z}$.

It can be verified that a random vector has multivariate normal distribution if and only if every linear combination of its components is univariate normal. Further, all the marginal and conditional distributions of the multivariate normal distribution are multivariate normal of appropriate dimension.

For the multivariate normal distribution the conditional distributions are linear functions of the subset of variables in the condition. More precisely, the following proposition can be proved.

Proposition B.1.2 *Let* $(\mathbf{X}^T, \mathbf{Y}^T)^T \sim \mathcal{N}_{p+q}(\boldsymbol{\mu}, \boldsymbol{C})$ *be a random vector, where the expectation* $\boldsymbol{\mu}$ *and the covariance matrix* $\boldsymbol{C}$ *are partitioned (with block sizes p and q) in the following way:*

$$\boldsymbol{\mu} = \begin{pmatrix} \boldsymbol{\mu}_{\mathbf{X}} \\ \boldsymbol{\mu}_{\mathbf{Y}} \end{pmatrix}, \quad \boldsymbol{C} = \begin{pmatrix} \boldsymbol{C}_{\mathbf{XX}} & \boldsymbol{C}_{\mathbf{XY}} \\ \boldsymbol{C}_{\mathbf{YX}} & \boldsymbol{C}_{\mathbf{YY}} \end{pmatrix}.$$

Here $\boldsymbol{C}_{\mathbf{XX}}$, $\boldsymbol{C}_{\mathbf{YY}}$ *are covariance matrices of* $\mathbf{X}$ *and* $\mathbf{Y}$, *whereas* $\boldsymbol{C}_{\mathbf{YX}} = \boldsymbol{C}_{\mathbf{XY}}^T$ *is the cross-covariance matrix. Assume that* $\boldsymbol{C}_{\mathbf{XX}}$, $\boldsymbol{C}_{\mathbf{YY}}$ *and* $\boldsymbol{C}$ *are regular. Then the conditional distribution of the random vector* $\mathbf{Y}$ *conditioned on* $\mathbf{X}$ *is* $\mathcal{N}_q(\boldsymbol{C}_{YX}\boldsymbol{C}_{\mathbf{XX}}^{-1}(\mathbf{X} - \boldsymbol{\mu}_{\mathbf{X}}) + \boldsymbol{\mu}_{\mathbf{Y}}, \boldsymbol{C}_{\mathbf{YY}|\mathbf{X}})$ *distribution, where*

$$\boldsymbol{C}_{\mathbf{YY}|\mathbf{X}} = \boldsymbol{C}_{\mathbf{YY}} - \boldsymbol{C}_{\mathbf{YX}}\boldsymbol{C}_{\mathbf{XX}}^{-1}\boldsymbol{C}_{\mathbf{XY}}.$$

The conditional expectation of $\mathbf{Y}$ conditioned on $\mathbf{X}$ is the expectation of the above conditional distribution:

$$\mathbb{E}(\mathbf{Y}|\mathbf{X}) = \boldsymbol{C}_{\mathbf{YX}}\boldsymbol{C}_{\mathbf{XX}}^{-1}(\mathbf{X} - \boldsymbol{\mu}_{\mathbf{X}}) + \boldsymbol{\mu}_{\mathbf{Y}}$$

which is a linear function of the coordinates of $\mathbf{X}$. In the $p = q = 1$ case, it is called the *regression line*, while in the $q = 1, p > 1$ case, *regression plane*. Summarizing, in the case of multidimensional Gaussian distribution the regression functions are linear functions of the variables in the condition, which fact has important consequences in multivariate statistical analysis. Note that multivariate normality can often be assumed due to the *multidimensional central limit theorem*, to be introduced right now.

Theorem B.1.3 *Let* $\mathbf{X}_1, \mathbf{X}_2, \dots \sim \mathcal{N}_p(\boldsymbol{\mu}, \boldsymbol{C})$ *be an i.i.d. sample, with the regular covariance matrix* $\boldsymbol{C}$. *Then the sequence of the standardized partial sums*

$$\frac{1}{\sqrt{n}}\boldsymbol{C}^{-1/2}\left(\sum_{i=1}^{n}\mathbf{X}_i - n\boldsymbol{\mu}\right)$$

converges to the p-dimensional standard normal distribution as $n \to \infty$.

The convergence is understood in distribution, which means the convergence of the cumulative multivariate distribution functions.

Definition B.1.4 *The empirical and corrected empirical covariance matrices based on the* $\mathbf{X}_1, \dots, \mathbf{X}_n$ *i.i.d. sample are*

$$\hat{\boldsymbol{C}} = \frac{1}{n}\boldsymbol{S} \quad \text{and} \quad \hat{\boldsymbol{C}}^* = \frac{1}{n-1}\boldsymbol{S},$$

where $\boldsymbol{S} = \sum_{i=1}^{n}(\mathbf{X}_i - \bar{\mathbf{X}})(\mathbf{X}_i - \bar{\mathbf{X}})^T$ *with* $\bar{\mathbf{X}} = \frac{1}{n}\sum_{i=1}^{n}\mathbf{X}_i$.

The corrected empirical covariance matrix is unbiased, while the empirical one is the asymptotically unbiased estimator of the true covariance matrix $\boldsymbol{C}$.

The following proposition applies to deterministic matrices, still, in the proof, they are treated as covariance matrices of random vectors.

Proposition B.1.5 *Let $\boldsymbol{A}$ and $\boldsymbol{B}$ be $n \times n$ symmetric, positive semidefinite matrices. Then the matrix $\boldsymbol{C}$ of entries $c_{ij} = a_{ij} b_{ij}$ $(i, j = 1, \ldots n)$ is also symmetric, positive semidefinite.*

Proof. The symmetry of $\boldsymbol{C}$ is trivial. Any symmetric, positive semidefinite matrix is a Gram-matrix, and hence, can be considered as the covariance matrix of an n-dimensional random vector. For simplicity, let $\mathbf{X} = (X_1, \ldots, X_n)^T \sim \mathcal{N}_n(\mathbf{0}, \boldsymbol{A})$ and $\mathbf{Y} = (Y_1, \ldots, Y_n)^T \sim \mathcal{N}_n(\mathbf{0}, \boldsymbol{B})$ be independent Gaussian random vectors. We define the random vector $\mathbf{Z}$ in the following way:

$$\mathbf{Z} := (X_1 Y_1, \ldots, X_n Y_n)^T.$$

It is easy to see that $\mathbb{E}(\mathbf{Z}) = \mathbf{0}$, since $\mathbb{E}(X_i Y_i) = \text{Cov}(X_i, Y_i) + \mathbb{E}(X_i) \cdot \mathbb{E}(Y_i) = 0$, $i = 1, \ldots, n$. If we verify that the covariance matrix of $\mathbf{Z}$ is $\boldsymbol{C}$ then we are ready as a covariance matrix is always positive semidefinite. Indeed, the ij-th entry of $\mathbb{E}(\mathbf{Z}\mathbf{Z}^T)$ is

$$\mathbb{E}(X_i Y_i X_j Y_j) = \mathbb{E}([X_i X_j] \cdot [Y_i Y_j]) = \mathbb{E}(X_i X_j) \cdot \mathbb{E}(Y_i Y_j) = a_{ij} \cdot b_{ij} = c_{ij},$$

where we again used that the expectation of the product of the two independent random variables in the brackets is the product of their expectations.

We remark that $\mathbf{Z}$ is not multivariate Gaussian, and it is not necessary that $\mathbf{X}$ and $\mathbf{Y}$ be so, just because of their independence, the above calculations are valid (as any component of $\mathbf{X}$ is independent of any component of $\mathbf{Y}$, their covariances are also zeros). However, if we assume multivariate normality, we can even construct an $\mathbf{X}$ and $\mathbf{Y}$ with covariance matrices $\boldsymbol{A}$ and $\boldsymbol{B}$, in the following way. Let $\boldsymbol{A} = \boldsymbol{U}\boldsymbol{\Lambda}\boldsymbol{U}^T$ and $\boldsymbol{B} = \boldsymbol{V}\boldsymbol{\Delta}\boldsymbol{V}^T$ be SD's, where the diagonal entries of the diagonal matrices $\boldsymbol{\Lambda}$ and $\boldsymbol{\Delta}$ are nonnegative. Then $\mathbf{X} = \boldsymbol{U}\boldsymbol{\Lambda}^{1/2}\mathbf{e}$ and $\mathbf{Y} = \boldsymbol{V}\boldsymbol{\Delta}^{1/2}\mathbf{f}$ have covariance matrices $\boldsymbol{A}$ and $\boldsymbol{B}$, where $\mathbf{e}$ and $\mathbf{f}$ are i.i.d. n-dimensional standard Gaussian vectors.

We also remark that the above kind of matrix product is called Hadamard product or Schur product (see Petz (2008); Bhatia (1997)) and denoted by $\boldsymbol{A} \circ \boldsymbol{B}$, whereas Theorem B.1.5 is frequently referred to as Schur's theorem, with a purely algebraic proof, see e.g., Horn and Johnson (1985).

B.2 Random matrices

The first random matrix of the history is the Wishart matrix, defined in the 1930s as the sample covariance matrix of a multivariate Gaussian sample.

Definition B.2.1 *Let $\mathbf{Z}_1, \ldots, \mathbf{Z}_n \sim \mathcal{N}_p(\mathbf{0}, \boldsymbol{I}_p)$ be i.i.d. p-dimensional normal sample. The distribution of the $p \times p$ matrix $\widetilde{\mathbf{W}} = \sum_{i=1}^n \mathbf{Z}_i \mathbf{Z}_i^T$ is called standard Wishart with parameters p, n (n is also called degree of freedom as in the $p = 1$ case this is the*

χ^2 distribution of degree of freedom n). More generally, let $\mathbf{X}_1, \ldots, \mathbf{X}_n \sim \mathcal{N}_p(\mathbf{0}, \boldsymbol{C})$ be an i.i.d. p-dimensional normal sample. The distribution of the $p \times p$ matrix $\boldsymbol{W} = \sum_{i=1}^{n} \mathbf{X}_i \mathbf{X}_i^T$ is called (central) Wishart with parameters p, n, and $\boldsymbol{C}$.

In view of the relation between the $\mathbf{Z}_i$'s and the $\mathbf{X}_i$'s in the Definition B.1.1, the relation $\boldsymbol{W} = \boldsymbol{C}^{1/2}\widetilde{\boldsymbol{W}}\boldsymbol{C}^{1/2}$ also holds, and it can be proved that n times the empirical covariance matrix of a p-dimensional normal sample, that is, the matrix $\boldsymbol{S}$ in Definition B.1.4, is a (central) Wishart matrix with parameters p, $n-1$ and the common covariance matrix. The Wishart matrix is symmetric and positive semidefinite, and in the $p < n$ case it is positive definite, with probability 1. The Wishart entries are far from being independent (in and above the main diagonal, otherwise the matrix is symmetric); the joint density of the entries (the density of the matrix) is revisited by means of easy transformations in Olkin (2002). The author of this paper also derived the distribution of eigenvalues and singular values of matrices arising from matrix factorizations, for fixed n and p, but he did not investigate asymptotics for the singular values of the $p \times n$ data matrix, whose entries are usually not independent.

The Wigner type matrices were introduced later, in the 1950s, the eigenvalues of which modelled energy levels in slow nuclear reactions. These symmetric matrices are more easy to treat as their diagonal- and upper-diagonal entries are independent random variables, however, the distribution of them need not be defined uniquely. Therefore, there are many variants of theorems applying to the limiting behavior of the eigenvalues of Wigner type matrices depending on the assumptions for the entries and the kind of the convergence.

In accord with the description in Alon, Krivelevich and Vu (2002); König (2005); Soshnikov (1999), first we formulate the famous *Wigner Semicircle Law* for the bulk spectrum of expanding symmetric matrices in a more general form than stated in the original papers Wigner (1955, 1958), see also Arnold (1967).

Theorem B.2.2 *Let a_{ij} $(i \le j)$ be independent real-valued random variables with the following properties.*

- *The distribution of a_{ij}'s is symmetric.*
- *All moments are finite. In view of symmetry, all odd moments vanish, especially, $\mathbb{E}(a_{ij}) = 0$ $(i \le j)$.*
- *$\mathbb{E}(a_{ij}^2) = \sigma^2$ $(i < j)$ and $\mathbb{E}(a_{ii}^2) < C$ $(i = 1, 2, \ldots)$ with $0 < \sigma, C < \infty$ constants.*
- *$\mathbb{E}(a_{ij}^{2k}) \le (Ck)^k$ $(k = 1, 2, \ldots)$ with constant $0 < C < \infty$ (called sub-Gaussian moments).*

Define a_{ij} for $i > j$ by $a_{ij} = a_{ji}$. Let $\lambda_1^{(n)} \ge \lambda_2^{(n)} \ge \cdots \ge \lambda_n^{(n)}$ be the spectrum of the random symmetric matrix $\boldsymbol{A}^{(n)} = (a_{ij})_{i,j=1}^{n,n}$. Denoting by

$$\tilde{\lambda}_i^{(n)} = \frac{1}{2\sigma\sqrt{n}}\lambda_i^{(n)}, \quad i = 1, \ldots, n,$$

the rescaled eigenvalues of $\mathbf{A}^{(n)}$, their empirical distribution function

$$F_n(x) = \frac{1}{n}|\{i : \tilde{\lambda}_i^{(n)} \leq x, \quad i = 1, \ldots, n\}$$

converges to a non-random limit $F(x) = \int_{-\infty}^{x} f(t)\, dt$, which is the c.d.f. corresponding to the semicircle density

$$f(x) = \begin{cases} \frac{2}{\pi}\sqrt{1 - x^2} & \text{if} \quad |x| \leq 1, \\ 0 & \text{if} \quad |x| > 1. \end{cases} \tag{B.2}$$

The convergence is understood almost surely (with probability 1) if the entries of all matrices $\mathbf{A}^{(n)}$ $(n = 1, 2, \ldots)$ are defined on the same probability space.

We remark that in the original theorem the convergence was understood in distribution; further, the entries were assumed to have an identical distribution in, and another identical distribution above, the main diagonal. Note that in the case of Gaussian entries, because of the zero expectation and equal variances, the upper diagonal entries are, indeed, identically distributed. The original proof of the Semicircle Law used the method of moments. For Gaussian distributed entries one can find the marginal distribution of the eigenvalues' joint distribution which is obtained from the joint distribution of the entries of $\mathbf{A}^{(n)}$ by means of the Jacobian transformation containing the derivatives of the matrix with respect to its eigenvalues and eigenvectors (with an appropriate parameterization). The first factor of this determinant is

$$\prod_{i<j} |\lambda_i^{(n)} - \lambda_j^{(n)}|, \tag{B.3}$$

supposing that there are no multiple eigenvalues (which is a zero probability event), while the second one depends on the eigenvectors (and follows Haar distribution), but in view of the independence of the two, this last one will be eliminated after integrating with respect to the eigenvectors to obtain the joint density of the eigenvalues (more precisely, it will only contribute to the constant term of the density). This derivation of the proof together with other generalizations can be found in Mehta (1990).

In König (2005) it is concluded that on the conditions of the above theorem, all eigenvalues of $\mathbf{A}^{(n)}$ lie in the interval $2\sigma\sqrt{n}[1 - \varepsilon, 1 + \varepsilon]$ for any $\varepsilon > 0$ with overwhelming probability, whereas the spacings between subsequent eigenvalues are of order $n^{-1/2}$ in the bulk of the spectrum and much larger close to the edge. However, the Semicircle Law applies to the bulk spectrum, and it provides limited information about the asymptotic behavior of any particular eigenvalue. In the last decades several finer results have been obtained for the largest eigenvalues and the spectral radius of $\mathbf{A}^{(n)}$, sometimes under mildly modified conditions.

Tracy and Widom (1994, 1996) characterized the limiting distribution of the first k eigenvalues (for any fixed $k \geq 1$) of the so-called Gaussian Orthogonal Ensemble (GOE), corresponding to the case when the upper diagonal entries are i.i.d. normally distributed random variables. The limiting distributions are called Tracy–Widom type distributions. Later, it turned out that the same distributions are obtained under the conditions of the Semicircle Law. This so-called universality at the edge of the spectrum of Wigner random matrices was proved in Soshnikov (1999) in the following sense. After proper rescaling, the first, second, third, etc. eigenvalues of a random symmetric matrix, meeting the conditions of Theorem B.2.2, converge to the distributions established by Tracy and Widom for the GOE, hence giving a generalization of their results.

To investigate the spectral radius of a Wigner type matrix, other authors relaxed the condition that the entries have symmetric distribution and hence, zero expectation. In Juhász (1981) the author considered the case when the entries in and above the main diagonal are independent, but the diagonal entries have an identical, and the off-diagonal ones another identical, distribution with a positive expectation. The following more general statement of Füredi and Komlós (1981) contains the aforementioned result as a special case.

Theorem B.2.3 *Let a_{ij} $(i \leq j)$ be independent (not necessarily identically distributed) random variables bounded with a common bound K, that is, $|a_{ij}| \leq K$, $\forall i, j$. Assume that for $i < j$, $\mathbb{E}(a_{ij}) = \mu$ and $\mathrm{Var}(a_{ij}) = \sigma^2$, further that $\mathbb{E}(a_{ii}) = \nu$. Define a_{ij} for $i > j$ by $a_{ij} = a_{ji}$. Let $\lambda_1^{(n)} \geq \lambda_2^{(n)} \geq \cdots \geq \lambda_n^{(n)}$ be the eigenvalues of the random symmetric matrix $\mathbf{A}^{(n)} = (a_{ij})_{i,j=1}^{n,n}$. The numbers K, μ, σ^2, ν will be kept fixed as $n \to \infty$.*

- *If $\mu > 0$, then the distribution of $\lambda_1^{(n)}$ can be approximated in order $\frac{1}{\sqrt{n}}$ by a normal distribution of expectation*

$$(n-1)\mu + \nu + \frac{\sigma^2}{\mu}$$

and variance $2\sigma^2$. Further,

$$\max_{2 \leq i \leq n} |\lambda_i^{(n)}| < 2\sigma\sqrt{n} + \mathcal{O}(n^{1/3} \ln n)$$

in probability (i.e., with probability tending to 1 as $n \to \infty$).

- *If $\mu = 0$, then*

$$\max_{1 \leq i \leq n} |\lambda_i^{(n)}| = 2\sigma\sqrt{n} + \mathcal{O}(n^{1/3} \ln n)$$

in probability.

Note that $\mathcal{O}(n^{1/3} \ln n)$ is also $o(\sqrt{n})$, therefore the second term in the above formulas is negligible compared to the first one. Theorem B.2.3 also implies that the largest absolute value eigenvalues of a Wigner type matrix $\boldsymbol{A}^{(n)}$ (all entries have zero expectation) is exactly of order $2\sigma\sqrt{n}$. However, if the common non-zero expectation of the off-diagonal entries is $\mu > 0$, then there is a large eigenvalue having an asymptotic normal distribution with bounded variance around its expectation of order

n. The result can easily be extended to the $\mu < 0$ case, therefore the spectral radius is of order n too. In fact, it is the shift in the expectation which puts off the edge of the spectrum. In Chapter 3, we show that in more general situations (with multiple shifts) there can be more than one eigenvalue of order n.

The following sharp concentration result of Alon, Krivelevich and Vu (2002) applies to general random symmetric matrices with uniformly bounded entries.

Theorem B.2.4 *Let a_{ij} $(1 \le i \le j \le n)$ be independent random variables with absolute value at most 1. Define a_{ij} for $1 \le j < i \le n$ by $a_{ij} = a_{ji}$. Let $\lambda_1 \ge \lambda_2 \ge \cdots \ge \lambda_n$ be the eigenvalues of the random symmetric matrix $\mathbf{A} = (a_{ij})_{i,j=1}^{n,n}$. Then for every positive integer $1 \le i \le \frac{n}{2}$, the probability that λ_i deviates from its median by more than t is at most $e^{-\frac{t^2}{32i^2}}$. The same estimate holds for the probability that λ_{n-i+1} deviates from its median by more than t.*

Observe that the estimate given in the above theorem does not depend on n (of course, the median does depend on it). The statement – apart from a constant factor in the exponent – remains valid if a_{ij}'s are uniformly bounded with some constant K. The authors of Theorem B.2.4 prove that the eigenvalues are also highly concentrated on their own expectations, since λ_i and its median are $\mathcal{O}(i)$ apart. Indeed, let m_i denote the median of λ_i. Then

$$|\mathbb{E}(\lambda_i) - m_i| \le \mathbb{E}|\lambda_i - m_i| = \int_0^\infty \mathbb{P}(|\lambda_i - m_i| > t)\, dt$$
$$\le \int_0^\infty e^{-\frac{t^2}{32i^2}}\, dt = 8\sqrt{2\pi}i.$$

Therefore, for all $t \gg i$ we have

$$\mathbb{P}(|\lambda_i - \mathbb{E}(\lambda_i)| > t) \le e^{-\frac{(1-o(1))t^2}{32i^2}} \quad \text{when} \quad 1 \le i \le \frac{n}{2}, \tag{B.4}$$

and the same estimate holds for the probability $\mathbb{P}(|\lambda_{n-i+1} - \mathbb{E}(\lambda_{n-i+1})| > t)$.

The authors also show that their estimate is sharp for the deviation of λ_1. For this purpose, they consider the following $n \times n$ adjacency matrix of an Erdős–Rényi random graph (see Erdős and Rényi (1960)): the diagonal entries are zeros, while the upper diagonal ones are independent Bernoulli distributed random variables with parameter $\frac{1}{2}$ (vertex pairs are connected with probability $\frac{1}{2}$, independently of each other), in which case, by Theorem B.2.3, $\lambda_1 = \frac{n}{2} + o(1)$. The concentration provided by Theorem B.2.4 for larger values of i is weaker than that provided for $i = 1$. The authors also note that for the adjacency matrix of a random graph, when the entries are in the [0,1] interval, the estimate of their Theorem B.2.4 can be improved to $4e^{-\frac{t^2}{8i^2}}$.

The proof of Theorem B.2.4 is based on the Talagrand inequality (see Talagrand (1996)) which is an efficient large deviation tool for product spaces. In fact, this technique is only applicable when the entries are uniformly bounded. Possibly, with other techniques, similar sharp concentration results can be obtained in the case of

Gaussian entries or just under the conditions of the Semicircle Law. For example, under the conditions of Theorem B.2.2, the individual eigenvalues are highly concentrated on the expected values of the corresponding order statistics based on the limiting distribution as follows.

Proposition B.2.5 *With the notation of Theorem B.2.2,*

$$\lim_{n\to\infty} \max_{1\le i\le n} |\tilde{\lambda}_i^{(n)} - \bar{\lambda}_i^{(n)}| = 0$$

almost surely, where $\bar{\lambda}_i^{(n)} = F^{-1}(\frac{2i-1}{2n})$, $i = 1, \ldots, n$, *and* F *is the c.d.f. of the semicircle density* (B.2).

This issue together with other generalizations for the singular values of square Gaussian matrices are discussed in the survey of Davidson and Szarek (2001).

Analogous estimates for the concentration of different norms of random rectangular matrices of complex entries were obtained in Meckes (2004) also using the Talagrand inequality. Achlioptas and McSherry (2007) generalized Theorem B.2.3 to rectangular matrices in the following way.

Theorem B.2.6 *Let* $\mathbf{A}$ *be a random* $m \times n$ *matrix with independent, uniformly bounded entries* a_{ij}*'s such that* $|a_{ij}| \le K$, $\mathbb{E}(a_{ij}) = 0$, *and* $\mathrm{Var}(a_{ij}) \le \sigma^2$ *with some constants* $0 < K, \sigma^2 < \infty$ *for* $i = 1, \ldots, m$; $j = 1, \ldots, n$. *Under these conditions, for any* $\alpha > \frac{1}{2}$, *if*

$$K < \sigma\sqrt{m+n}(7\alpha \ln(m+n))^{-3},$$

then

$$\mathbb{P}\left(\|\mathbf{A}\| > \frac{7}{3}\sigma\sqrt{m+n}\right) < (m+n)^{\frac{1}{2}-\alpha}.$$

The theorem implies that the spectral norm (largest singular value) of the above type $m \times n$ random rectangular matrix is of order $\sqrt{m+n}$ in probability, that is, with probability tending to 1 as $m, n \to \infty$.

Going back to the Wishart matrices, in the density of which the factor (B.3) also appears, the first asymptotic result for the bulk spectrum with appropriate normalization, analogous to the Wigner's theorem, is due to Marchenko and Pastur (1967), where the limiting distribution depends on the ratio $\gamma = p/n$ when $p, n \to \infty$. Yin *et al.* (1988) gave the exact almost certain limit (depending on γ) for the largest eigenvalue of Gram sequences under special conditions. The limiting distribution of the largest eigenvalue of the sample covariance matrix based on a multivariate normal distribution was also characterized in Johnstone (2001) who proved that the distribution of the appropriately centered and scaled largest eigenvalue of the standard Wishart matrix with parameters p and n approaches the Tracy–Widom law when $p, n \to \infty$ in such a way that $\frac{n}{p} = \gamma > 1$ is a fixed constant.

References

Achlioptas D and McSherry F 2007 Fast computation of low-rank matrix approximations. *J. ACM* **54** (2), Article 9.

Alon N, Krivelevich M and Vu VH 2002 On the concentration of eigenvalues of random symmetric matrices. *Isr. J. Math.* **131**, 259–267.

Arnold, L 1967 On the asymptotic distribution of the eigenvalues of random matrices. *J. Math. Anal. Appl.* **20**, 262–268.

Bhatia R 1997 *Matrix Analysis*, Springer.

Davidson KR and Szarek SJ 2001 Local operator theory, random matrices and Banach spaces, in *Handbook of the Geometry of Banach spaces* (eds Johnson WB and Lindenstrauss J), Vol. 1, Chapter 8, Elsevier Science, Amsterdam, pp. 317–366.

Erdős P and Rényi A 1960 On the evolution of random graphs. *Publ. Math. Inst. Hung. Acad. Sci.* **5**, 17–61.

Füredi Z and Komlós J 1981 The eigenvalues of random symmetric matrices. *Combinatorica* **1**, 233–241.

Horn RA and Johnson CR 1985 *Matrix Analysis*, Cambridge University Press, Cambridge.

Johnstone IM 2001 On the distribution of the largest eigenvalue in principal components analysis. *Ann. Stat.* **29**, 295–327.

Juhász F 1981 On the spectrum of random graphs. In *Algebraic Methods in Graph Theory, Coll. Math. Soc. J. Bolyai* (ed. Lovász L), North–Holland, Amsterdam, vol. 25, pp. 313–316.

König W 2005 Orthogonal polynomial ensembles in probability theory. *Probab. Surv.* **2**, 385–447.

Marchenko VA and Pastur LA 1967 Distribution of eigenvalues for some set of random matrices. *Mat. USSR-Sbornik* **1**, 457–486.

Meckes M 2004 Concentration of norms and eigenvalues of random matrices. *J. Funct. Anal.* **211** (2), 508–524.

Mehta LM 1990 *Random Matrices*, 2nd edn Academic Press.

Olkin I 2002 The 70th anniversary of the distribution of random matrices: A survey. *Linear Algebra Appl.* **354**, 231–243.

Petz D 2008 *Quantum Information Theory and Quantum Statistics*, Springer.

Soshnikov A 1999 Universality at the end of the spectrum in Wigner random matrices. *Commun. Math. Phys.* **207**, 697–733.

Talagrand M 1996 Concentration of measure and isoperimetric inequalities in product spaces. *Publ. Math. I.H.E.S.* **81**, 73–205.

Tracy CA and Widom H 1994 Level-spacing distributions and the Airy kernel. *Commun. Math. Phys.* **159**, 151–174.

Tracy CA and Widom H 1996 On orthogonal and symplectic matrix ensembles. *Commun. Math. Phys.* **177**, 727–754.

Wigner EP 1955 Characteristic vectors of bordered matrices with infinite dimensions. *Ann. Math.* **62**, 548–564.

Wigner EP 1958 On the distribution of the roots of certain symmetric matrices. *Ann. Math.* **67**, 325–328.

Yin YQ, Bai ZD and Krishnaiah PR 1988 On the limit of the large-dimensional sample covariance matrix. *Probab. Theory Relat. Fields* **78**, 509–521.

Appendix C

Multivariate statistical methods

C.1 Principal component analysis

In multivariate data analysis, usually there are strong dependencies between the coordinates of the underlying random vector. With an appropriate transformation we want to describe its covariance structure by means of independent variables. In practical applications, there may be linear or near linear dependencies between the components of multidimensional data. To reduce the dimensionality, the first step is to simplify the covariance structure. If the underlying distribution is multivariate Gaussian (which is frequently the case due to the multidimensional central limit theorem B.1.3), after subtracting the means it suffices to treat the empirical covariance matrix of the observations, but in other situations (if the data are from a multivariate, absolutely continuous distribution) we can also be confined to the covariances. In such cases, instead of independence, we will speak of uncorrelatedness of the components.

Let $\mathbf{X}$ be a p-dimensional random vector with expectation $\boldsymbol{\mu}$ and covariance matrix $\boldsymbol{C}$. The *principal component transformation* assigns the following p-dimensional random vector $\mathbf{Y}$ to $\mathbf{X}$:

$$\mathbf{Y} = \boldsymbol{U}^T(\mathbf{X} - \boldsymbol{\mu}),$$

where $\boldsymbol{C} = \boldsymbol{U}\boldsymbol{\Lambda}\boldsymbol{U}^T$ is the SD of the positive definite (possibly, semidefinite) covariance matrix $\boldsymbol{C}$, see (A.10). It is easy to see that the random vector $\mathbf{Y}$ has expectation $\mathbf{0}$ and covariance matrix $\boldsymbol{\Lambda}$. As $\boldsymbol{\Lambda}$ is a diagonal matrix, the components of $\mathbf{Y}$ are uncorrelated (in the Gaussian case also independent) with variances $\lambda_1 \geq \cdots \geq \lambda_p \geq 0$, the diagonal entries of $\boldsymbol{\Lambda}$, that is, the eigenvalues of $\boldsymbol{C}$. Denoting by $\mathbf{u}_1, \ldots, \mathbf{u}_p$ the

Spectral Clustering and Biclustering: Learning Large Graphs and Contingency Tables, First Edition.
Marianna Bolla.

corresponding unit-norm eigenvectors, the ith component of $\mathbf{Y}$, called the ith *principal component* is

$$Y_i = \mathbf{u}_i^T(\mathbf{X} - \boldsymbol{\mu}), \quad i = 1, \ldots, p.$$

In fact, Y_i is a linear combination of the components of $\mathbf{X}$ normalized such that $\|\mathbf{u}_i\| = 1$. The sum of the variances of the principal components is equal to the sum of the variances of X_i's, since

$$\sum_{i=1}^{p} \mathrm{Var}(Y_i) = \sum_{i=1}^{p} \lambda_i = \mathrm{tr}(\boldsymbol{C}) = \sum_{i=1}^{p} \mathrm{Var}(X_i)$$

and

$$\mathrm{Var}(Y_1) \geq \mathrm{Var}(Y_2) \geq \cdots \geq \mathrm{Var}(Y_p) \geq 0.$$

Thus, we may say that the set of the principal components explains the total variation of the original random vector's components, in decreasing order. If $r = \mathrm{rank}(\boldsymbol{C}) < p$, then the principal components $Y_{r+1}, \ldots, Y_p$ have no relevance.

By Proposition A.3.11, the principal components can also be obtained as solutions of the following sequential maximization task:

$$\max_{\mathbf{v}\in\mathbb{R}^p,\ \|\mathbf{v}\|=1} \mathrm{Var}(\mathbf{v}^T(\mathbf{X} - \boldsymbol{\mu})) = \max_{\mathbf{v}\in\mathbb{R}^p,\ \|\mathbf{v}\|=1} \mathbf{v}^T\boldsymbol{C}\mathbf{v} = \lambda_1$$

and the maximum is attained with the choice $\mathbf{v} = \mathbf{u}_1$ (uniquely if $\lambda_1 > \lambda_2$). Hence, the first principal component $Y_1 = \mathbf{u}_1^T(\mathbf{X} - \boldsymbol{\mu})$ is obtained as the maximum variance, normalized linear combination of the components of $\mathbf{X}$. This was the $k = 1$ case. Further, for $k = 2, 3, \ldots, r$, in view of Proposition A.3.11,

$$\max_{\substack{\mathbf{v}\in\mathbb{R}^p,\ \|\mathbf{v}\|=1 \\ \mathrm{Cov}(\mathbf{v}^T\mathbf{X}, Y_i)=0\,(i=1,\ldots,k-1)}} \mathrm{Var}(\mathbf{v}^T(\mathbf{X} - \boldsymbol{\mu})) = \max_{\substack{\mathbf{v}\in\mathbb{R}^p,\ \|\mathbf{v}\|=1 \\ \mathbf{v}^T\mathbf{u}_i=0\,(i=1,\ldots,k-1)}} \mathbf{v}^T\boldsymbol{C}\mathbf{v} = \lambda_k$$

and the maximum is attained with the choice $\mathbf{v} = \mathbf{u}_k$ (uniquely if $\lambda_k > \lambda_{k+1}$). Hence, the kth principal component $Y_k = \mathbf{u}_k^T(\mathbf{X} - \boldsymbol{\mu})$ is obtained as the maximum variance, normalized linear combination of the components of $\mathbf{X}$ under the condition of its uncorrelatedness with the preceding principal components $Y_1, \ldots, Y_{k-1}$.

A more general statement is also true. For a fixed positive integer $k \leq r$, such that $\lambda_k > \lambda_{k+1}$, the first k principal components provide the best rank k approximation of $\mathbf{X}$ in the following sense:

$$\min_{\substack{\boldsymbol{A}\text{ is } p\times p \\ \mathrm{rank}(\boldsymbol{A})=k}} \mathbb{E}\|\mathbf{X} - \boldsymbol{A}\mathbf{X}\|^2 = \mathbb{E}\|\mathbf{X} - \boldsymbol{P}\mathbf{X}\|^2,$$

where $\boldsymbol{P}$ is the orthogonal projection onto $\mathrm{Span}\{\mathbf{u}_1, \ldots, \mathbf{u}_k\}$.

In fact, it is the ratio $\sum_{i=1}^{k} \lambda_i / \sum_{i=1}^{p} \lambda_i$ which tells us the proportion of $\mathbf{X}$'s total variation explained by the first k principal components. Therefore, it suffices to retain only the first k principal components if there is a noticeable gap in the spectrum of $\boldsymbol{C}$

between λ_k and λ_{k+1}. Based on a statistical sample $\mathbf{X}_1, \dots, \mathbf{X}_n \sim \mathcal{N}(\boldsymbol{\mu}, \boldsymbol{C})$, where $n \gg p$, for $k = 0, \dots, r$ we test the hypothesis that the last $p - k$ eigenvalues of $\boldsymbol{C}$ are equal, until it is accepted. The likelihood ratio test statistic is based on the spectrum of the empirical covariance matrix.

Note that there are more sophisticated models for reducing the dimensionality. For example, the following model of factor analysis, for given k, looks for the k-dimensional standard Gaussian vector $\mathbf{f} = (f_1, \dots, f_k)^T$ of common factors and p-dimensional Gaussian vector $\mathbf{e}$ of independent individual factors, where $\mathbf{f}$ and $\mathbf{e}$ are also independent of each other, such that

$$\mathbf{X} = \boldsymbol{A}\mathbf{f} + \mathbf{e} + \boldsymbol{\mu} \tag{C.1}$$

with some $p \times k$ matrix $\boldsymbol{A}$, whose entries are called *factor loadings*. There is a wide literature about the optimum choice of k for which the decomposition (C.1), or a good approximation for it, exists. If the above discussed sequential hypothesis testing of the principal component analysis stops at an integer k, then the $p \times k$ matrix $(\sqrt{\lambda_i}\mathbf{u}_1, \dots, \sqrt{\lambda_k}\mathbf{u}_k)$ is a good candidate for $\boldsymbol{A}$ (this is called principal factor analysis). For further details, see e.g., Mardia *et al.* (1979).

C.2 Canonical correlation analysis

Now the coordinates of our random vector are partitioned into two parts, and we want to find maximally correlated linear combinations of them. Let $(\mathbf{X}^T, \mathbf{Y}^T)^T$ be a $(p+q)$-dimensional random vector. For simplicity, we assume that the coordinates have zero expectation. The covariance matrix $\boldsymbol{C}$ is partitioned (with block sizes p and q) accordingly:

$$\boldsymbol{C} = \begin{pmatrix} \boldsymbol{C}_{\mathbf{XX}} & \boldsymbol{C}_{\mathbf{XY}} \\ \boldsymbol{C}_{\mathbf{YX}} & \boldsymbol{C}_{\mathbf{YY}} \end{pmatrix}$$

where $\boldsymbol{C}_{\mathbf{XX}}, \boldsymbol{C}_{\mathbf{YY}}$ are covariance matrices of $\mathbf{X}$ and $\mathbf{Y}$, whereas $\boldsymbol{C}_{\mathbf{YX}} = \boldsymbol{C}_{\mathbf{XY}}^T$ is the cross-covariance matrix of them. Assume that $\boldsymbol{C}_{\mathbf{XX}}, \boldsymbol{C}_{\mathbf{YY}}$ and $\boldsymbol{C}$ are regular.

Consider the following successive maximization problem. In the first step we look for maximally correlated linear combination of the $\mathbf{X}$- and $\mathbf{Y}$-coordinates. Obviously, the problem is equivalent to finding unit variance linear combinations with maximum covariance as follows:

$$\begin{aligned} \max_{\mathbf{a}\in\mathbb{R}^p,\, b\in\mathbb{R}^q} \operatorname{Corr}(\mathbf{a}^T\mathbf{X}, \mathbf{b}^T\mathbf{Y}) &= \max_{\operatorname{Var}(\mathbf{a}^T\mathbf{X})=\operatorname{Var}(\mathbf{b}^T\mathbf{Y})=1} \operatorname{Cov}(\mathbf{a}^T\mathbf{X}, \mathbf{b}^T\mathbf{Y}) \\ &= \max_{\substack{\mathbf{a}^T\boldsymbol{C}_{\mathbf{XX}}\mathbf{a}=1 \\ \mathbf{b}^T\boldsymbol{C}_{\mathbf{YY}}\mathbf{b}=1}} \mathbf{a}^T\boldsymbol{C}_{\mathbf{XY}}\mathbf{b}. \end{aligned}$$

With the notation

$$\tilde{\mathbf{a}} = \boldsymbol{C}_{\mathbf{XX}}^{1/2}\mathbf{a}, \quad \tilde{\mathbf{b}} = \boldsymbol{C}_{\mathbf{XX}}^{1/2}\mathbf{b}, \quad \boldsymbol{B} = \boldsymbol{C}_{\mathbf{XX}}^{-1/2}\boldsymbol{C}_{\mathbf{XY}}\boldsymbol{C}_{\mathbf{YY}}^{-1/2} \tag{C.2}$$

the above maximization task is equivalent to that of Proposition A.3.9. Let $\boldsymbol{B} = \boldsymbol{VSU}^T$ be the SVD of the $p \times q$ matrix $\boldsymbol{B}$ of rank $r \leq \min\{p, q\}$ as described in Theorem A.3.5 and Equation (A.12). Then

$$\max_{\substack{\mathbf{a}^T \boldsymbol{C}_{\mathbf{XX}}\mathbf{a}=1 \\ \mathbf{b}^T \boldsymbol{C}_{\mathbf{YY}}\mathbf{b}=1}} \mathbf{a}^T \boldsymbol{C}_{\mathbf{XY}}\mathbf{b} = \max_{\substack{\tilde{\mathbf{a}}\in\mathbb{R}^p, \tilde{\mathbf{b}}\in\mathbb{R}^q \\ \|\tilde{\mathbf{a}}\|=1, \|\tilde{\mathbf{b}}\|=1}} \tilde{\mathbf{a}}^T \boldsymbol{B}\tilde{\mathbf{b}} = s_1$$

and it is attained with the choice $\tilde{\mathbf{a}} = \mathbf{v}_1$ and $\tilde{\mathbf{b}} = \mathbf{u}_1$ (uniquely if $s_1 > s_2$), where $\mathbf{v}_1, \mathbf{u}_1$ is the first singular vector pair with corresponding singular value s_1. The first canonical vector pair is obtained by back transformation:

$$\mathbf{a}_1 = \boldsymbol{C}_{\mathbf{XX}}^{-1/2}\mathbf{v}_1, \quad \mathbf{b}_1 = \boldsymbol{C}_{\mathbf{YY}}^{-1/2}\mathbf{u}_1.$$

The pair of linear combinations $\mathbf{a}_1^T\mathbf{X}$, $\mathbf{b}_1^T\mathbf{Y}$ is called the first canonical variable pair, and their correlation s_1 is the *first canonical correlation.*

This was the $k = 1$ case. For $k = 2, 3, \ldots, r$ we are looking for maximally correlated linear combinations $\mathbf{a}^T\mathbf{X}$ and $\mathbf{b}^T\mathbf{Y}$ which are uncorrelated with the first $k - 1$ canonical variables $\mathbf{a}_i^T\mathbf{X}$ and $\mathbf{b}_i^T\mathbf{Y}$ ($i = 1, \ldots, k - 1$), respectively. Since the correlations are again viewed as covariances of the unit-variance variables, and also making use of the relations

$$\mathrm{Cov}\big(\mathbf{a}^T\mathbf{X}, \mathbf{a}_i^T\mathbf{X}\big) = \mathbf{a}^T \boldsymbol{C}_{\mathbf{XX}}\mathbf{a}_i, \quad \mathrm{Cov}\big(\mathbf{b}^T\mathbf{Y}, \mathbf{b}_i^T\mathbf{Y}\big) = \mathbf{b}^T \boldsymbol{C}_{\mathbf{YY}}\mathbf{b}_i,$$

our maximization problem, making use of (C.2), has the following form and solution:

$$\max_{\substack{\mathbf{a}^T \boldsymbol{C}_{\mathbf{XX}}\mathbf{a}=1 \\ \mathbf{b}^T \boldsymbol{C}_{\mathbf{YY}}\mathbf{b}=1 \\ \mathbf{a}^T \boldsymbol{C}_{\mathbf{XX}}\mathbf{a}_i=0\,(i=1,\ldots,k-1) \\ \mathbf{b}^T \boldsymbol{C}_{\mathbf{YY}}\mathbf{b}_i=0\,(i=1,\ldots,k-1)}} \mathbf{a}^T \boldsymbol{C}_{\mathbf{XY}}\mathbf{b} = \max_{\substack{\tilde{\mathbf{a}}\in\mathbb{R}^p, \tilde{\mathbf{b}}\in\mathbb{R}^q \\ \|\tilde{\mathbf{a}}\|=1, \|\tilde{\mathbf{b}}\|=1 \\ \tilde{\mathbf{a}}^T\mathbf{u}_i=0\,(i=1,\ldots,k-1) \\ \tilde{\mathbf{b}}^T\mathbf{v}_i=0\,(i=1,\ldots,k-1)}} \tilde{\mathbf{a}}^T \boldsymbol{B}\tilde{\mathbf{b}} = s_k$$

and the maximum is attained with the choice $\tilde{\mathbf{a}} = \mathbf{v}_k$ and $\tilde{\mathbf{b}} = \mathbf{u}_k$ (uniquely if $s_k > s_{k+1}$).

The kth *canonical vector pair* is obtained by back transformation:

$$\mathbf{a}_k = \boldsymbol{C}_{\mathbf{XX}}^{-1/2}\mathbf{v}_k, \quad \mathbf{b}_k = \boldsymbol{C}_{\mathbf{YY}}^{-1/2}\mathbf{u}_k.$$

The pair of linear combinations $\mathbf{a}_k^T\mathbf{X}$, $\mathbf{b}_k^T\mathbf{Y}$ is called kth *canonical variable pair*, and their correlation s_k is the kth *canonical correlation.*

Since the canonical correlations are, in fact, correlations, for the singular values of the matrix $\boldsymbol{B}$, the relation

$$1 \geq s_1 \geq s_2 \geq \cdots \geq s_r > 0$$

holds. In the case of a multivariate Gaussian sample, a sequence of hypotheses can be tested for the number of canonical correlations significantly explaining the relation between $\mathbf{X}$ and $\mathbf{Y}$. The likelihood ratio statistic is based on the singular values of the matrix $\hat{\boldsymbol{B}}$ calculated from the empirical covariances in the following way:

$$\hat{\boldsymbol{B}} = \hat{\boldsymbol{C}}_{\mathbf{XX}}^{-1/2} \hat{\boldsymbol{C}}_{\mathbf{XY}} \hat{\boldsymbol{C}}_{\mathbf{YY}}^{-1/2}.$$

C.3 Correspondence analysis

Here we have two-way classified data for two categorical variables taking on, say, m and n different values, respectively. We want to find independent, maximally correlated factors of the $m \times n$ contingency table $\boldsymbol{C}$ of nonnegative integer counts, where $\sum_{i=1}^{m} \sum_{j=1}^{n} c_{ij} = N$ is the sample size. The table is normalized such that the entries of the new table $\boldsymbol{W}$ are $w_{ij} = c_{ij}/N$, hence $\sum_{i=1}^{m} \sum_{j=1}^{n} w_{ij} = 1$. Therefore, we can speak in terms of the joint distribution $\mathbb{W}$ of the two underlying categorical variables. Denoting by $p_i = \sum_{j=1}^{n} w_{ij}$ $(i = 1, \dots, m)$ and $q_j = \sum_{i=1}^{m} w_{ij}$ $(j = 1, \dots, n)$ the row- and column sums of $\boldsymbol{W}$, the marginal distributions are $\mathbb{P} = \{p_1, \dots, p_m\}$ and $\mathbb{Q} = \{q_1, \dots, q_n\}$, respectively. We assume that p_i's and q_j's are all positive, that is, there are no identically zero rows or columns in the contingency table.

Our purpose is to find a ψ, ϕ pair taking on values with respect to the distributions $\mathbb{P}$ and $\mathbb{Q}$ that solve

$$\max_{\mathbb{E}_{\mathbb{P}}^2 \psi = \mathbb{E}_{\mathbb{Q}}^2 \phi = 1} \mathbb{E}_{\mathbb{W}}(\psi\phi) = \max_{\substack{\sum_{i=1}^{m} \psi^2(i) p_i = 1 \\ \sum_{j=1}^{n} \phi^2(j) q_j = 1}} \sum_{i=1}^{m} \sum_{j=1}^{n} w_{ij} \psi(i) \phi(j)$$

where the vectors $\boldsymbol{\psi} = (\psi(1), \dots, \psi(m))^T$ and $\boldsymbol{\phi} = (\phi(1), \dots, \phi(n))^T$ contain the possible values of ψ and ϕ in their coordinates, respectively. With the notation

$$\boldsymbol{P} = \text{diag}(p_1, \dots, p_m), \quad \boldsymbol{Q} = \text{diag}(q_1, \dots, q_n),$$

further, with

$$\mathbf{x} = \boldsymbol{P}^{1/2} \boldsymbol{\psi}, \quad \mathbf{y} = \boldsymbol{Q}^{1/2} \boldsymbol{\phi}, \quad \boldsymbol{B} = \boldsymbol{P}^{-1/2} \boldsymbol{W} \boldsymbol{Q}^{-1/2},$$

the maximization task above is equivalent to

$$\max_{\|\mathbf{x}\| = \|\mathbf{y}\| = 1} \mathbf{x}^T \boldsymbol{B} \mathbf{y}.$$

In view of Proposition A.3.9, the maximum is the largest singular value of $\boldsymbol{B}$. The matrix $\boldsymbol{B}$ is called *normalized contingency table* or *correspondence matrix* belonging to the contingency table $\boldsymbol{C}$. It can easily be seen that the maximum singular value of $\boldsymbol{B}$ is $s_0 = 1$, since by the Cauchy-Schwarz inequality, $\mathbb{E}_{\mathbb{W}}(\psi\phi) \le \sqrt{\mathbb{E}_{\mathbb{P}}^2 \psi \cdot \mathbb{E}_{\mathbb{Q}}^2 \phi} = 1$, but with the constant random variables $\psi_0 = 1$ and $\phi_0 = 1$, equality can be attained. The ψ_0, ϕ_0 pair is called the *trivial correspondence factor pair*. If $\boldsymbol{B}$ (or equivalently, $\boldsymbol{C}$) is non-decomposable (see Definition A.3.28), then the 1 is a single singular value. We will assume this in the sequel.

Then for $k = 1, 2, \dots, r-1$, where $r = \text{rank}(\boldsymbol{B}) = \text{rank}(\boldsymbol{C}) \le \min\{m, n\}$, we are looking for the kth *correspondence factor pair* ψ_k, ϕ_k such that they solve the following maximization task:

$$\max_{\substack{\mathbb{E}_{\mathbb{P}}^2\psi=\mathbb{E}_{\mathbb{Q}}^2\phi=1 \\ \mathbb{E}_{\mathbb{P}}(\psi\psi_\ell)=\mathbb{E}_{\mathbb{Q}}(\phi\phi_\ell)=0\,(\ell=0,1,\dots,k-1)}} \mathbb{E}_{W}(\psi\phi) \tag{C.3}$$

where ψ_ℓ, ϕ_ℓ denotes the ℓth correspondence factor pair ($\ell = 0, 1, \dots, k-1$) already found.

The conditions $\mathbb{E}_{\mathbb{P}}(\psi\psi_0) = 0$ and $\mathbb{E}_{\mathbb{Q}}(\phi\phi_0) = 0$ imply that $\mathbb{E}_{\mathbb{P}}\psi = 0$ and $\mathbb{E}_{\mathbb{Q}}\phi = 0$, and in view of this, the above optimization problem can be reformulated as follows.

Let $\boldsymbol{B} = \boldsymbol{V}\boldsymbol{S}\boldsymbol{U}^T$ denote the SVD of the $m \times n$ matrix $\boldsymbol{B}$ as in Equation (A.12). Then the solution of (C.3) in the kth step is

$$\max_{\substack{\mathbb{E}_{\mathbb{P}}\psi=0,\ \text{Var}_{\mathbb{P}}\xi=1 \\ \mathbb{E}_{\mathbb{Q}}\phi=0,\ \text{Var}_{\mathbb{Q}}\phi=1 \\ \text{Cov}_{\mathbb{P}}\xi\xi_\ell=0\,(i=1,\dots,k-1) \\ \text{Cov}_{\mathbb{Q}}\xi\xi_\ell=0\,(i=1,\dots,k-1)}} \text{Cov}_{\mathbb{W}}(\psi, \phi) = \max_{\substack{\mathbf{x}\in\mathbb{R}^m,\ \mathbf{y}\in\mathbb{R}^n \\ \|\mathbf{x}\|=1,\ \|\mathbf{y}\|=1 \\ \mathbf{x}^T\mathbf{v}_i=0\,(i=1,\dots,k-1) \\ \mathbf{y}^T\mathbf{u}_i=0\,(i=1,\dots,k-1)}} \mathbf{x}^T\boldsymbol{B}\mathbf{y} = s_k$$

and the maximum is attained with the choice $\mathbf{x} = \mathbf{v}_k$ and $\mathbf{y} = \mathbf{u}_k$, where $\mathbf{v}_k, \mathbf{u}_k$ is the singular vector pair corresponding to the singular value s_k of $\boldsymbol{B}$ (uniquely if $s_k > s_{k+1}$).

The $\mathbb{P}$- and $\mathbb{Q}$-distributed random variables ψ_k and ϕ_k taking on values that are the coordinates of the vectors $\boldsymbol{P}^{-1/2}\mathbf{v}_k$ and $\boldsymbol{Q}^{-1/2}\mathbf{u}_k$, are called kth *correspondence factor pair*, and their correlation is s_k. If the underlying contingency table is non-decomposable, then its non-zero singular values, being correlations, are

$$1 = s_0 > s_1 \ge s_2 \ge \cdots \ge s_{r-1} > 0.$$

Let $\boldsymbol{B}_k = (b_k(i, j))$ denote the best rank k approximation of $\boldsymbol{B}$ in the sense of Theorem A.3.21. Then the entries of $\boldsymbol{B}$ are approximated by

$$b_k(i, j) = \sum_{\ell=0}^{k-1} s_\ell u_\ell(i) v_\ell(j)$$

and consequently, in the kth step we get the following approximation of the entry w_{ij}:

$$p_i q_j \left(1 + \sum_{\ell=1}^{k-1} s_\ell \psi_\ell(i)\phi_\ell(j)\right).$$

In the $k = 1$ case this gives the approximation of the table with a rank 1 table, which contains the products of the marginals. It can be shown (see Benzécri *et al.* (1980)) that $N\|\boldsymbol{B} - \boldsymbol{B}_1\|_2 = N\sum_{\ell=1}^{r-1} s_\ell^2$ is the χ^2 statistic testing the independence of the two underlying categorical variables, where N is the sample size. Likewise,

$$N\|\boldsymbol{B} - \boldsymbol{B}_k\|_2 = N\sum_{\ell=k}^{r-1} s_\ell^2$$

and hence, it is the ratio $\sum_{\ell=1}^{k-1} s_\ell^2 / \sum_{\ell=1}^{r-1} s_\ell^2$ which gives the proportion of the χ^2 statistic explained by the first k correspondence factors.

We remark that, in view of Equation (A.11), between the left and right correspondence factors the following transition formulas work:

$$\psi_k(i) = \frac{1}{s_k p_i} \sum_{j=1}^{n} w_{ij} \phi_k(j), \quad i = 1, \ldots, m; \ k = 1, \ldots, r-1$$

$$\phi_k(j) = \frac{1}{s_k q_j} \sum_{i=1}^{m} w_{ij} \psi_k(i), \quad j = 1, \ldots, n; \ k = 1, \ldots, r-1.$$

In the $k = 1$ case, the above coordinates are sometimes interpreted as optimum scoring of the row and column categories, see Mardia *et al.* (1979).

The above decomposition also gives rise to the following k-dimensional representation of the rows and columns:

$$(s_0\psi_0(i), s_1\psi_1(i), \ldots, s_{k-1}\psi_{k-1}(i)), \quad i = 1, \ldots, m$$
$$(s_0\phi_0(j), s_1\phi_1(j), \ldots, s_{k-1}\phi_{k-1}(j)), \quad j = 1, \ldots, n.$$

In fact, this is a $(k-1)$-dimensional representation, since we can disregard the first, all-1 coordinates.

In Benzécri *et al.* (1980) and Greenacre (1984) it is proved that the squared Euclidean distance of two different row representatives is the so-called χ^2-distance of the corresponding rows in the best approximating rank k table (with respect to the marginal measures). In case of a binary table $\boldsymbol{C}$, if the first k correspondence factors explain a large proportion of the χ^2 statistic, then the k-dimensional representatives have the following allocation properties: representatives of two row categories are 'close' to each other if they occur together in 'many' columns, and vice versa, representatives of two column categories are 'close' to each other if they occur together in 'many' rows. Moreover, representatives of row categories are 'close' to representatives of column categories which 'frequently' occur together. On this basis, cluster analysis techniques for the representatives are applicable.

C.4 Multivariate regression and analysis of variance

As we have discussed in Section B.1, the general regression problem is solved by taking conditional expectation. In case of multivariate normally distributed data, the regression function is linear. By the multidimensional central limit theorem B.1.3 the multivariate normality can often be assumed, and even if not, we may look for the best linear dependence between the target and predictor variables. Here we only consider the following linear model: the expectation of the target variable Y depends linearly on p predictor variables, the values of which are deterministic (they are given measurement points, the unknown linear combination of which determines the expected value of Y), and a random measurement error is added. We have $n > p$ measurements $Y_1, \ldots, Y_n$, components of the random vector $\mathbf{Y}$. The random, so-called homoscedastic errors are the i.i.d. $\varepsilon_1, \dots, \varepsilon_n \sim \mathcal{N}(0, \sigma^2)$ variables, components of the random vector $\boldsymbol{\varepsilon} \sim \mathcal{N}_n(\mathbf{0}, \sigma^2 \boldsymbol{I}_n)$. With the measurement points collected in the $n \times p$ data matrix $\mathbf{X}$, the linear model is formulated like

$$\mathbf{Y} = \mathbf{X}\mathbf{a} + \boldsymbol{\varepsilon}$$

where $\mathbf{a} \in \mathbb{R}^p$ is the vector of the unknown parameters (coefficients of the regression function) and for simplicity, the constant term is omitted (we assume that $\bar{\mathbf{Y}} = \mathbf{0}$). We are looking for the least square estimator of the parameter vector $\mathbf{a}$ that minimizes $\|\mathbf{Y} - \mathbf{X}\mathbf{a}\|^2$. By the argument of Gauss, denoting by $\hat{\mathbf{a}}$ the optimum $\mathbf{a}$, the vector $\mathbf{X}\hat{\mathbf{a}}$ is the projection of $\mathbf{Y}$ onto the subspace spanned by the column vectors of $\mathbf{X}$. The $n \times n$ matrix of this orthogonal projection is

$$\boldsymbol{P} = \mathbf{X}(\mathbf{X}^T\mathbf{X})^{-1}\mathbf{X}^T$$

provided that $\mathbf{X}$ is of full rank: $\mathrm{rank}(\mathbf{X}) = p$ (otherwise generalized inverse is to be used). Therefore, $\hat{\mathbf{a}}$ is the solution of the system of Gaussian normal equations:

$$\mathbf{X}^T\mathbf{X}\mathbf{a} = \mathbf{X}^T\mathbf{Y}.$$

This system always has a solution, since the random vector $\mathbf{X}^T\mathbf{Y}$ is within the subspace spanned by the columns of $\mathbf{X}^T$ (rows of $\mathbf{X}$), and the same subspace is spanned by the columns of $\mathbf{X}^T\mathbf{X}$. If $\mathrm{rank}(\mathbf{X}) = p$, then the unique solution is

$$\hat{\mathbf{a}} = (\mathbf{X}^T\mathbf{X})^{-1}\mathbf{X}^T\mathbf{Y}.$$

Otherwise, there are infinitely many solutions obtained by generalized inverses. For example, using the Moore–Penrose inverse (see Definition A.3.8),

$$\hat{\mathbf{a}} = (\mathbf{X}^T\mathbf{X})^{+}\mathbf{X}^T\mathbf{Y}$$

is a possible solution in the $\mathrm{rank}(\mathbf{X}) < p$ case.

The Gauss–Markov theory states that the above $\hat{\mathbf{a}}$ is a best linear unbiased estimator (BLUE) in the following sense. $\mathbb{E}(\hat{\mathbf{a}}) = \mathbf{a}$, the coordinates of $\hat{\mathbf{a}}$ are linear functions of the coordinates of $\mathbf{Y}$. Further, among other linear, unbiased estimators (say $\tilde{\mathbf{a}}$), $\hat{\mathbf{a}}$ is the best, that is, more efficient in the sense that for the covariance matrices

$$\mathrm{Var}(\hat{\mathbf{a}}) \leq \mathrm{Var}(\tilde{\mathbf{a}})$$

holds, meaning that the difference of the right- and left-hand sides is positive semidefinite (also positive definite when $\hat{\mathbf{a}} \neq \tilde{\mathbf{a}}$). At the same time, the unbiased estimate

$$\widehat{\sigma^2} = \frac{1}{n-p}\|\mathbf{Y} - \mathbf{X}\hat{\mathbf{a}}\|^2$$

of σ^2 is obtained.

The following model of the one-way analysis of variance (ANOVA) completely fits into the framework of linear models. Here we have univariate normal measurements in k different groups. Based on $n_1, \dots, n_k$ independent homoscedastic observations (with the same variance σ^2 and identically distributed within the groups), we want to decide whether the expectations of the groups are equal or not. Let $Y_{ij} \sim \mathcal{N}(\mu + a_i, \sigma^2)$ $(j = 1, \dots, n_i)$ denote the sample entries within the group i $(i = 1, \dots, k)$, components of the n-dimensional random vector $\mathbf{Y}$ in this order, where $n = \sum_{i=1}^{k} n_i$. Here μ and $\mathbf{a} = (a_1, \dots, a_k)^T$ are unknown parameters such that $\sum_{i=1}^{k} a_i = 0$ (not a restriction as μ is just introduced to compensate for this). With these, we have the linear model

$$\mathbf{Y} = \mu \mathbf{1}_n + \mathbf{X}\mathbf{a} + \boldsymbol{\varepsilon}$$

where $\boldsymbol{\varepsilon} \sim \mathcal{N}_n(\mathbf{0}, \sigma^2 \boldsymbol{I}_n)$, $\mathbf{1}_n$ is the constantly 1 vector of $\mathbb{R}^n$, and the structure matrix $\mathbf{X}$ is destined for decomposing the components of $\mathbf{Y}$ as their expectation plus an $\mathcal{N}(0, \sigma^2)$ error term. The columns of $\mathbf{X}$ are so-called *partition vectors* having 0 or 1 coordinates depending on the group memberships. More precisely, let $C_1 = \{1, \dots, n_1\}$, $C_2 = \{n_1 + 1, \dots, n_1 + n_2\}, \dots, C_k = \{\sum_{i=1}^{k-1} n_i + 1, \dots, n\}$ be the index sets corresponding to the measurements in the k groups. Then the entries of the rank k matrix $\mathbf{X}$ are the following

$$x_{ij} = \begin{cases} 1 & \text{if} \quad i \in C_j \\ 0 & \text{otherwise} \end{cases} \qquad i = 1, \dots, n; \; j = 1, \dots, k.$$

After this, the parameter estimation is the same as in the usual linear model using the Gauss normal equations. To test the null-hypothesis $a_1 = a_2 = \cdots = a_k = 0$, we need to decompose the empirical variance of the sample into *between- and within-group variances*. The Fisher–Cochran theorem is applicable, which provides us with the χ^2-distributed statistics producing the F-distributed test statistic, see Rao (1973) for more details. Here we do not discuss the test procedure, however, the decomposition of variances and the equivalent decomposition of the centering matrix into projections will play a crucial role in the subsequent sections and also in spectral clustering. The following formula provides the decomposition of the complete sample's empirical variance (better to say, n times this empirical variance) into between- and within-cluster parts:

$$\sum_{i=1}^{k} \sum_{j=1}^{n_i} (Y_{ij} - \bar{Y}_{..})^2 = \sum_{i=1}^{k} n_i (\bar{Y}_{i.} - \bar{Y}_{..})^2 + \sum_{i=1}^{k} \sum_{j=1}^{n_i} (Y_{ij} - \bar{Y}_{i.})^2$$

or briefly,

$$T = B + W$$

where $\bar{Y}_{..} = \frac{1}{n} \sum_{i=1}^{k} \sum_{j=1}^{n_i} Y_{ij}$ is the sample mean, while $\bar{Y}_{i.} = \frac{1}{n_i} \sum_{j=1}^{n_i} Y_{ij}$ is the mean of the group i ($i = 1, \dots, k$). With some linear algebra, the total-, between- and within-cluster variances can be written as quadratic forms corresponding to special projections:

$$T = \mathbf{Y}^T (\boldsymbol{I}_n - \mathbf{1}_n \mathbf{1}_n^T) \mathbf{Y}, \quad B = \mathbf{Y}^T \boldsymbol{P} \mathbf{Y}, \quad W = \mathbf{Y}^T \boldsymbol{A} \mathbf{Y}$$

where the so-called *centering matrix* $\boldsymbol{I}_n - \mathbf{1}_n \mathbf{1}_n^T$ projects onto the orthogonal complementary subspace of $\mathbf{1}_n$ in $\mathbb{R}^n$, $\boldsymbol{P} = \mathbf{X}(\mathbf{X}^T\mathbf{X})^{-1}\mathbf{X}^T$ projects onto the subspace spanned by the columns (partition vectors) of $\mathbf{X}$, and $\boldsymbol{A} = \boldsymbol{A}_1 \oplus \cdots \oplus \boldsymbol{A}_k$ is the Kronecker-sum (see Definition A.3.34) of the *centering matrices* $\boldsymbol{A}_i = \boldsymbol{I}_{n_i} - \mathbf{1}_{n_i} \mathbf{1}_{n_i}^T$ ($i = 1, \dots, k$).

The model of the one-way multivariate analysis of variance (MANOVA) works analogously, with measurements in k different groups, but here we have p-variate measurements $\mathbf{Y}_{ij} \sim \mathcal{N}_p(\boldsymbol{\mu} + \mathbf{a}_i, \boldsymbol{C})$ $(j = 1, \ldots, n_i;\ i = 1, \ldots, k)$, where $\sum_{i=1}^k \mathbf{a}_i = \mathbf{0}$ is assumed. We will only need the decomposition

$$\boldsymbol{T} = \boldsymbol{B} + \boldsymbol{W}$$

of n times the $p \times p$ sample covariance matrix into *between- and within-group covariance matrices* in the following way:

$$\begin{aligned}
\boldsymbol{T} &= \sum_{i=1}^{k} \sum_{j=1}^{n_i} (\mathbf{Y}_{ij} - \bar{\mathbf{Y}}_{..})(\mathbf{Y}_{ij} - \bar{\mathbf{Y}}_{..})^T \\
\boldsymbol{B} &= \sum_{i=1}^{k} n_i (\bar{\mathbf{Y}}_{i.} - \bar{\mathbf{Y}}_{..})(\bar{\mathbf{Y}}_{i.} - \bar{\mathbf{Y}}_{..})^T \\
\boldsymbol{W} &= \sum_{i=1}^{k} \sum_{j=1}^{n_i} (\mathbf{Y}_{ij} - \bar{\mathbf{Y}}_{i.})(\mathbf{Y}_{ij} - \bar{\mathbf{Y}}_{i.})^T
\end{aligned} \tag{C.4}$$

where $\bar{\mathbf{Y}}_{..} = \frac{1}{n} \sum_{i=1}^k \sum_{j=1}^{n_i} \mathbf{Y}_{ij}$ is the sample mean vector, while $\bar{\mathbf{Y}}_{i.} = \frac{1}{n_i} \sum_{j=1}^{n_i} \mathbf{Y}_{ij}$ is the mean vector of group i $(i = 1, \ldots, k)$.

Another extension of the one-way ANOVA is the two-way one when the continuous measurements are grouped in two different aspects, providing a biclustering of them in the form of a contingency table (there may be more than one measurements within the cells), see Mardia *et al.* (1979); Rao (1973) for details.

C.5 The k-means clustering

Here we only consider one method of cluster analysis for finding groups of data points in a finite dimensional Euclidean space. Given the points $\mathbf{x}_1, \ldots, \mathbf{x}_n \in \mathbb{R}^d$ and an integer $1 < k < n$, we are looking for the k-partition of the index set $\{1, \ldots, n\}$ (or equivalently, the clustering of the points into k disjoint non-empty subsets) which minimizes the following *k-variance* of the points over all possible k-partitions $P_k = (C_1, \ldots, C_k)$:

$$S_k^2(\mathbf{x}_1, \ldots, \mathbf{x}_n) = \min_{P_k} S_k^2(P_k; \mathbf{x}_1, \ldots, \mathbf{x}_n) = \min_{P_k} \sum_{a=1}^{k} \sum_{j \in C_a} \|\mathbf{x}_j - \mathbf{c}_a\|^2 \tag{C.5}$$

where $\mathbf{c}_a = \frac{1}{|C_a|} \sum_{j \in C_a} \mathbf{x}_j$ is the center of cluster a $(a = 1, \ldots, k)$.

In general, $d \le k$, and they are much less than n. In fact, the above k-variance corresponds to the trace of the within-cluster term $\boldsymbol{W}$ in (C.4). Since the total variance is fixed, minimizing $\frac{\mathrm{tr}(\boldsymbol{W})}{\mathrm{tr}(\boldsymbol{B})} = \frac{\mathrm{tr}(\boldsymbol{W})}{\mathrm{tr}(\boldsymbol{T}) - \mathrm{tr}(\boldsymbol{W})}$ is equivalent to minimizing $\mathrm{tr}(\boldsymbol{W})$, that is, S_k^2.

To find the global minimum is NP-complete, but the iteration of the k-means algorithm, first described in Steinhaus (1956), is capable to find a local minimum in polynomial time. The vectors $\mathbf{c}_1, \ldots, \mathbf{c}_k$ are usually referred to as the *centroids* of the clusters, and in a more abstract formulation of the above optimization task, for example, MacQueen (1967) also looked for them. Roughly speaking, starting with an initial clustering, the iteration of the simple k-means algorithm consists of the following two alternating steps. In the first step, fixing the clustering of the points, it finds the cluster centers (they will be the baricenters by the Steiner's theorem).

In the second one, the algorithm relocates the points in such a way that it assigns a point to the cluster, the center of which is the closest to it (in case of ambiguity the algorithm chooses the smallest index such cluster). If there exists a well-separated k-clustering of the points (even the largest intra-cluster distance is smaller than the smallest inter-cluster one) the convergence of the algorithm to the global minimum is proved in Dunn (1973, 1974), with a convenient starting. Under relaxed conditions, the speed of the algorithm is increased by a filtration in Kanungo *et al.* (2002). The algorithm runs faster if the separation between the clusters increases and an overall running time of $\mathcal{O}(kn)$ can be guaranteed.

Sometimes the points $\mathbf{x}_1, \dots, \mathbf{x}_n$ are endowed with the positive weights $d_1, \dots, d_n$, where without loss of generality $\sum_{i=1}^{n} d_i = 1$ can be assumed. In such cases the *weighted k-variance* of the points

$$\tilde{S}_k^2(\mathbf{x}_1, \dots, \mathbf{x}_n) = \min_{P_k} \tilde{S}_k^2(P_k; \mathbf{x}_1, \dots, \mathbf{x}_n) = \min_{P_k} \sum_{a=1}^{k} \sum_{j \in C_a} d_j \|\mathbf{x}_j - \mathbf{c}_a\|^2 \qquad \text{(C.6)}$$

is minimized by the weighted k-means algorithm, where $\mathbf{c}_a = \frac{1}{\sum_{j \in C_a} d_j} \sum_{j \in C_a} d_j \mathbf{x}_j$ is the weighted center of cluster a ($a = 1, \dots, k$). The k-means algorithm can be easily adapted to this situation (see Subsection 5.2.1 for the pseudocode). Note that $\tilde{S}_k^2(\mathbf{x}_1, \dots, \mathbf{x}_n)$ corresponds to the k-variance with respect to the distribution $d_1, \dots, d_n$. In this context, $S_k^2(\mathbf{x}_1, \dots, \mathbf{x}_n)$ is the special case when this law is uniform. Likewise, instead of L^2-distances, other kind of distance functions in the objective function may be used.

A well-known drawback of the k-means clustering is that the clusters need to be convex in order to achieve satisfactory results. That is, the k-means algorithm forms spherical clusters whether or not the underlying data distribution obeys this form. Otherwise, our data can be mapped into a feature space and we apply k-means clustering for the mapped data which already have this spherical structure. With the notation of Section 1.5, we need not actually map our data, but can find the squared Euclidean distance between a feature point $\phi(\mathbf{x}_\ell)$ and the center $\mathbf{c}$ of its cluster C via the kernel K in the following way:

$$\begin{aligned} &\|\phi(\mathbf{x}_\ell) - \mathbf{c}\|^2 \\ &= \langle \phi(\mathbf{x}_\ell), \phi(\mathbf{x}_\ell) \rangle + \frac{1}{|C|^2} \sum_{i \in C} \sum_{j \in C} \langle \phi(\mathbf{x}_i), \phi(\mathbf{x}_j) \rangle - \frac{2}{|C|} \sum_{i \in C} \langle \phi(\mathbf{x}_i), \phi(\mathbf{x}_\ell) \rangle \\ &= K(\mathbf{x}_\ell, \mathbf{x}_\ell) + \frac{1}{|C|^2} \sum_{i \in C} \sum_{j \in C} K(\mathbf{x}_i, \mathbf{x}_j) - \frac{2}{|C|} \sum_{i \in C} K(\mathbf{x}_i, \mathbf{x}_\ell). \end{aligned}$$

C.6 Multidimensional scaling

For clustering purposes, sometimes we have abstract objects which are basically not in a metric space, and all we have are their pairwise distances. In a broad sense, the notion of a *distance matrix* in the upcoming definition allows us to define distances between object subjectively, possibly as monotonous decreasing functions of their pairwise similarities.

Definition C.6.1 *The $n \times n$ matrix $\boldsymbol{D}$ is called distance matrix if it is symmetric and*

$$d_{ii} = 0 \quad (i = 1, \dots, n), \quad \text{and} \quad d_{ij} \geq 0 \quad (i \neq j).$$

This definition allows the entries of the distance matrix not to obey the triangle inequalities (see Definition A.1.1), and even if they do so, the metric defined by the distances is not necessarily the Euclidean one.

Our purpose is to find a dimension d and points $\mathbf{x}_1, \dots, \mathbf{x}_n \in \mathbb{R}^d$ such that their pairwise Euclidean distances approach the entries of the distance matrix as much as possible. Of course, if the distances are real distances of objects in a physical space, the method of multidimensional scaling to be introduced is assumed to find the dimension and configuration of the hidden points, at least, up to translation, rotation, and reflection. For example, if someone provides us with the pairwise distances of cities (not too far apart), we will be able to reconstruct their mutual position. Even in this case, our measurements may be subject to error, therefore, we want to find a solution to the problem, which is able to give a good approximation in any of the above cases.

The situation when the distances can exactly be realized in a Euclidean space is defined now.

Definition C.6.2 *The $n \times n$ distance matrix is Euclidean if there is a positive integer d and points $\mathbf{x}_1, \dots, \mathbf{x}_n \in \mathbb{R}^d$ such that*

$$\|\mathbf{x}_i - \mathbf{x}_j\| = d_{ij}, \quad i, j = 1, \dots, n.$$

The following theorem gives a necessary and sufficient condition for a distance matrix to be Euclidean. Intuitively, one needs somehow to eliminate the translation invariance, therefore makes use of the already defined *centering matrix* $\boldsymbol{C}_n = \boldsymbol{I}_n - \frac{1}{n}\mathbf{1}_n\mathbf{1}_n^T$ (we will drop the index n in the sequel).

Theorem C.6.3 *Given an $n \times n$ distance matrix $\boldsymbol{D}$, the matrix $\boldsymbol{A} = (a_{ij})$ is defined as $a_{ij} = -\frac{1}{2}d_{ij}^2$ $(i, j = 1, \dots, n)$. The matrix $\boldsymbol{D}$ is Euclidean if and only if the symmetric matrix $\boldsymbol{CAC}$ is positive semidefinite.*

For the proof see Mardia *et al.* (1979). We just remark that the construction for the point configuration, under the conditions of Theorem C.6.3 , is the following. The dimension will be the rank of $\boldsymbol{CAC}$, that is, $d = \text{rank}(\boldsymbol{CAC})$. Let $\lambda_1 \geq \cdots \geq \lambda_d > 0$ be the strictly positive eigenvalues of $\boldsymbol{CAC}$ with corresponding unit-norm eigenvectors

$\mathbf{u}_1, \ldots, \mathbf{u}_d$. The points will be row vectors of the $n \times d$ matrix $(\sqrt{\lambda_1}\mathbf{u}_1, \ldots, \sqrt{\lambda_d}\mathbf{u}_d)$. In fact, this idea is inspired by Equation (A.13) in the context of Gram-matrices. This system is unique apart from translation, rotation, and reflection, not to mention the indeterminacy due to possible multiple eigenvalues and the triviality that the coordinates can be inflated with any number of zero coordinates. Note that the construction gives a hint how to find an approximate solution when $\boldsymbol{D}$ is not Euclidean, but not 'far' from that, that is, the matrix $\boldsymbol{CAC}$ has some slightly negative eigenvalues. Then we omit those and use the eigenvectors corresponding to the positive ones in the above construction. Even if $\boldsymbol{D}$ is Euclidean, the rank of $\boldsymbol{CAC}$ may be so large that we want a smaller dimensional configuration that reconstructs the distances with a tolerable error. For this purpose, we retain the largest eigenvalues is the analysis, better to stay, we look for a gap in the spectrum so that the eigenvalues behind this gap are negligible compared to those before the gap. Then the number of the outstanding eigenvalues will be the dimension of the points we look for.

C.7 Discriminant analysis

For the time being, except the regression (where we had a target variable depending on the predictor ones), we have discussed methods of the so-called *unsupervised learning*, when we retrieved information from our data without any preliminary assumption. Now, another method of *supervised learning* is introduced, when we already have some preliminary knowledge about the classification of our data (made by an expert) and we want to reproduce this classification based merely on multivariate measurements. In such situations we have a learning sample to build the artificial intelligence, and we test it on on the same or on another sample. In this way, so-called expert systems are constructed, and in case of good performance, they can be used (with care) for automatic classification.

At the beginning, we have a p-dimensional sample of independent observations, but they are not identically distributed, rather they form a mixture of k multivariate distributions, which are clearly distinguished by an expert in the so-called learning sample. For example, we have p clinical measurements of patients coming from k different diagnostic groups. If the measurements have something to do with the diagnosis, there is a hope that with some algorithm we are able to assign a patient to one of the groups merely based on his/her measurements. Based on the classes of the learning sample and some intuition, we are provided with the following knowledge:

- p-variate densities $f_1(\mathbf{x}), \ldots, f_k(\mathbf{x})$ of the classes (usually these are multivariate Gaussian with estimated parameters);
- the prior probabilities $\pi_1, \ldots, \pi_k$ of a randomly selected object belonging to the classes, $\sum_{i=1}^{k} \pi_i = 1$ (they are usually proportional to the sample sizes, but can as well correspond to the expert's intuition).

Our purpose is to find a partition $\mathcal{X}_1, \ldots, \mathcal{X}_k$ of the p-dimensional sample space so that the obtained classes would, as much as possible, coincide with the original ones. Equivalently, we have to find a decision rule which decides the membership of an object based on its measurement $\mathbf{x} \in \mathbb{R}^p$. Our algorithm minimizes the following average loss function:

$$L = \sum_{i=1}^{k} \pi_i L_i.$$

Here the average loss L_i is due to misclassifying objects of $\mathcal{X}_i$, defined as

$$L_i = \sum_{j=1}^{k} \int_{\mathcal{X}_j} r_{ij} f_i(\mathbf{x})\, d\mathbf{x}$$

where the risk $r_{ij} \geq 0$ of classifying an object of class i into class j is given for $i, j = 1, \ldots, k$. After some simple calculation

$$L = -\sum_{j=1}^{k} \int_{\mathcal{X}_j} S_j(\mathbf{x})\, d\mathbf{x},$$

where $S_j(\mathbf{x}) = -\sum_{i=1}^{k} \pi_i r_{ij} f_i(\mathbf{x})$ is called jth *discriminant informant*, and for given $\mathbf{x}$, we want to maximize

$$\sum_{j=1}^{k} \int_{\mathcal{X}_j} S_j(\mathbf{x})\, d\mathbf{x} \tag{C.7}$$

over the set of k-partitions of the sample space. A simple lemma guarantees that the maximum is attained with the following k-partition $\mathcal{X}_1^*, \ldots, \mathcal{X}_k^*$: an object with measurement $\mathbf{x}$ is classified into $\mathcal{X}_i^*$ for which, $i = \text{argmax}_j S_j(\mathbf{x})$ (such an i is not necessarily unique, but any choice, say the smallest such i, will give the same maximum of (C.7).

Now, let us make the following simplification: $r_{ij} = 1$ for $i \neq j$ and of course, $r_{ii} = 0$ ($j = 1, \ldots, k$). This assumption is quite natural: all misclassifications have the same risk, and there is no risk of a correct classification. By this, the discriminant informant simplifies to

$$S_j(\mathbf{x}) = -\sum_{i \neq j} \pi_i f_i(\mathbf{x}) = -\sum_{i=1}^{k} \pi_i f_i(\mathbf{x}) + \pi_j f_j(\mathbf{x}) = c + \pi_j f_j(\mathbf{x})$$

where the constant c does not depend on j, therefore instead $S_j(\mathbf{x})$ we can as well maximize $\pi_j f_j(\mathbf{x})$. That is, an object with measurement $\mathbf{x}$ is placed into the group j for which $\pi_j f_j(\mathbf{x})$ is maximum. Observe that this is nothing else but a *Bayesian decision rule*. Indeed, let Y denote the cluster membership, and $\mathbf{X}$ is the underlying p-variate random vector. Then for a randomly selected object with measurement $\mathbf{x}$ the following conditional probability is maximized in j:

$$P(Y = j \,|\, \mathbf{X} = \mathbf{x}) = \frac{\pi_j f_j(\mathbf{x})}{\sum_{i=1}^{k} \pi_i f_i(\mathbf{x})}$$

where we used the Bayes rule. The maximization is equivalent to maximizing the numerator with respect to $j = 1, \ldots, k$. Further, if all the prior probabilities are equal, for given $\mathbf{x}$, we maximize $\mathbf{f}_j(\mathbf{x})$ which is just the *maximum likelihood discrimination rule*.

We can further simplify the maximization if the distribution of class j is $\mathcal{N}_p(\boldsymbol{\mu}_j, \boldsymbol{C}_j)$ with positive definite covariance matrix $\boldsymbol{C}_j$ $(j = 1, \ldots, k)$. Adapting Equation (B.1) for the density of the classes, for given $\mathbf{x}$, instead of $\pi_j f_j(\mathbf{x})$, we can maximize its natural logarithm. After leaving out the terms which do not depend on j, one can easily see that the following *quadratic informant* (quadratic function of the coordinates of $\mathbf{x}$) has to be maximized with respect to j:

$$Q_j(\mathbf{x}) = -\frac{1}{2}\ln|\boldsymbol{C}_j| - \frac{1}{2}(\mathbf{x} - \boldsymbol{\mu}_j)^T \boldsymbol{C}_j^{-1}(\mathbf{x} - \boldsymbol{\mu}_j) + \ln \pi_j.$$

In the case of $\boldsymbol{C}_1 = \cdots = \boldsymbol{C}_k = \boldsymbol{C}$, we can disregard the terms which do not depend on j, and the following *linear informant* will decide the group memberships:

$$L_j(\mathbf{x}) = \boldsymbol{\mu}_j^T \boldsymbol{C}_j^{-1}\mathbf{x} - \frac{1}{2}\boldsymbol{\mu}_j^T \boldsymbol{C}_j^{-1}\boldsymbol{\mu}_j + \ln \pi_j.$$

If $k = 2$, then we put an object with measurement $\mathbf{x}$ into the first group if $L_1(\mathbf{x}) \geq L_2(\mathbf{x})$ and to the second one, otherwise. That is, the sample space is separated into two parts by means of a hyperplane. In Mardia *et al.* (1979), it is shown that in case of k groups, $k - 1$ hyperplanes will do this job.

We remark that the sample means and covariance matrices are usually estimated from the sub-samples after checking for multivariate normality.

In another approach, R. A. Fisher looked for the linear function $\mathbf{a}^T\mathbf{x}$, the coefficients of which maximize the ratio of the between-groups sum of squares to the within-groups sum of squares. That is, with the notation of (C.4),

$$\frac{\mathbf{a}^T \boldsymbol{B}\mathbf{a}}{\mathbf{a}^T \boldsymbol{W}\mathbf{a}} \tag{C.8}$$

has to be maximized with respect to $\mathbf{a}$. Since the scale of $\mathbf{a}$ does not affect the above ratio, $\|\mathbf{a}\| = 1$ can be assumed. Mardia *et al.* (1979) and Rao (1973) prove that the vector $\hat{\mathbf{a}}$ which maximizes the above ratio is the unit-norm eigenvector corresponding to the largest eigenvalue of the matrix $\boldsymbol{W}^{-1}\boldsymbol{B}$, which is of rank at most $k - 1$ (since $\text{rank}(\boldsymbol{B}) = k - 1$ in the general case). The function $\hat{\mathbf{a}}^T\mathbf{x}$ is called *Fisher's linear discriminant function* or the *first canonical variate*. Based on this, we allocate $\mathbf{x}$ into the group i if

$$\|\hat{\mathbf{a}}^T\mathbf{x} - \hat{\mathbf{a}}^T\bar{\mathbf{x}}_i\| \leq \|\hat{\mathbf{a}}^T\mathbf{x} - \hat{\mathbf{a}}^T\bar{\mathbf{x}}_j\| \quad \forall j \neq i$$

where $\bar{\mathbf{x}}_j$ is the sample mean of group j. In the $k = 2$ case, this rule is identical to that given by the linear informants, where $\boldsymbol{\mu}_1$ and $\boldsymbol{\mu}_2$ are estimated by the group means $\bar{\mathbf{x}}_1$ and $\bar{\mathbf{x}}_2$, respectively. Note that this is not true in the $k > 2$ case. Remark that successively, number of $\text{rank}(\boldsymbol{W}^{-1}\boldsymbol{B})$ canonical variates can be computed, which gives rise to differentiate between the groups in dimension $k - 1$. Canonical variates also have important relation to the canonical correlations, see Mardia *et al.* (1979); Rao (1973) for details.

In practice, first we process the discrimination on the learning sample, and in case of good performance, we can apply the algorithm for a test sample of new-coming objects. In the absence of a test sample, we can randomly select objects from the same learning sample with some resampling method, called *bootstrapping*. The performance itself is evaluated by the cross-classification of the objects: we calculate the $k \times k$ *confusion matrix* the ij-th entry of which is the number of objects classified into class i by the expert, and into class j by the algorithm.

References

Benzécri JP *et al.* 1980 *Pratique de l'Analyse des Données, 2. L'Analyse des Correspondances*, Dunod, Paris.

Dunn JC 1973 A fuzzy relative of the ISODATA Process and its use in detecting compact well-separated clusters. *J. Cybernetics* **3**, 32–57.

Dunn JC 1974 Well-separated clusters and optimal fuzzy partitions. *J. Cybernetics* **4**, 95–104.

Greenacre MJ 1984 *Theory and Applications of Correspondence Analysis*, Academic Press, New York.

Kanungo T, Mount DM, Netanyahu NS *et al.* 2002 An efficient k-means clustering algorithm: Analysis and implementation. *IEEE Trans. Pattern Anal. Mach. Intell.* **24** (7), 881–892.

MacQueen JB 1967 Some methods for classification and analysis of multivariate observations, in *Proc. 5th Berkeley Symposium on Mathematical Statistics and Probability* (eds Cam L and Neyman J), University of California Press, Berkeley, USA, Vol. 1, pp. 281–297.

Mardia KV, Kent JT and Bibby JM 1979 *Multivariate Analysis*, Academic Press, London.

Rao CR 1973 *Linear Statistical Inference and its Applications*, Wiley.

Steinhaus H 1956 Sur la division des corp materiels en parties. *Bull. Acad. Polon. Sci. Cl. III* **4**, 801–804.

Index

Spectral Clustering and Biclustering: Learning Large Graphs and Contingency Tables, First Edition.
Marianna Bolla.

www.ingramcontent.com/pod-product-compliance
Lightning Source LLC
Chambersburg PA
CBHW060101010325
22684CB00003B/9

* 9 7 8 1 1 1 8 3 4 4 9 2 7 *